Climate Uncer

Climate Uncertainty and Risk

Rethinking Our Response

By Judith A. Curry

ANTHEM PRESS

Anthem Press
An imprint of Wimbledon Publishing Company
www.anthempress.com

This edition first published in UK and USA 2023
by ANTHEM PRESS
75–76 Blackfriars Road, London SE1 8HA, UK
or PO Box 9779, London SW19 7ZG, UK
and
244 Madison Ave #116, New York, NY 10016, USA

British Library Cataloguing-in-Publication Data
A catalogue record for this book is available from the British Library.

Library of Congress Cataloging-in-Publication Data
A catalog record for this book has been requested.
2023930881

ISBN-13: 978-1-78527-816-7 (Hbk)
ISBN-10: 1-78527-816-9 (Hbk)

ISBN-13: 978-1-83998-925-4 (Pbk)
ISBN-10: 1-83998-925-4 (Pbk)

This title is also available as an e-book.

CONTENTS

LIST OF FIGURE AND TABLES

Figure

Tables

DESCRIPTION

World leaders have made a forceful statement that climate change is the greatest challenge facing humanity in the twenty-first century. However, little progress has been made in implementing policies to address climate change in a meaningful way. In *Climate Uncertainty and Risk*, eminent climate scientist Judith Curry shows how we can break through this stalemate. This book helps us rethink the climate change problem, the risks we are facing, and our response. It helps us strategize on how we can best engage with our environment and support human well-being, while responding to climate change.

Climate Uncertainty and Risk provides a comprehensive framework for understanding the climate change debate. It shows how both the climate change problem and its solutions have been oversimplified. It explains how understanding uncertainty helps us to better assess the risks. It describes how uncertainty and disagreement can be part of the decision-making process. It provides a road map for formulating pragmatic solutions that can improve our well-being in the twenty-first century.

Judith Curry brings a unique perspective to the debate on climate change. She is a distinguished climate scientist who has engaged extensively with decision makers in both the private and public sectors on a range of issues related to weather and climate. She interacts with scientists, activists and politicians on both sides of the climate change debate. In her search for wisdom on the challenge of climate change, she incorporates the philosophy and sociology of science, ethics, risk management and politics. *Climate Uncertainty and Risk* is essential reading for those concerned about the environment, professionals dealing with climate change, and our national leaders.

ACKNOWLEDGMENTS

My journey in developing the perspectives presented in this book has been enriched by endless conversations with Peter Webster. Special thanks to Peter for this and for his ongoing support. The Denizens of my blog *Climate Etc.* have provided a wide range of expertise and perspectives on the climate change challenge, broadly defined, which has expanded my thinking on these topics. CFAN's clients have anchored me in the real world of decision-making on weather- and climate-related risks.

Specifically with regard to the preparation of this book, extensive comments from Peter Webster, Paul Klipfel, Peter Hartley, Meredith Whelan, Bruce Moran, and Margaret Tippett are greatly appreciated. Comments from Tim Palmer, Larry Kummer, Ingrid Vogel, Suellen Knopick, Bob Grossman, Carole Vogel and Mike Tippett were very helpful. Very special thanks to Mark Jelinek for his extensive efforts in technical editing of the book, including fact-checking, quote-checking, paraphrase-checking, reference hunting, footnoting, and overall document preparation.

I am also very appreciative of Anthem Press for the rigor of their review process and support in producing this book.

This book is dedicated to the next generations, particularly my daughter Meredith and granddaughter Clara.

AUTHOR'S FOREWORD

"The major problems in the world are the result of the difference between how nature works and the way people think."

—Anthropologist Gregory Bateson[1]

World leaders have made a forceful statement that climate change is the greatest challenge facing humanity in the twenty-first century. While the majority of people are worried by climate change, most are unwilling to follow the call from the United Nations for "rapid, far-reaching and unprecedented changes in all aspects of society."[2] Further, many of the technologies needed to effectively transition the world's economy away from fossil fuels are not ready for large-scale deployment. As a result, there is acrimonious worldwide political debate on implementing climate policies, which even if successful, have little chance of improving the climate or human well-being in the twenty-first century.

How did we come to be between a rock and hard place on the issue of climate change? This book shows how the narrow and politicized framing of the climate debate has resulted in an oversimplification of both the scientific problem and its solutions. My personal journey in navigating the climate debate provides insights into the problem and ways forward for finding solutions.

Prior to about 2003, it was fashionable in academic circles to be skeptical about the highly confident conclusions being issued in the Intergovernmental Panel on Climate Change (IPCC) assessment reports on human-caused climate change. I became concerned about the way these assessment reports were treating uncertainty and confidence levels in their conclusions.[3] Apart from reading the IPCC Reports, I was mostly oblivious to the public debate and controversies surrounding climate change.

I inadvertently entered the public debate on climate change on September 14, 2005. The American Association for the Advancement of Science organized a press conference for a paper I co-authored that described a substantial increase in the global proportion of category 4 and 5 hurricanes. The unplanned and uncanny timing of publication of this paper was three weeks after Hurricane Katrina had devastated New Orleans. Our "15 minutes" stretched into days, weeks, and months, as Hurricane Katrina became a major focusing event for the global warming debate. I was treated like a rock star by the environmental movement. But on the other side of the debate, I was targeted as a global warming alarmist, capitalizing on this tragedy to increase research funding and for personal publicity, and a threat to capitalism and the American way of life.

I had lost my bearings during this episode; I took a step back and tried to understand all this craziness and learn from it. I wrote a journal article on "Mixing Politics and Science in Testing the Hypothesis that Greenhouse Warming is Causing a Global Increase in Hurricane Intensity."[4] I made my first major foray into the blogosphere, checking in at all the blogs where the paper was being discussed. I landed on the "skeptic" blog *Climate Audit*,[5] where the commenters had some good questions about the statistical analyses and wanted to see the raw data.

When the IPCC Fourth Assessment Report was published in 2007, I joined the consensus in supporting this document as authoritative; I was convinced by the rigors of the process. While I didn't personally agree with everything in the document (I still had nagging concerns about the treatment of uncertainty and overconfidence), I bought into the meme of "don't trust what one scientist says, listen to the IPCC." I was becoming increasingly concerned by the political activism of people involved in the IPCC and by policies that didn't make sense to me. But after all, "don't trust what one scientist says," and I continued to substitute the IPCC assessment for my own personal judgments regarding the public debate on climate change.

My perspective on all this changed as a result of Climategate,[6] which began in November 2009 with the hacking and unauthorized release of emails from the Climatic Research Unit at the University of East Anglia. These email exchanges between climate scientists and authors of the IPCC were released to various internet locations several weeks before the Copenhagen Summit on climate change. When I read the Climategate emails, they confirmed concerns and suspicions that I had been developing about politics and personal agendas encroaching on the IPCC assessment process. Attempting to help calm the waters with the climate skeptics, I wrote an essay entitled "On the Credibility of Climate Research" that was published on the blog *Climate Audit*,[7] where the Climategate story broke. Several days later, my essay "An open letter to graduate students and young scientists in fields related to climate research" was published by the *New York Times*.[8]

A big part of my reaction to events unfolding after Climategate was the concern that I had somehow been duped into substituting the judgments of the IPCC for my own in my public statements on climate change. I started my own blog *Climate Etc.* in 2010,[9] which launched a series on *Climate Science and the Uncertainty Monster*. As people in other fields became increasingly interested in my essays at *Climate Etc.*, I redefined and broadened my academic network to include open-minded individuals from the fields of physics, philosophy, economics, social psychology, law, engineering, management, communications, artificial intelligence, statistics, and political and policy sciences.

As I drifted away from the mainstream narrative about climate change, academia increasingly seemed like "wrong trousers" for me, with its disciplinary compartmentalization, culture of climate consensus enforcement, and freedom of speech issues. I resigned my tenured faculty position in 2017.[10]

As a result of the politicization of climate science, I found that I had lost my love of science in the context of the academic ecosystem. I left academia with the hope of regaining my love of scientific research and its applications. I am now employed in the private sector as President of Climate Forecast Applications Network (CFAN).[11] I am working on new and better ways to apply weather and climate data, weather forecast information, and regional climate change scenarios to support real-world decision-making to manage weather- and climate-related risks. Most importantly, I am exhilarated by the lack of political and peer constraints in conducting my research, serving the clients of CFAN, and in my public statements about climate change.

By engaging with decision makers in both the private and public sector on issues related to weather and climate variability, I have learned about the complexity of different decisions that depend, at least in part, on weather and climate information. I have learned the importance of careful determination and conveyance of the uncertainty associated with a prediction. I have found that the worst outcome is a prediction issued with a high level of confidence that turns out to be wrong; a close second is missing the possibility of an extreme event. Simply put, I have "skin in the game" in terms of my predictions and assessments of risk; I lose clients and damage the reputation of my company if I issue an overconfident prediction that turns out to be wrong.

It is this perspective that I bring to the debate on climate change—integration of a strong academic reputation in climate science with extensive real-world engagement with public- and private-sector decision makers. My transition away from being a supporter of the mainstream climate consensus has led to wide-ranging reflections on the climate debate that incorporates the philosophy and sociology of science, ethics, and politics. My perspective has been developed and shared in academic publications, numerous posts at my blog *Climate Etc.*, reports to the clients of CFAN, and testimonies presented in US Congressional Hearings.[12]

The goal of *Climate Uncertainty and Risk* is to stimulate new ideas and broader thinking about the climate change challenge and its solutions. This is accomplished by understanding the deep uncertainties surrounding the issue, new methods of generating regional scenarios for twenty-first century climate variability and change, and the principles of risk management and decision-making under deep uncertainty. The point of considering these ideas is to increase the range of policy options and enlarge the landscape for decision makers on issues surrounding climate change. My hope in writing this book is to lay out a path for eventually acquiring wisdom as to how we can best engage with our environment and support human well-being and thrivability while responding to climate change.

The climate change problem provides much scope for disagreement among reasonable and intelligent people. This book makes no attempt to arbitrate specific scientific disputes or to prescribe policy preferences. The book is designed to acknowledge and engage with the complete spectrum of perspectives, from people that are convinced to those that are unconvinced by the existing paradigm of climate change. It is aimed at those concerned about the environment, professionals dealing with climate change, and our national leaders.

This book was started during 2020—the first year of COVID-19. The scientific and public debates surrounding COVID-19 have become a major topic at my blog *Climate Etc.*[13] While the climate debate has unfurled over the course of decades, with COVID-19 we are seeing the same complex interplay of scientific research, politics, and risk governance unfurl on the timescale of months. COVID-19 is challenging how we think about uncertainty and risk, both personally and societally. The comparisons and contrasts between climate change and COVID-19 enrich our understanding of both risks.

Judith Curry
Reno, Nevada, USA
August 20, 2022

Notes

1 *An Ecology of Mind*, 2011, http://www.anecologyofmind.com/.
2 António Guterres, "Secretary-General's Remarks at Climate Finance Ministerial Meeting [as Delivered] Secretary-General," United Nations Secretary General (United Nations, October 13, 2018), https://www.un.org/sg/en/content/sg/statement/2018-10-13/secretary-generals-remarks-climate-finance-ministerial-meeting.
3 Judith Curry, "Some Thoughts on Uncertainty: Applying Lessons to the CCSP Synthesis and Assessment Products," Judith Curry's Home Page—Climate

Change (Georgia Institute of Technology, October 21, 2003), https://curry.eas.gatech.edu/climate/index.htm.

4 J. A. Curry et al., "Mixing Politics and Science in Testing the Hypothesis That Greenhouse Warming Is Causing a Global Increase in Hurricane Intensity," *Bulletin of the American Meteorological Society* 87, no. 8 (August 1, 2006): 1025–1038, https://doi.org/10.1175/bams-87-8-1025.

5 "Climate Audit," Climate Audit, accessed August 17, 2022, https://climateaudit.org/.

6 "Climategate," Nature News (Nature Publishing Group), accessed August 17, 2022, https://www.nature.com/collections/synrzkgmlf.

7 Judith Curry, "Curry: On the Credibility of Climate Research," Climate Audit, November 22, 2009, https://climateaudit.org/2009/11/22/curry-on-the-credibility-of-climate-research-2/.

8 Andrew C. Revkin, "A Climate Scientist Who Engages Skeptics" (The New York Times, November 27, 2009), https://archive.nytimes.com/dotearth.blogs.nytimes.com/2009/11/27/a-climate-scientist-on-climate-skeptics/.

9 "Climate Etc.," Climate Etc., accessed August 17, 2022, https://judithcurry.com/.

10 Scott Waldman, "Judith Curry Retires, Citing 'Craziness' of Climate Science," Politico Pro, January 4, 2017, https://subscriber.politicopro.com/article/eenews/1060047798.

11 "CFAN Climate Forecast Applications Network," CFAN Climate Forecast Applications Network, accessed August 17, 2022, http://www.cfanclimate.net/.

12 "About," Climate Etc., accessed August 17, 2022, https://judithcurry.com/about/.

13 "Covid Search Results," Climate Etc., accessed August 17, 2022, https://judithcurry.com/?s=Covid.

Part One

THE CLIMATE CHANGE CHALLENGE

"[M]y most fundamental objective is to urge a change in the perception and evaluation of familiar data."

—Philosopher of science Thomas Kuhn[1]

Change in the Earth's climate and its adverse effects have always been a common concern of humankind. The current challenge of climate change is typically formulated as:

- The Earth's climate is warming.
- A warming climate is dangerous.
- We are causing the warming by emitting carbon dioxide (CO_2) from burning fossil fuels.
- We need to prevent dangerous climate change by rapidly reducing and then eliminating our CO_2 emissions.

In spite of the perceived urgency of the problem and international climate treaties and agreements that were first signed in 1992, global CO_2 emissions continue to increase while targets and deadlines continue to be missed.[2]

Most people feel that climate change is a very serious issue. Depending on your perspective and values, there will be much future loss and damage from either climate change itself, or from the policies designed to prevent climate change. Conflicts surrounding climate change have been exacerbated by oversimplifying both the problem and its solutions.

Acknowledging disagreement is not the same as rejecting climate change as an important problem. In the context of the international treaties and agreements on climate change, both the problem of climate change and its solutions are framed as a global issue. This framing of the central challenge that focuses on reducing global carbon emissions has allowed technical fixes such as geo-engineering and low-carbon energy to take center-stage. This focus has come at the expense of a host of wider visions for social, economic, and political change, particularly at the national and local levels.

Part One describes how the challenge of climate change has evolved in the context of a complex interplay among scientists, the organizations that support research, government-sponsored assessments of climate research, national and international climate policy, politics, and the needs and desires of peoples and nations in a rapidly changing world. Polarization has deepened in a fog of confusion about what we know versus what we do not know and what we cannot know. A populace that is trying to understand climate change is left confused by international and national policies and commitments that do not seem doable or politically feasible.

To assess objectively the risks from climate change and the policies designed to mitigate it, we need to step back from the current debate and broaden our framework for thinking about climate change. Part One provides a framework for clarifying our thinking about these challenges.

Notes

1 Thomas S. Kuhn, "The Structure of Scientific Revolution," in *The Structure of Scientific Revolution* (Chicago, IL: University Of Chicago Press, 1962), x.
2 Jeff Tollefson, "COP26 Climate Summit: A Scientists' Guide to a Momentous Meeting," Nature news (Nature Publishing Group, October 25, 2021), https://www.nature.com/immersive/d41586-021-02815-w/index.html.

Chapter One

INTRODUCTION

"We are drowning in information while starving for wisdom."
—American biologist E. O. Wilson[1]

Within the public domain, there is a widespread narrative of climate catastrophe if we do not urgently reduce or eliminate emissions from burning fossil fuels. Example quotes are provided below from United Nations (UN) officials and national leaders:

"The clock is ticking towards climate catastrophe." (Ban Ki-moon, UN Secretary-General, 2015)[2]
"We face a direct existential threat." (Antonio Guterres, UN Secretary-General, 2018)[3]
"There's one issue that will define the contours of this century more dramatically than any other, and that is the urgent threat of a changing climate." (US President Barack Obama, 2014)[4]
"We are killing our planet. Let's face it, there is no planet B." (Emmanuel Macron, President of France, 2018)[5]

In the 1990s, the world's nations embarked on a path to prevent dangerous human-caused climate change by stabilizing the concentrations of atmospheric greenhouse gases, especially carbon dioxide (CO_2). These efforts were codified by the 1992 United Nations Framework Convention on Climate Change (UNFCCC) treaty.[6]

The Intergovernmental Panel on Climate Change (IPCC)[7] plays a primary role in legitimizing UNFCCC policies. The IPCC prepares periodic assessment reports that are formulated around identifying human influences on climate, adverse environmental and socioeconomic impacts of climate change, and stabilization of CO_2 concentrations in the atmosphere.

How concerned should we be about climate change? The IPCC Assessment Reports do not support the concept of imminent global catastrophe associated with global warming. However, a minority of scientists, some very vocal, believe that catastrophic scenarios are more realistic than the IPCC's *likely* scenarios. There is also a very vocal contingent among journalists and politicians that supports the catastrophe narrative.

At the same time, there are other scientists that do not view climate change to be a serious threat. Many of these adopt the lukewarmer perspective,[8] which expects warming to be on the lower end of the IPCC *likely* range and do not expect the impacts to be alarming or catastrophic. Some politicians and industrialists reject the solutions put forward by the international climate treaties in favor of near-term economic development.

This chapter lays out the contours of the climate change problem: ambiguities surrounding the definition of climate change, what we know with confidence, what we do not know and cannot know, and whether climate change is dangerous.

1.1 What Is "Climate Change"?

"Well, brother, you mustn't be too hard upon me; but, to tell the truth, I didn't remark the elephant." (Ivan Krylov, author of *Krilof and His Fables*, 1815)[9]

A changing climate has been the norm throughout the Earth's 4.6-billion-year history. The Earth's temperature and weather patterns change naturally over timescales ranging from decades to millions of years. Natural variations in the surface climate originate in two ways. Internal climate fluctuations associated with circulations in the atmosphere and ocean produce exchanges of energy, water, and carbon between the atmosphere, oceans, land, and ice. External influences on the climate system include variations in the energy received from the sun and the effects of volcanic eruptions. Human activities influence climate through changing land use and land cover. Humans are also changing atmospheric composition by increasing the emissions of CO_2 and other greenhouse gases and by altering the concentrations of aerosol particles in the atmosphere.

Over the past several decades, the definition of climate change has shifted away from this broader interpretation. Article 1 of the UNFCCC defines climate change as

"a change of climate which is attributed directly or indirectly to *human activity* that alters the composition of the global atmosphere and which is in addition to natural climate variability observed over comparable time periods."[10]

The UNFCCC thus makes a distinction between climate *change*—attributable to human activities altering the atmospheric composition (mainly CO_2) and climate *variability*—attributable to natural causes. This redefinition of "climate change" to refer only to human-caused changes to the atmospheric composition has effectively eliminated natural climate change from the public discussion—the common parlance refers to "climate change," with no mention of natural climate variability. Any change that is observed over the past century is now implicitly assumed to be caused by human emissions to the atmosphere. This assumption leads to connecting every unusual weather or climate event to human-caused climate change from fossil fuel emissions.

The term "climate change" does not just connote the science of human-caused global warming, but also an entire worldview of society. Geographer Mike Hulme identifies *climate reductionism* as a form of analysis and prediction in which the interdependencies that shape human life within the physical world are correlated with climate change.[11] Human-caused climate change is then elevated to the role of the dominant predictor of societal change. Multiple possibilities of the future are effectively closed off as climate predictions assert their determinative influence over food production, health, tourism and recreation, human migration, violent conflict, and so on. Other environmental, economic, and social factors that influence these societal problems then become marginalized.

The ever-expanding narrative of climate change is entraining a range of social values into the proposed solutions. The momentum of the climate change narrative leads to claims that there is a solution to many other societal problems within the climate change cause—an example is social justice in the context of the US Green New Deal.[12] This link acts to energize both causes and leverages the climate change narrative to blame or attack those opposed to the separate cause.

Climate change has thus become a grand narrative in which human-caused climate change has become a dominant cause of societal problems. This perspective was highlighted on the cover of a recent issue of *Time Magazine* with the title *Climate Is Everything*.[13] Everything that goes wrong reinforces the conviction that there is only one thing we can do to prevent societal problems—stop burning fossil fuels. This grand narrative leads us to think that if we solve the problem of burning fossil fuels, then these other problems would also be solved. This belief leads us away from a deeper investigation of the true causes of these other problems. The end result is narrowing of the viewpoints and policy options that we are willing to consider in dealing with complex issues such as public health, water resources, weather disasters, and national security.

The closing statement in Hulme's paper: "And so the future is reduced to climate."[14]

1.2 What We Know with Confidence

"We understand a lot of the physics in its basic form. We don't understand the emergent behavior that results from it." (Climate scientist William Collins)[15]

The foundation of our understanding of the processes of climate variability and change rests on fundamental laws of physics such as Newton's laws of motion, the first and second laws of thermodynamics, ideal gas laws, gravitation, and conservation of mass and energy. These fundamental laws are incorporated into numerous theories of complex processes that contribute to our understanding of climate processes. These include the theories of rotating fluids, boundary layers, and radiative transfer. These theories are widely accepted. The theory of greenhouse warming of the climate system is a metatheory that incorporates many hypotheses and theories about how components of the Earth system work and interact.

Science is a process for understanding how nature works, whereby we explore new ideas to find new representations of the world that explain what is observed. Part of science is to conduct experiments, make observations, do calculations and make predictions. But another part of science asks deep questions about how nature works.

What constitutes evidence in climate science? Scientific evidence is generally regarded to consist of observations and experimental results. In complex natural systems, the epistemic status of observations is not straightforward. There are data homogenization adjustments, model assimilation of observations, retrieval algorithms to interpret voltages measured by satellites, and paleoclimate proxies. As a result, many climate data records are not without controversy and there are ongoing revisions to many data records.

Scientific investigations of the dynamics of the climate system have more in common with systems of biology and economics than with laboratory physics and chemistry, owing to the inherent complexity of the system and the inability to conduct controlled experiments. Complexity is not the same thing as complicated. Complicated systems have many parts but simple chains of causation. Complexity of the climate system arises from the chaotic behavior and nonlinearity of the equations for motions in the atmosphere and ocean, and the feedbacks between subsystems for the atmosphere, oceans, land surface, and glacier ice.

Climate change associated with increasing concentrations of atmospheric CO_2 is a theory in which the basic mechanism is well understood, but whose magnitude is highly uncertain.

Here are the incontrovertible facts about global warming:

- Average global surface temperatures have overall increased since about 1860.
- CO_2 has infrared emission spectra, and thus acts to warm the planet.
- Humans have been adding CO_2 to the atmosphere via emissions from burning fossil fuels.

The above facts are strongly supported by scientific evidence, and there is no significant disagreement in the scientific community on these points. However, these three facts, either individually or collectively, do not tell us much about the most consequential issues associated with climate change:

1. Whether and to what extent CO_2 and other human-caused emissions have dominated over natural climate variability as the cause of the recent warming.
2. How much the climate can be expected to change over the twenty-first century.
3. Whether warming is dangerous.
4. Whether radically reducing CO_2 emissions will improve human well-being in the twenty-first century.

The first two points are in the realm of science, requiring logical arguments, model simulations, and expert judgment to assess "whether" and "how much." The issue of "dangerous" is an issue of societal values, about which science has little to say. Whether reducing CO_2 emissions will improve human well-being is an issue of economics and technology, as well as being contingent on the relative importance of natural climate variability versus human-caused global warming for the twenty-first century.

In the 1970s, an international conference in Stockholm assessed the main scientific problems to be solved before reliable climate forecasting could be possible.[16] The conference identified a number of problems, but focused on two. The first concerned an inability to simulate the amount and character of clouds in the atmosphere. Clouds are important because they govern the balance between solar heating and infrared cooling of the planet, and thereby are an important control on Earth's temperature. The second concerned an inability to forecast the behavior of oceans, which is important for transporting heat, changing atmospheric circulation patterns that influence clouds, and the storage of carbon. The wide differences among climate model simulations of clouds and ocean circulations continue to be primary sources of uncertainties in the current generation of climate models.

A 2011 article in *Nature* interviewed prominent climate scientists for their opinions on "The Real Holes in Climate Science."[17] The article highlighted four topics: (i) the inability of climate models to simulate regional climate change; (ii) wide divergence in the climate model simulations of global rainfall patterns; (iii) the interactions of atmospheric aerosol particles with clouds, which influences how cloud interact with the Earth's solar heating and infrared cooling; and (iv) paleoclimate reconstructions of past climates from tree rings.

At a 2014 Workshop sponsored by the American Physical Society,[18] I stated that I regarded the following to be the most important gaps in current understanding:

- Solar impacts on climate, including indirect effects beyond solar heating
- Multi-decadal and century-scale natural internal variability associated with large-scale ocean circulations
- Mechanisms of vertical heat transfer in the ocean
- Fast thermodynamic feedbacks (water vapor, clouds, atmospheric lapse rate) that determine the climate sensitivity to increases in atmospheric greenhouse gases

From the perspective of 2021 and many more recent investigations, I would add the following:

- Earth's carbon budget and carbon cycle
- Ice sheet dynamics
- Geothermal heat transfer under the oceans and ice sheets

Many of these issues are discussed in Part Two of this book. Other scientists would undoubtedly come up with different issues that they regard as key gaps in understanding climate change. It is noteworthy that when asked, individual scientists tend to cite uncertainties in their own field of expertise, while apparently accepting the consensus views on topics that are further away from their own expertise.[19]

1.3 Is Global Warming Dangerous?

"Who are you gonna believe, me or your lying eyes?" (Comedian Richard Pryor)[20]

The 1992 UN Framework Convention on Climate Change Treaty states as its objective: "stabilization of greenhouse gas concentrations in the atmosphere at a level that would prevent *dangerous* anthropogenic interference with the climate system."[21]

Actions to reduce emissions and otherwise mitigate global warming presuppose that warming is dangerous. However, there is no truly objective determination of the level at which climate change becomes dangerous, or how we should compare this risk with others. How we perceive and evaluate risks and dangers from global warming is a complex issue, which is discussed more completely in Chapter Ten.

Public support or opposition to climate policies (e.g., treaties, regulations, taxes, subsidies) is greatly influenced by perceptions of the risks and dangers posed by global warming. The time horizon for "danger" matters greatly—now, versus 30 years from now, versus the twenty-second century.

1.3.1 The Goldilocks Dilemma

> "What's needed is an approach to these challenges that is just right—not too hot, and not too cold." (Meteorologist and policy analyst William Hooke)[22]

Before addressing a definition of "dangerous" in the context of climate change, we need to confront the "Goldilocks dilemma."

The Goldilocks principle states that something desirable must fall within certain margins, as opposed to reaching extremes.[23] The Goldilocks principle is derived from the children's story *The Three Bears*. Each bear has their own preference of food, beds, and room size. After testing each of the three items, Goldilocks determines that one of them is always too much in one extreme (too hot, too large), one is too much in the opposite extreme (too cold, too small), and one is "just right."

In planetary science, the "Goldilocks zone" is terminology for the band around a sun where temperatures are neither too hot nor too cold for liquid water to exist.[24] However, when it comes to planet Earth, we have adopted a much narrower definition of the Goldilocks zone for climate.[25]

The IPCC and UNFCCC have implicitly adopted the premise that our climate was "just right" prior to human interference, measuring dangerous warming as the temperature change since preindustrial times. Preindustrial is generally regarded as prior to 1750 and the start of the industrial revolution.[26]

So, which climate do we want? Few would choose the preindustrial climate of the eighteenth century, which occurred during the Little Ice Age and was one of the coldest centuries of the last millennia.[27] The Little Ice Age was associated with viciously cold winters in North America, Europe, and China. Mount Vernon, the site of George Washington's home, has posted an article entitled *Washington's Winters*, characterizing winters in the late eighteenth century as "frozen rivers, knee-deep snows, sleet, frigid temperatures, and other winter miseries."[28]

Defining dangerous relative to a baseline during the cold Little Ice Age does not relate well to people's weather preferences. America's domestic migration patterns reflect an appreciation of warm winters, and that the effects of climate change since the 1970s are perceived overall to have been an improvement.[29] Analysis of the change in weather preferences in China from 1971 to 2013 showed that China's weather conditions were perceived to have improved during this period.[30]

Climate change is expected to have impacts that are unequally distributed geographically. Apart from hypothesized regions of worsening climate, areas of the world that currently cannot easily support populations and agriculture (e.g., Northern China, Siberia, northern Canada) may become more desirable in a future warmer climate regime. In short, we can expect both winners and losers with any global or regional climate change. The Goldilocks dilemma asks: dangerous for whom, where and when?

1.3.2 Defining "Dangerous"

"Nothing is so firmly believed as that what we least know." (French Renaissance philosopher Michel de Montaigne)[31]

Despite the 1992 UNFCCC Treaty aimed at preventing dangerous anthropogenic interference with the climate system, the UNFCCC has avoided and then struggled to provide a definition of dangerous. So, what actually constitutes dangerous climate change?

It was not until 2010 that clarification of dangerous was provided by UN international negotiators: "In 2010, governments agreed that emissions need to be reduced so that global temperature increases are limited to below two degrees Celsius."[32] The two degree Celsius (2°C) target is relative to preindustrial temperatures, which presupposes that the warming observed since the mid-nineteenth century is contributing to climate danger, despite the Goldilocks dilemma. The 2°C limit is used politically to motivate the urgency of action to reduce CO_2 emissions.[33] At a recent UN Climate Summit, Secretary-General Ban Ki-moon warned: "Without significant cuts in emissions by all countries, and in key sectors, the window of opportunity to stay within less than two degrees will soon close forever."[34]

The scientific validity of the two-degree target has been questioned. Sustainability scientist Carl Jaeger describes how the two-degree limit has evolved in a somewhat ad hoc and contradictory fashion: policy makers have treated it as a scientific finding, and scientists treat it as a political issue. It has been presented as a threshold separating a domain of safety from one of catastrophe, and as an optimal strategy that balances costs and benefits.[35]

To avoid making value judgments, the IPCC does not define climate change in context of dangerous; the IPCC Assessment Reports refer to "reasons for concern." The 5th Assessment Report (AR5) (2014) articulates five integrative reasons for concern[36]:

- Risk to unique and threatened systems (e.g., coral reefs, tropical mountain glaciers, endangered species)
- Risk of extreme weather events (e.g., heat waves, floods, droughts, wildfires, hurricanes)
- Disparities of impacts and vulnerabilities (e.g., disproportionate harm to developing countries and the poor)
- Aggregate damages (i.e., net global market damages)
- Risks of large-scale discontinuities ("tipping points" resulting in rapid sea-level rise, ocean acidification, and strong amplifiers of warming)

Tipping points—abrupt or nonlinear transition to a different state—become likely to occur once some threshold has been crossed, with regional or global consequences that are largely uncontrollable and beyond our management. In other words, tipping points are points of no return, at least on the century timescale. The IPCC AR5 considered a number of potential tipping points, including ice sheet collapse, collapse of the Atlantic overturning circulation, and carbon release from permafrost thawing. Every single catastrophic scenario considered by the IPCC AR5 has a rating of *very unlikely* or *exceptionally unlikely* and/or has *low confidence*.[37] The only tipping point that the IPCC AR5 considers *likely* in the twenty-first century is disappearance of Arctic sea ice during summer, which reforms each winter even under extreme warming scenarios.

There is no actual large-scale threshold (or tipping point) in the climate that has been clearly linked to 2°C global warming. Global average warming is not the only kind of climate change that is potentially dangerous, and greenhouse gases such as CO_2 are not the only cause of dangerous climate change. Some potential thresholds cannot be meaningfully linked to global temperature change and are more regional in nature; others are sensitive to rates of climate change, and some are most sensitive to spatial gradients of temperature change.

Among the greatest concerns about climate change are its impacts on extreme events such as floods, droughts, wildfires, and hurricanes. However, there is little evidence that the recent warming has worsened such events (see Chapters Seven and Nine for a more thorough discussion). The first half of the twentieth century had more extreme weather than the second half, when human-caused global warming is claimed to have been mainly responsible for observed climate change.[38] The disconnect between historical data for the past 100 years and climate model-based projections of worsening extreme weather events presents a real conundrum regarding the appropriate basis on which to assess risk and make policies when theory and historical data are in such disagreement.

With regard to the perception that severe weather events seem more frequent and more severe over the past decade, there are several factors in play. The first is increasing vulnerability and exposure associated with increasing population and concentration of wealth in coastal and other disaster-prone regions. The second factor is natural climate variability. Many extreme weather events have documented relationships with natural climate variability. In the United States, extreme weather events (e.g., droughts, heat waves, and hurricanes) were significantly worse in the 1930s and 1950s.[39] A recent analysis summarizing many studies finds no evidence to support claims that any part of the overall increase in global economic losses from weather and climate disasters can be attributed to global warming.[40]

1.3.3 The Catastrophe Narrative

"The climate change debate has entered what we might call the Campfire Phase, in which the goal is to tell the scariest story." (Journalist and policy analyst Oren Cass)[41]

In spite of the challenges associated with the Goldilocks dilemma and in providing a meaningful definition of dangerous climate change, there is a widespread narrative within the public domain of near-term climate catastrophe (absent deep emissions cuts). Climate change is now routinely referred to as the "climate crisis." This catastrophe narrative is not supported by mainstream science, but rather propagates via emotive engagement.[42] Climate catastrophe narratives are emboldened by climate reductionism (Section 1.1), whereby climate change is regarded as the dominant cause of current and future societal problems.

There have been many speculations about a climate apocalypse, which is a hypothetical scenario involving the global civilization and potential human extinction as a direct or indirect result of global warming. Under such scenarios, some or all of the Earth may be rendered uninhabitable as a result of extreme temperatures, severe weather events, inability to grow crops, and so on. Several examples are provided here:

- In his 2019 BBC documentary *Climate Change—The Facts*, English broadcaster and natural historian Sir David Attenborough warns that dramatic action needs to be taken against climate change within the next decade to avoid irreversible damage to the natural world and the collapse of human societies.[43]
- Journalist David Wallace-Wells published a book in 2019 entitled *The Uninhabitable Earth*. His article in the *New York Magazine* with the same title has

the subtitle: "Famine, economic collapse, a sun that cooks us: What climate change could wreak—sooner than you think."[44]

• Journalist Jonathan Franzen in 2019 writes in the *New Yorker*: "The climate apocalypse is coming. If you're younger than sixty, you have a good chance of witnessing the radical destabilization of life on earth—massive crop failures, apocalyptic fires, imploding economies, epic flooding, hundreds of millions of refugees fleeing regions made uninhabitable by extreme heat or permanent drought. If you're under thirty, you're all but guaranteed to witness it."[45]

Such apocalyptic statements rightfully draw criticism for leading people to believe that these scenarios are inevitable, rather than implausible. Dystopian climate fiction is the place to explore such scenarios. Alarm about apocalyptic climate scenarios is often referred to as "doomism." Doomism is regarded by climate activists as being as pernicious as denial in terms of leading us down a path of inaction.[46]

That said, there is an important role in risk assessment for articulating plausible worst-case scenarios (see Chapter Nine). Is it possible that something really dangerous could happen to the Earth's climate during the twenty-first century? Yes, it is possible, but natural climate variability (perhaps in conjunction with human-caused climate change) may be a more likely source of undesirable change than human causes alone. Here is the best articulation of potential danger from climate change that I have come across:

> "[N]on-linear response processes of natural ecosystems transmitted through complex cause-effect chains would lead to a sudden upward shift in the level of climate related damages and disasters that finally result in civil unrest in some regions of the world as those societies lost their capacity to deal with the additional climate risk(s)."[47]

1.3.4 Vulnerability to Climate Change

> "Every truth that we may think complete will prove itself untruth at the moment of shipwreck." (Psychiatrist and philosopher Karl Jaspers)[48]

Catastrophes are major large-scale events that produce great and sudden harm. Extreme events such as landfalling major hurricanes, floods, wildfires, heat waves, and droughts can have catastrophic impacts. While such events are not unexpected, their frequency or severity may increase in the future and they are a surprise to the individual locations that are impacted by a specific event. Natural events become catastrophes when the population is large and unprepared, infrastructure is exposed, and humans have tampered with ecosystems that provide a natural safety barrier (e.g., deforestation, draining wetlands, destroying coastal mangroves).

Social vulnerability to extreme weather and climate events is local to regional in scale (see Chapter Thirteen). A recent study quantified the changes in socioeconomic vulnerability, expressed as fatalities over exposed population and economic losses, to climate-related hazards between 1980 and 2016.[49] A clear decreasing trend in both human and economic vulnerability was found, with global average mortality and economic loss rates dropping by 6.5 and nearly five times, respectively, relative to their values in 1980. There was a clear negative correlation between vulnerability to weather/climate hazards and wealth, with the greatest vulnerability at the lowest income levels. That is, the poor suffer the most from weather and climate disasters.

Notes

1 E. O. Wilson, *Consilience—The Unity of Knowledge* (New York: Vintage Books, 1999), 294.
2 Ban Ki-moon, "Secretary General Opening Remarks SG/SM/17396-ENV/ DEV/1614" (Speech, Twenty-first Conference of the Parties to the United Nations Framework Convention on Climate Change, Paris, France, December 7, 2015, https://www.un.org/press/en/2015/sgsm17396.doc.htm).
3 António Guterres, "Secretary-General's Remarks on Climate Change" (Speech, United Nations Headquarters, September 10, 2018, https://www.un.org/sg/ en/content/sg/statement/2018-09-10/secretary-generals-remarks-climate-change-delivered).
4 Barack Obama, "Remarks by the President at United Nations Climate Change Summit" (Speech, United Nations Headquarters, September 23, 2014, https://obamawhitehouse.archives.gov/the-press-office/2014/09/23/remarks-president-un-climate-change-summit).
5 Emmanuel Macron, "Speech of the President of the Republic, Emmanuel Macron. Before the Congress of the United States of America" (Speech, United States Congress, April 25, 2018, https://www.elysee.fr/en/emmanuel-macron/2018/04/25/speech-by-the-president-of-the-republic-emmanuel-macron-at-the-congress-of-the-united-states-of-america).
6 United Nations, "United Nations Framework Convention on Climate Change" (New York: United Nations, General Assembly, 1992), https://unfccc.int/resource/ docs/convkp/conveng.pdf.
7 "Intergovernmental Panel on Climate Change," https://www.ipcc.ch/.
8 Patrick J. Michaels and Chip Knappenberger, *Lukewarming: The New Climate Science that Changes Everything* (Washington, DC: Cato Institute, 2016).
9 Ivan Krylov, "An Inquisitive Man," In *Krilof and His Fables*, ed. W. R. S. Ralston (London: Strahan and Company, 1869), 42–44.

10 United Nations, "United Nations Framework Convention on Climate Change."

11 Mike Hulme, "Reducing the Future to Climate: A Story of Climate Determinism and Reductionism," *Osiris* 26, no. 1 (2011): 245–266, https://doi.org/10.1086/661274.

12 U.S. Congress, House of Representatives, *Recognizing the duty of the Federal Government to create a Green New Deal*, 116th Congress, 2019–2020, H.Res.109, https://www.congress.gov/bill/116th-congress/house-resolution/109/text.

13 *The Story behind TIME's 'Climate Is Everything' Cover, Time* (Time, 2021), https://time.com/5954495/story-behind-climate-is-everything-cover/.

14 Hulme, "Reducing the Future to Climate," 245–266.

15 American Physical Society, "Climate Change Statement Review Workshop—Transcript of Proceedings," January 8, 2014, 36, https://www.aps.org/policy/statements/upload/climate-seminar-transcript.pdf.

16 United Nations, "Report of the United Nations Conference on the Human Environment" (New York: United Nations, 1973), https://www.un.org/ga/search/view_doc.asp?symbol=A/CONF.48/14/REV.1.

17 Quirin Schiermeier, "The Real Holes in Climate Science," *Nature* 463, no. 7279 (2010): 284–287, https://doi.org/10.1038/463284a.

18 American Physical Society, "Climate Change Statement Review Workshop."

19 Quirin Schiermeier, "The Real Holes in Climate Science."

20 *Richard Pryor Live on the Sunset Strip*, 1982.

21 United Nations, "United Nations Framework Convention on Climate Change," 4.

22 William Hooke, "A Goldilocks Approach to Climate Change," Living on the Real World, March 29, 2017, https://www.livingontherealworld.org/a-goldilocks-approach-to-climate-change/.

23 Kenneth E. Boulding, *Evolutionary Economics* (Beverly Hills, CA: Sage, 1981).

24 Brian Dunbar, "The Goldilocks Zone," NASA (NASA, October 7, 2003), https://www.nasa.gov/vision/earth/livingthings/microbes_goldilocks.html.

25 Peter J. Webster, "The Role of Hydrological Processes in Ocean-Atmosphere Interactions," *Reviews of Geophysics* 32, no. 4 (1994): 427, https://doi.org/10.1029/94rg01873.

26 Andrew P. Schurer et al., "Importance of the Pre-Industrial Baseline for Likelihood of Exceeding Paris Goals," *Nature Climate Change* 7, no. 8 (2017): 563–567, https://doi.org/10.1038/nclimate3345.

27 Philipp Blom, *Nature's Mutiny: How the Little Ice Age of the Long Seventeenth Century Transformed the West and Shaped the Present* (New York, NY: Liveright Publishing Corporation, a division of W.W. Norton & Company, 2020).

28 "Washington's Winters," George Washington's Mount Vernon, accessed September 24, 2021, https://www.mountvernon.org/george-washington/so-hard-a-winter/.

29 Patrick J. Egan and Megan Mullin, "Recent Improvement and Projected Worsening of Weather in the United States," *Nature* 532, no. 7599 (2016): 357–360, https://doi.org/10.1038/nature17441.

30 Zihang Fang et al., "Will Climate Change Make Chinese People More Comfortable? A Scenario Analysis Based on the Weather Preference Index," *Environmental Research Letters* 15, no. 8 (July 2020): 084028, https://doi.org/10.1088/1748-9326/ab9965.

31 John Bartlett, *Familar Quotations: A Collection of Passages, Phrases, and Proverbs Traced to Their Sources in Ancient and Modern Literature* (Boston, MA: Little, Brown and Company, 1911), 775.

32 "The United Nations Framework Convention on Climate Change (UNFCCC)," environnet, July 13, 2016, http://www.environnet.in.th/en/archives/1678.

33 "Two Degrees: The History of Climate Change's Speed Limit," Carbon Brief, December 8, 2014, https://www.carbonbrief.org/two-degrees-the-history-of-climate-changes-speed-limit/.

34 "UN Climate Summit: Ban Ki-Moon Final Summary," Unfccc.int, September 25, 2014, https://unfccc.int/news/un-climate-summit-ban-ki-moon-final-summary.

35 Carlo C. Jaeger and Julia Jaeger, "Three Views of Two Degrees," *Regional Environmental Change* 11, no. S1 (August 2010): 15–26, https://doi.org/10.1007/s10113-010-0190-9.

36 Rajendra K. Pachauri and Leo Meyer, *Climate Change 2014: Synthesis Report* (Geneva, CH: IPCC, 2015).

37 Matthew Collins, Reto Knutti et al., "Long-Term Climate Change: Projections, Commitments and Irreversibility Pages 1029 to 1076," *Climate Change 2013—The Physical Science Basis*, n.d., 1029–1136, https://doi.org/10.1017/cbo9781107415324.024.

38 Michael J. Kelly, "Trends in Extreme Weather Events since 1900—an Enduring Conundrum for Wise Policy Advice," *Journal of Geography & Natural Disasters* 06, no. 01 (2016), https://doi.org/10.4172/2167-0587.1000155.

39 Judith Curry, "Statement to the Committee on Environment and Public Works of the United States Senate," U.S. Congress, Senate, Committee on Environment and Public Works, *Hearing on Review of the President's Climate Action Plan*, 113th Cong., 2nd sess., January 14, 2014, https://judithcurry.com/wp-content/uploads/2014/01/curry-senatetestimony-2014-final.pdf.

40 Roger Pielke, "Economic 'Normalisation' of Disaster Losses 1998–2020: A Literature Review and Assessment," *Environmental Hazards* 20, no. 2 (May 2020): 93–111, https://doi.org/10.1080/17477891.2020.1800440.

41 Oren Cass (@oren_cass), "The Climate Change Debate Has Entered What We Might Call the 'Campfire Phase', in Which the Goal Is to Tell the Scariest Story. /1," Twitter, July 12, 2017, https://twitter.com/oren_cass/status/885164438436925440.

42 "The Catastrophe Narrative," Climate etc., November 15, 2018, https://judithcurry.com/2018/11/14/the-catastrophe-narrative/.

43 *Climate Change—The Facts, BBC IPlayer* (BBC, 2019), https://www.bbc.co.uk/iplayer/episode/m00049bl/climate-change-the-facts.

44 David Wallace-Wells, "When Will the Planet Be Too Hot for Humans? Much, Much Sooner than You Imagine," Intelligencer, July 10, 2017, https://nymag.com/intelligencer/2017/07/climate-change-earth-too-hot-for-humans.html.

45 Jonathan Franzen and Rachel Riederer, "What If We Stopped Pretending the Climate Apocalypse Can Be Stopped?," *The New Yorker*, September 8, 2019, https://www.newyorker.com/culture/cultural-comment/what-if-we-stopped-pretending.

46 Jonathan Watts, "Climatologist Michael E. Mann: 'Good People Fall Victim to Doomism. I Do Too Sometimes'," *The Guardian* (Guardian News and Media, February 27, 2021), https://www.theguardian.com/environment/2021/feb/27/climatologist-michael-e-mann-doomism-climate-crisis-interview.

47 M. Obersteiner et al., "Managing Climate Risk," Welcome to IIASA PURE (IR-01-051, December 1, 2001), http://pure.iiasa.ac.at/id/eprint/6471/.

48 Karl Jaspers et al., *Tragedy Is Not Enough* (London: Gollancz, 1953).

49 Giuseppe Formetta and Luc Feyen, "Empirical Evidence of Declining Global Vulnerability to Climate-Related Hazards," *Global Environmental Change* 57 (2019): 101920, https://doi.org/10.1016/j.gloenvcha.2019.05.004.

Chapter Two

CONSENSUS, OR NOT?

"I know that most men, including those at ease with problems of the greatest complexity, can seldom accept even the simplest and most obvious truth, if it would oblige them to admit the falsity of conclusions which they have delighted in explaining to colleagues, which they have proudly taught to others, and which they have woven, thread by thread, into their lives."

—Russian writer Leo Tolstoy[1]

While the public may understand little about climate science, nearly everyone has been exposed to the statement that there is a consensus among scientists regarding dangerous climate change. This chapter explores the history and consequences of the scientific consensus building activities undertaken by the Intergovernmental Panel on Climate Change (IPCC).

For genuinely well-established scientific theories, the concept of consensus is irrelevant. For example, there is no point in discussing a consensus that the Earth orbits the sun, or that the hydrogen molecule has less mass than the nitrogen molecule. While a consensus may arise surrounding a specific scientific hypothesis or theory, the existence of a consensus is not itself the evidence.[2]

There is a key difference between a "scientific consensus" and a "consensus of scientists."[3] A scientific consensus is a relatively stable paradigm that structures and organizes scientific knowledge. By contrast, a consensus of scientists represents a deliberate expression of collective judgment by a scientific institution or a group of scientists, often at the official request of a government or other organization.[4]

Under the auspices of the IPCC, the international climate community has worked for the past 30 years to establish a scientific consensus on human-caused climate change. The IPCC has codified consensus seeking into its assessment procedures: "In taking decisions, drawing conclusions, and adopting reports, the IPCC Plenary and Working Groups shall use all best endeavours to reach consensus." The IPCC consensus has been described as a "manufactured consensus" (or a consensus of scientists), arising from an intentional consensus building process.[5]

Among the best indicators to nonexperts about climate change is the existence of a consensus among experts. Messaging on the climate consensus went viral with this 2013 tweet from US President Obama:

"Ninety-seven percent of scientists agree: #climatechange is real, man-made and dangerous."[6]

President Obama's tweet linked to a paper by Cook et al. that analyzed the abstracts of almost 12,000 climate-related papers.[7] Their analysis showed that 97.1 percent of these papers either supported or assumed that humans are causing climate change. While this paper and its methodology were widely criticized, the important point is not whether the number is 97 percent or some other number. The key scientific issue is not *whether* anthropogenic greenhouse gases have caused any increase in global temperature—the issue is *how much* global warming has been caused by humans, about which there is disagreement among scientists.

"Consensus entrepreneurs"[8] include politicians, social scientists, and activists who defend and extend the climate consensus for political purposes. Consensus entrepreneurs extend the scope of the climate consensus well beyond its scientific basis and often without regard for the actual substance of the scientific consensus. While notions of dangerous are outside the domain of science, the concept of dangerous is typically included in public statements of the IPCC consensus. President Obama's tweet about the 97 percent paper was an example of consensus entrepreneurship, with an unsupported extension of the consensus. The consensus in the Cook et al. paper referred to abstracts of published papers, not to scientists agreeing about climate change; nor did the paper refer to any agreement about "dangerous" climate change.

The utility of consensus in the public debate on climate change is described in a *Scientific American* article that interviewed John Cook, lead author of the 97 percent paper:

"A part of what we were doing was closing that consensus gap, and the consensus gap is delaying climate action. We wanted it to have a tangible impact."[9]

Once the climate consensus evolved into a consensus with flexible and ever-expanding objects, it became a way of exerting power over the public discourse on climate change. The IPCC consensus is used as an appeal to authority in the representation of scientific results as the basis for urgent policy making:

"If we feel that a policy question deserves to be informed by scientific knowledge, then we have no choice but to ask, what is the consensus of experts on this matter? [...] If there is a consensus of experts—as there is today over the reality of anthropogenic climate change—then we have a case for moving forward with relevant action."[10]

This statement by historian Naomi Oreskes reflects "speaking truth to power" as a strategy for science to guide policy making. The key question is whether existing scientific knowledge is certain enough to compel action. Oreskes argues that we should not expect logically indisputable proof, but rather a robust consensus of experts.

Social scientist Jeroen Van der Sluijs argues that the IPCC has adopted a "speaking consensus to power" approach that sees uncertainty and dissent as problematic and attempts to mediate these into a consensus.[11] The speaking consensus to power strategy acknowledges that available knowledge is inconclusive and uses consensus as a proxy for truth through a negotiated interpretation of the inconclusive body of scientific evidence. The consensus to power strategy reflects a specific vision of how politics deals with scientific uncertainties.

2.1 The Problem of Overconfidence

"We are all confident idiots." (Social psychologist David Dunning)[12]

In a 2007 speech before the UN Commission on Sustainable Development following publication of the IPCC 4th Assessment Report (AR4), Gro Harlem Brundtland, the UN Secretary-General's Special Envoy on Climate Change, said:

> "So what is it that is new today? What is new is that *doubt has been eliminated*. The report of the International Panel on Climate Change is clear [...] It is irresponsible, reckless, and deeply immoral to question the seriousness of the situation. The time for diagnosis is over. Now it is time to act."[13]

Apart from political motivations, attempts to eliminate doubt about human-caused global warming are directly related to overconfidence in the relevant scientific findings. Specifically, with regards to the IPCC, the consensus building process and emphasis on expert judgment produces a petri dish that grows overconfidence.[14]

Climate assessments inflate the confidence of their conclusions through defining confidence levels in a manner that defies common understanding. The words used to describe "High" confidence include "moderate evidence" and "medium consensus," which are more descriptive of the common understanding of "Medium" confidence. The words used to describe "Medium" confidence include: "a few sources, limited consistency, models incomplete, methods emerging [...] competing schools of thought," that are more descriptive of the common understanding of "Low" confidence.[15] Such misleading terminology contributes to overconfidence in the conclusions and reflects acceptance of lower standards of evidence on what is believed to be true.

An article on *How the Experts Messed Up on Covid* provides some important insights regarding overconfidence of experts.[16] A true expert knows when to be uncertain, and is aware of the provisional nature of our knowledge. Trustworthiness does not just come from being right, but from communicating the limits of the evidence and regularly updating one's view in light of new data and analysis. Overconfidence from experts, coupled with a willingness to denigrate those who publicly dissent, can make it more difficult to change course and diminishes trust in experts.

2.2 Why Scientists Disagree

"Science is always sold as facts, and it's not, it's process. And that process is mainly arguing." (Climate scientist Tamsin Edwards)[17]

The use of expert judgment in creating consensus statements for the IPCC reports raises several important questions and concerns. Which experts and how many? By what method is the expert judgment formulated? What biases might enter into the expert judgment? Expert judgment encompasses a wide variety of techniques, ranging from a single undocumented opinion, to preference surveys, to formal elicitation with external validation.[18]

Relative to the broad space of disagreement about aspects of climate change, there is apparently little disagreement *within* the IPCC. This general agreement arises from the IPCC's consensus seeking strategy and the individual experts selected to serve as lead authors.

As an example of disagreement among a large number of experts, consider projections of twenty-first century sea level rise for the high emissions scenario (RCP8.5). The IPCC AR5 (2013) made the following assessment:

"Global mean sea level rise for 2081–2100 will *likely* be in the range of 0.45–0.82 meters for RCP8.5. (*medium confidence*)"[19]

At around the same time as the AR5, Horton et al.[20] report on an expert elicitation of 90 experts actively publishing on sea level, regarding their expectations for sea level change by the year 2100. For the high warming scenario of 4.5°C by

2100 (equivalent to warming driven by RCP8.5), the median values of the *likely* range as reported by each polled expert are 0.7–1.2 meters by 2100. However, 5 of the 90 respondents cited values exceeding three meters, with the highest value exceeding six meters.

Why do experts disagree about climate change and its impacts? How is the IPCC's *medium confidence* in a narrow range of sea level rise projections justified in the face of such disagreement among experts? Each scientist has a vision of reality that is the best they have found so far, and there may be substantial disagreement among individual scientists. Disagreement among experts becomes a political issue for a policy-relevant issue such as climate change, when such disagreement is perceived as diluting the will to act.

The most fundamental source of disagreement regarding the theory of human-caused climate change is the importance of natural climate variability. Why do climate scientists disagree on the relative importance of natural versus human-caused climate change? The historical data is sparse and inadequate, particularly in the oceans. There is disagreement about the value of different classes of evidence, notably the value of global climate model simulations and paleoclimate reconstructions from geologic data. There is also disagreement about the appropriate logical framework for linking and assessing the evidence. And finally, there is little acknowledgment that some climate processes are poorly understood or even unknown.

Disagreement among climate scientists was explored in a 2014 Workshop organized by the American Physical Society to consider its statement on climate change.[21] A committee of eminent physicists, each with no particular expertise in climate science, selected three climate scientists that support the IPCC consensus findings plus three climate scientists who have challenged aspects of it (including myself). We were each asked to address the same list of questions with reference to the IPCC AR5.

While each of the six scientists agreed on most of the primary scientific evidence, each had a unique perspective on how to reason about the evidence, what conclusions could be drawn and with what level of certainty. The format effectively required circumspection, justification, and candid admissions of lack of knowledge. The transcript of this Workshop is a remarkable document—it provides, in my opinion, the most accurate portrayal of the scientific debates surrounding climate change, making clear what the real scientific disputes are about.[22] From reading the full transcript of the Workshop, Australian policy analyst and journalist Rupert Darwall concluded that climate scientists are much less certain than they tell the public.[23]

2.3 Biases Caused by a Consensus Building Process

"Like a magnetic field that pulls iron filings into alignment, a powerful cultural belief is aligning multiple sources of scientific bias in the same direction." (Policy scientist Daniel Sarewitz)[24]

Consensus is viewed as a proxy for truth in many discussions of science. A consensus formed by the independent and free deliberations of many is a strong indicator of truth. However, a scientific argument can evolve prematurely into a ruling theory if cultural forces are sufficiently strong and aligned in the same direction.[25] Premature theories enforced by an explicit consensus building process can harm scientific progress because of the questions that do not get asked and the investigations that are not undertaken.

If the objective of scientific research is to obtain truth and minimize error, how might a consensus seeking process introduce bias into the science and increase the chances for error? "Confirmation bias" is a well-known psychological principle that connotes the seeking or interpretation of evidence in ways that are partial to existing beliefs, expectations, or an existing hypothesis. Confirmation bias usually refers to unwitting selectivity in the acquisition and interpretation of evidence. Philosopher Michael Kelly's "invisible hand" process tends to bestow a certain competitive advantage to our prior beliefs with respect to confirmation and disconfirmation.[26]

Confirmation bias can become even stronger when people confront questions that trigger moral emotions and concerns about group identity. People's beliefs become stronger and more extreme when they're surrounded by like-minded colleagues. They come to assume that their opinions are not only the norm but also the truth—creating what social psychologist Jonathan Haidt calls a "tribal-moral community," with its own sacred values about what's worth studying and what's taboo.[27] Such biases can lead to widely accepted claims that reflect the scientific community's blind spots more than they reflect justified scientific conclusions.[28]

Communication theorist Jean Goodwin has examined the manufacture of consensus by the IPCC. She asserts that there is no mechanism from within science for establishing that a scientific consensus exists. She contends that the IPCC's consensus claim is primarily aimed at non-scientists, and constitutes an appeal to authority. She argues that once the consensus claim was made, scientists involved in the ongoing IPCC process had reasons not just to consider the scientific evidence, but also to consider the possible effect of their statements on their ability to defend the consensus claim.[29]

Social psychologist Lee Jussim points to another relevant factor, describing how the academic and journalism ecosystems can reach conclusions that are detached from the evidence.[30] This is possible because of selectivity in reporting and describing results. Data laundering is the process by which data with high uncertainty is transformed to justify clear and compelling conclusions.[31] The path from dissertation to journal article to press release is often one of data laundering and idea laundering that produces more exciting outcomes.

The phenomenon described by Jussim is evident in the IPCC assessment reports. An analysis of the IPCC Synthesis Summaries showed that linguistic devices amplified the strength of statements, as part of transferring messages effectively from the scientific context to a policy-maker audience.[32] The style and tone of the IPCC Summaries were found to contribute to important absences and imbalances in emphasis. Alarming media reports are based upon the less-alarming Synthesis Summary and Summary for Policy Makers, which have oversimplified the complex findings in the main text of the Report. In short, what the public consumes as scientific information about climate change is carefully crafted and laundered "spin."[33]

Economist Richard Tol has been involved in the IPCC Assessment Reports since 1994 and served as a Convening Lead Author on the IPCC AR5 Working Group III (WGIII) on Mitigation of Climate Change. In an interview with UK journalist Roger Harrabin, Tol made the following statement:

"I think that the Fifth Assessment Report in the initial drafts of the Summary for Policymakers that actually came through I think very loud is that the new insight […] is that many of the more dramatic impacts of climate change are really symptoms of mismanagement and poverty and can be controlled if we had better governance and more development."[34]

Tol describes how the IPCC shifted its Summary for Policymakers away from this conclusion and towards the traditional doom and gloom message that would justify greenhouse gas emissions reduction. Tol withdrew his name from the Summary for Policy Makers.[35]

What are the implications of these concerns for the IPCC's consensus on human-caused climate change? Cognitive biases in the context of an institutionalized consensus-building process have arguably resulted in the consensus becoming increasingly confirmed in a self-reinforcing way. The IPCC consensus has become canonized through a political process, bypassing the long and complex validation process for such a complex topic as to whether the conclusions are actually true. An extended group of scientists derive their confidence in the consensus in a second-hand manner from the institutional authority of the IPCC and the emphatic way in which the consensus is portrayed.

2.4 Heresy, Doubt, and Denial

"We can never be sure that the opinion we are endeavoring to stifle is a false opinion; and if we were sure, stifling it would be an evil still. [...] All silencing of discussion is an assumption of infallibility." —(Nineteenth-century philosopher John Stuart Mill)[36]

Consensus is a strong claim, and it opens up a wide space for arguments. By representing their work as a consensus, the IPCC essentially delegitimized the objections of those commonly labeled as skeptics or deniers, and committed themselves to an ongoing, decades-long process of either responding to them or attempting to marginalize them.

Consensus entrepreneurs have expended considerable efforts in demarcating those qualified to contribute to the climate change consensus from those who are not.[37] Their efforts have characterized skeptics as few in number,[38] extreme,[39] and scientifically and morally suspect.[40] Consensus entrepreneurship actively works to dismiss any skepticism about climate change as cranks or politically motivated deniers. Further, consensus entrepreneurship encourages illegitimate attacks on scientists whose views depart from the approved narrative in even the smallest detail.[41]

Numerous theories to explain climate denial have been put forward, including efforts to forestall action on climate change for venal motives.[42,43,44] However, complex theories of climate denial are not needed. Many scientists and members of the public simply find the claim for dangerous human-caused climate change to be inadequately justified at the present time. Such sentiments are expressed in three recently published books:

- Steven Koonin's *Unsettled: What Climate Science Tells Us. What it Doesn't, and Why it Matters*;[45]
- Bjorn Lomborg's *False Alarm: How Climate Panic Costs is Trillions, Hurts the Poor, and Fails to Fix the Planet*;[46] and
- Michael Schellenberger's *Apocalypse Never: Why Environmental Alarmism Hurts Us All*.[47]

The climate change issue has been complicated by concerns about how science and expert judgment are being mobilized in debates that are essentially political. In other words, some people are suspicious about how the different interests and values of public actors on the climate change issue align with their political goals.[48]

2.4.1 Scientific Skepticism

"The very best thing that you may do in your life is create a speck of intense irritation for someone whose views you vigorously dispute, around which a pearl of new intelligence may then accrete." (Professor of International Affairs Henry Farrell, in a review of *The Enigma of Reason* by Mercier and Sperber)[49]

Science is a process that seeks universally valid knowledge as a public good, which requires disinterestedness, skepticism, and originality. Science works just fine when there is more than one hypothesis to explain something. In fact, disagreement spurs scientific progress through creative tension and efforts to resolve the disagreement.

Minority perspectives have an important and respected role to play in advancing science, as a means for testing ideas, and in pushing the knowledge frontier forward. In context of the Mertonian norms of science, skepticism means that scientific claims must be exposed to critical scrutiny before being accepted.[50] Scientific skepticism is motivated by curiosity and reflects dissatisfaction with existing knowledge and concerns about the critical inferences. Scientific skepticism actively challenges the notion that we have completed the knowledge development process and have thoroughly vetted the evidence. Scientific skepticism is about the process, not the conclusions.

When philosophers speak of skepticism being a foundation of science, they are not referring to unbridled cynicism or prejudicial doubt. Scientific skepticism is about protecting the integrity of the scientific process. By contrast, social skepticism is driven by dislike of a particular conclusion rather than concern over the methods used to reach the conclusion.[51]

In the context of climate science, there are three categories of skeptics:

1. Research scientists challenging existing hypotheses and working to extend knowledge frontiers
2. Watchdog auditors, whose main concern is accountability, quality control and transparency of the science
3. Merchants of doubt, who distort and magnify scientific uncertainties as an excuse for inaction (social skeptics)

The populist notion that all climate skeptics are being paid by big oil or are right-wing ideologues (Category 3) is simply not correct. Category 1 includes a broad range of scientists, from Nobel Laureates to cranks. Category 2 is sociologically the most interesting—science auditing is a new movement that has been enabled by the internet and the blogosphere.

The climate auditor movement was initiated by Steve McIntyre, a semi-retired Canadian mining executive.[52] In 2002, McIntyre became suspicious of the so-called climate hockey stick—the paleoclimate reconstruction of temperatures for the past 1,000 years based on the research of Mann et al.[53,54] that was featured prominently in the IPCC Third Assessment Report (TAR) (2001).[55] McIntyre stated: "In financial circles, we talk about a hockey stick curve when some investor presents you with a nice, steep curve in the hope of palming something off on you."[56] From 2003 to 2005, Steve McIntyre and environmental economist Ross McKitrick published three high-profile papers that were critical of the Mann et al. papers.[57,58,59]

McIntyre started the blog *Climate Audit* (climateaudit.org) in 2005 so that he could defend himself against claims being made at the blog *RealClimate* (realclimate.org) by Mann and others with regard to his critique of the hockey stick. McIntyre's auditing became very popular not only with climate skeptics, but also with the progressive open-source community, and the number of such blogs grew. Internet voting by the public awarded *Climate Audit* the 2007 Weblog Best Science Blog award.[60]

With their focus on data quality and statistical analysis methods, it is very difficult to categorize the climate auditors as anti-science. A 2013 article published in *The Guardian* asked the question "Are climate skeptics the true champions of the scientific method?"[61] The climate audit movement initiated a new era of extended and unforgiving online peer review of scientific publications on climate change.

A recent example of blogospheric auditing occurred on my blog, *Climate Etc.* UK financier and independent climate scientist Nic Lewis audited a paper by Resplandy et al. published in *Nature* on the topic of ocean heat content.[62,63] This paper presented the startling and alarming conclusion that the oceans were absorbing much more heat than previously thought, and hence the Earth's climate was more sensitive to fossil fuel emissions. There was extensive coverage of this paper in the mainstream media. Lewis identified problems with the statistical analysis of trends in the paper, and also problems with the uncertainty analysis. Lewis made every effort to engage with the authors prior to publishing his critique. Lewis' critique that was published on my blog resulted in a correction to the paper being submitted by the authors. Ten months later, *Nature* retracted the original paper. The auditing of Resplandy et al. by Nic Lewis provided much needed probity to the paper, which was missed in the peer review process and the publicity following its publication.[64]

2.4.2 *Climate Heretics*

"One of the great achievements of the Enlightenment—the liberation of historical and scientific enquiry from dogma—is quietly being reversed." (Philosopher Edward Skidelsky)[65]

Given the important role of skepticism in science, how did skepticism about climate change come to be an accusation, with some scientific researchers in academia being branded as deniers,[66] heretics,[67] misinformers,[68] and anti-science?[69]

Virtually, all academic climate scientists are within the so-called 97 percent consensus regarding the existence of a human impact on warming of the Earth's climate. Climate heresy seems to be associated at least as much with the sociological aspects of the public debate on climate change as with the actual content of climate science.

Which scientists are ostracized and labeled as heretics or deniers? Independent thinkers, who are not supportive of the IPCC consensus, are suspect. Any criticism of the IPCC (its conclusions, its process, its leadership) can lead to ostracism. Failure to advocate for CO_2 mitigation policies leads to suspicion. The most reliable way to get labeled as a denier is to associate in any way with so-called enemies of the climate consensus and their preferred policies—oil companies, conservative think tanks, climate auditors such as Steve McIntyre, or even the wrong political party.[70]

Academic scientists appearing on activists' lists of climate skeptics, misinformers, and deniers[71,72] include Roger Pielke Jr.,[73] Roger Pielke Sr.,[74] Richard Lindzen,[75] Richard Muller,[76] and John Christy.[77] Each of these scientists has published extensively and has an impressive curriculum vitae, but among other "transgressions" has been invited by Republicans to testify before the US Congress.

The label neo-skeptic refers to a scientist who comments publicly about concerns that current climate policies may not have the desired impact on changing the climate and human welfare for the better and are not commensurate with the uncertainties in climate science and societal impacts.[78] Economist Bjorn Lomborg is the archetypal neo-skeptic.[79] Neo-skepticism is regarded by activists as something that needs to be combatted because it threatens the global imperative of carbon mitigation policies at any cost.

"Denial entrepreneurship" has arguably replaced "consensus entrepreneurship" among climate activists. Historian Naomi Oreskes published an article in *The Guardian* entitled "There is a new form of climate denialism to look out for—so do not celebrate yet."[80] Summary quote:

"There is also a new, strange form of denial that has appeared on the landscape of late, one that says that renewable sources can't meet our energy needs. Oddly, some of these voices include climate scientists, who insist that we must now turn to wholesale expansion of nuclear power."[81]

The particular denialist scientists referred to are James Hansen,[82] Tom Wigley,[83] Ken Caldeira,[84] and Kerry Emanuel.[85] Each of these individuals is a genuinely distinguished scientist. James Hansen's 1988 US Congressional testimony on climate change was instrumental in raising broad awareness of global warming.[86] Hansen is one of the world's leading advocates of action to avoid dangerous climate change and has been arrested multiple times in protest demonstrations.[87]

Individuals and governments can legitimately disagree about the matters of policy that concern the neo-skeptics and the broader aspects of climate politics. Therefore, the policy discussion should not be about believing or denying "The Science" that supposedly supports just a narrow range of policies. Nevertheless, researchers that question the favored climate change policies are referred to as deniers, even though they may generally accept consensus climate science.

2.4.3 Challenging the Consensus on COVID-19 Origins

"The 'party of science' had turned the spirit of inquiry into a religious dogma handed down by credentialed experts." (Journalist Rebecca Weisser)[88]

The early consensus on COVID-19 origins provides some interesting perspectives on the manufacture of a consensus by scientists and its political enforcement. The COVID-19 virus first appeared in Wuhan, China, where there is a laboratory that conducts research on bat coronaviruses. From the beginning, the possibility that this virus accidentally escaped from the lab was dismissed quite forcefully by prominent virologists.

The consensus that COVID-19 had an entirely natural origin was established by two op-eds in early 2020—*The Lancet* in February and *Nature Medicine* in March.[89,90] *The Lancet* op-ed stated, "We stand together to strongly condemn conspiracy theories suggesting that COVID-19 does not have a natural origin."[91] The pronouncements in these op-eds effectively shut down inquiry. The so-called fact checkers of *PolitiFact* used the preemptive declaration of scientific consensus in these op-eds to shut down any discussion of the lab leak hypothesis.[92] Articles in the mainstream press repeatedly stated that a consensus of experts had ruled lab escape to be out of the question or extremely unlikely.

The enormous gap between the actual state of knowledge in early 2020 and the confidence displayed in the two op-eds should have been obvious to anyone in the field of virology, or for that matter anyone with critical faculties. There were scientists from adjacent fields who said as much.

In May 2021, science reporter Nicholas Wade published a lengthy article in the *Bulletin of Atomic Scientists* that identified conflict of interests in the scientists writing *the Lancet* letter in hiding any links with the Wuhan lab.[93] Publication of Wade's article triggered a cascade of defections from scientists. Matthew Crawford describes the defections as "not simply from a consensus that no longer holds, but from a fake consensus that is no longer enforceable."[94] *Politifact* subsequently withdrew its Wuhan-Lab theory "fact check."[95]

What is concerning about this episode is not so much that a consensus has been overturned, but that a fake consensus was so easily enforced for more than a year. Scientists who understood that there was a great deal of uncertainty surrounding the origins of the virus did not speak up. Wade notes that in today's universities, challenging the consensus can be very costly. Careers can be destroyed for stepping out of line. Any virologist who challenges the community's declared views risked being labeled as a heretic, being "canceled" on social media, and having their next grant application turned down by the panel of fellow virologists that advises the government grant distribution agency.

2.5 Rethinking Consensus Messaging

> "'[S]cientific consensus,' when used as a rhetorical bludgeon, predictably excites reciprocally contemptuous and recriminatory responses by those who are being beaten about the head and neck with it." (Legal scholar Dan Kahan)[96]

Apart from concerns about the content of the climate change consensus and the role of "speaking consensus to power" in decision-making, the consensus approach used by the IPCC has received a number of criticisms from IPCC insiders. IPCC authors Michael Oppenheimer, Brian O'Neill, and Shardul Agrawala warn of the need to guard against overconfidence and argue that the IPCC consensus emphasizes expected outcomes, whereas it is equally important that policy makers understand the more extreme possibilities that consensus may exclude or downplay.[97] IPCC authors Arnulf Grübler and Nebojsa Nakicenovic opine that there is a danger that the consensus position might lead to a dismissal of uncertainty in favor of spuriously constructed expert opinion.[98]

A significant indicator that winds are shifting on the role of the climate consensus is the recent book *Discerning Experts: The Practices of Scientific Assessment for Environmental Policy*.[99] The authors include IPCC lead author Michael Oppenheimer as well as historian Naomi Oreskes. Oreskes has been a staunch defender of the IPCC consensus and consensus-seeking process. Surprisingly, this book makes a case *against* consensus seeking in climate change assessments.

The authors are concerned that scientists have been underestimating the rate of climate change. The book argues that scientists may consciously or unconsciously pull back from reporting on extreme outcomes.[100]

In defending scientific perspectives that say climate change "is worse than we thought," Oppenheimer et al. have opened a welcome can of worms with arguments against consensus seeking—these arguments are exactly in line with what climate science skeptics have been arguing for decades. Since publication of the IPCC AR5 in 2013, it has been clear that the conclusions of the IPCC Assessment Reports are too tame for some scientists and activists that are more alarmed about global warming than the IPCC consensus position. In fact, quoting the IPCC Reports has become a favored strategy of people who are labeled as skeptics or deniers.

The IPCC consensus building process arguably played a useful role in the early synthesis of the scientific knowledge on climate change. However, the ongoing process to negotiate a scientific consensus on uncertain science has had the unintended consequence of oversimplifying both the problem and its solution, introducing biases into the both the science and related decision-making processes.[101] Consensus seeking and enforcement by consensus entrepreneurs trivialize and politicize climate science and the public debate on climate change.

Scientific assessments do not need to be consensual to be authoritative. Speaking consensus to power acts to conceal uncertainties, ambiguities, dissent, and ignorance behind a scientific consensus. Greater openness about scientific uncertainties and ignorance, plus more transparency about dissent and disagreement, would provide policymakers with a more complete picture of climate science and its limitations.[102]

Notes

1 Leo Tolstoy, *What Is Art?* (Indianapolis, IN: Bobbs-Merrill, 1981).
2 J. A. Curry and P. J. Webster, "Climate Change: No Consensus on Consensus," *CAB Reviews: Perspectives in Agriculture, Veterinary Science, Nutrition and Natural Resources* 8, no. 001 (January 2013), https://doi.org/10.1079/pavsnnr20138001.
3 Stephen P. Turner, *The Politics of Expertise* (New York, NY: Taylor & Francis, 2013).
4 M Anthony Mills, "Manufacturing Consensus," The New Atlantis, 2021, https://www.thenewatlantis.com/publications/manufacturing-consensus.
5 Jean Goodwin, "The Authority of the IPCC First Assessment Report and the Manufacture of Consensus," (paper presented at the National Communication Association convention, Chicago, IL, November, 2009), https://lib.dr.iastate.edu/cgi/viewcontent.cgi?article=1004&context=engl_conf.
6 Barack Obama (@BarackObama), "Ninety-seven percent of scientists agree: #climate change is real, man-made and dangerous.," Twitter, May 16, 2013, https://twitter.com/BarackObama/status/335089477296988160?s=20.

7 John Cook et al., "Quantifying the Consensus on Anthropogenic Global Warming in the Scientific Literature," *Environmental Research Letters* 8, no. 2 (2013): 024024, https://doi.org/10.1088/1748-9326/8/2/024024.

8 Mike Hulme, "Scientists speaking with one voice—panacea or pathology?" (presented at Circling the Square 2, Nottingham, UK, June 22–23, 2015), https://blogs.nottingham.ac.uk/circlingthesquare/2015/06/24/mike-hulme-scientists-speaking-with-one-voice-panacea-or-pathology/.

9 Gayathri Vaidyanathan, "How to Determine the Scientific Consensus on Global Warming," Scientific American (Scientific American, July 24, 2014), https://www.scientificamerican.com/article/how-to-determine-the-scientific-consensus-on-global-warming/.

10 Naomi Oreskes, "Science and Public Policy: What's Proof Got to Do with It?," *Environmental Science and; Policy* 7, no. 5 (2004): 369–383, https://doi.org/10.1016/j.envsci.2004.06.002.

11 Jeroen P. van der Sluijs, "Uncertainty and Dissent in Climate Risk Assessment: A Post-Normal Perspective," *Nature and Culture* 7, no. 2 (January 2012): 174–195, https://doi.org/10.3167/nc.2012.070204.

12 David Dunning, "We Are All Confident Idiots," Pacific Standard (Pacific Standard, October 27, 2014), https://psmag.com/social-justice/confident-idiots-92793.

13 Gro Harlem Brundtland, "Speech at the UN Commission on Sustainable Development" (Speech, Fifteenth session of the UN Commission on Sustainable Development, United Nations, New York, April 30, 2007, https://www.regjeringen.no/en/topics/foreign-affairs/the-un/Brundtland_speech_CSD/id465906/).

14 J.A.Curry and P.J.Webster, "Climate Science and the Uncertainty Monster," *Bulletin of the American Meteorological Society* 92, no. 12 (2011): 1667–1682, https://doi.org/10.1175/2011bams3139.1.

15 Judith Curry, "National Climate Assessment: A Crisis of Epistemic Overconfidence," Climate etc., January 3, 2019, https://judithcurry.com/2019/01/02/national-climate-assessment-a-crisis-of-epistemic-overconfidence/.

16 Stuart Ritchie and Michael Story, "How the Experts Messed up on COVID," UnHerd, November 9, 2020, https://unherd.com/2020/10/how-the-experts-messed-up-on-covid/.

17 "BBC inside Science, the Perils of Explaining Science, Living to 500, What's Good for Your Teeth and The Future of Stargazing," BBC Radio 4 (BBC, January 12, 2017), https://www.bbc.co.uk/programmes/b087pf71.

18 Michael Oppenheimer, Christopher M. Little, and Roger M. Cooke, "Expert Judgement and Uncertainty Quantification for Climate Change," *Nature Climate Change* 6, no. 5 (2016): 445–451, https://doi.org/10.1038/nclimate2959.

19 Rajendra K. Pachauri and Leo Meyer, Climate Change 2014: Synthesis Report Summary for Policymakers (Geneva, CH: IPCC, 2015), 13.

20 Benjamin P. Horton et al., "Expert Assessment of Sea-Level Rise by Ad 2100 and AD 2300," *Quaternary Science Reviews* 84 (2014): 1–6, https://doi.org/10.1016/j.quascirev.2013.11.002.

21 American Physical Society, "Climate Change Statement Review Workshop—Transcript of Proceedings," January 8, 2014, 36, https://www.aps.org/policy/statements/upload/climate-seminar-transcript.pdf.

22 "Climate Change Statement Review Workshop," American Physical Society, January 8, 2014, https://www.aps.org/policy/statements/upload/climate-seminar-transcript.pdf.

23 Rupert Darwall, "A Veneer of Certainty Stoking Climate Alarm," Competitive Enterprise Institute, November 28, 2017, https://cei.org/studies/a-veneer-of-certainty-stoking-climate-alarm/.

24 Daniel Sarewitz, "Beware the Creeping Cracks of Bias," *Nature* 485, no. 7397 (2012): 149–149, https://doi.org/10.1038/485149a.

25 Nuzzo, "How Scientists Fool Themselves," 182–185.

26 Thomas Kelly, "The Epistemic Significance of Disagreement," 167–196.

27 Jonathan Haidt, "The Bright Future of Post-Partisan Social Psychology," (presented at Society for Personality and Social Psychology Annual Meeting, San Antonio, TX, January 27, 2011), https://www.youtube.com/watch?v=E34mlS3M7OA.

28 José L. Duarte et al., "Political Diversity."

29 Jean Goodwin, "The Authority of the IPCC."

30 Lee Jussim et al., "Interpretations and Methods: Towards a More Effectively Self-Correcting Social Psychology," *Journal of Experimental Social Psychology* 66 (2016): 116–133, https://doi.org/10.1016/j.jesp.2015.10.003.

31 Y. A. de Vries et al., "The Cumulative Effect of Reporting and Citation Biases on the Apparent Efficacy of Treatments: The Case of Depression," *Psychological Medicine* 48, no. 15 (February 2018): 2453–2455, https://doi.org/10.1017/s0033291718001873.

32 Kjersti Fløttum, Des Gasper, and Asuncion Lera St. Clair, "Synthesizing a Policy-Relevant Perspective from the Three IPCC 'Worlds'—a Comparison of Topics and Frames in the SPMs of the Fifth Assessment Report," *Global Environmental Change* 38 (2016): 118–129, https://doi.org/10.1016/j.gloenvcha.2016.03.007.

33 Ernest Hugh O'Boyle, George Christopher Banks, and Erik Gonzalez-Mulé, "The Chrysalis Effect," *Journal of Management* 43, no. 2 (October 2016): 376–399, https://doi.org/10.1177/0149206314527133.

34 "Changing Climate, The Science," BBC Radio 4 (BBC, November 16, 2015), https://www.bbc.co.uk/programmes/b06p7d29.

35 Ibid.

36 John Stuart Mill, *On Liberty* (London, UK: John W. Parker and Son, 1859), 34.

37 Jean Goodwin, "The Authority of the IPCC."

38 Naomi Oreskes, "The Scientific Consensus on Climate Change," *Science* 306, no. 5702 (February 2004): 1686–1686, https://doi.org/10.1126/science.1103618.

39 Klaus Hasselmann, "The Climate Change Game," *Nature Geoscience* 3, no. 8 (2010): 511–512, https://doi.org/10.1038/ngeo919.

40 W. R. Anderegg et al., "Expert Credibility in Climate Change," *Proceedings of the National Academy of Sciences* 107, no. 27 (2010): 12107–12109, https://doi.org/10.1073/pnas.1003187107.

41 Aaron M. McCright and Riley E. Dunlap, "Cool Dudes: The Denial of Climate Change among Conservative White Males in the United States," *Global Environmental Change* 21, no. 4 (2011): 1163–1172, https://doi.org/10.1016/j.gloenvcha.2011.06.003.

42 Stephan Lewandowsky et al., "Seepage: Climate Change Denial and Its Effect on the Scientific Community," *Global Environmental Change* 33 (2015): 1–13, https://doi.org/10.1016/j.gloenvcha.2015.02.013.

43 Kari Marie Norgaard, "Climate Denial: Emotion, Psychology, Culture, and Political Economy," *Oxford Handbooks Online*, 2011, https://doi.org/10.1093/oxfordhb/9780199566600.003.0027.

44 Peter J. Jacques, "A General Theory of Climate Denial," *Global Environmental Politics* 12, no. 2 (2012): 9–17, https://doi.org/10.1162/glep_a_00105.

45 Steven Koonin, "Unsettled: What Climate Science Tells Us, What it Doesn't, and Why it Matters," (2021) BenBella Books, 316 pp.

46 Bjorn Lomborg, "False Alarm: How Climate Change Costs Us Trillions, Hurts the Poor, and Fails to Fix the Planet," (2020), Basic Books, 422 pages.

47 Michael Schellenberger, "Apocalypse Never: Why Environmental Alarmism Hurts Us All," (2020), Harper, 430 pp.

48 Mike Hulme, accessed October 5, 2012, https://www.earth.columbia.edu/videos/view/exploring-trust-and-mistrust-in-climate-science-and-solutions.

49 Henry Farrell, "In Praise of Negativity," Crooked Timber, July 24, 2020, https://crookedtimber.org/2020/07/24/in-praise-of-negativity/.

50 Robert King Merton, "The Normative Structure of Science," in *The Sociology of Science: Theoretical and Empirical Investigations* (Chicago, IL: University of Chicago Press, 1973), 267–278.

51 "What Is Social Skepticism," The Ethical Skeptic, May 1, 2012, https://theethicalskeptic.com/tag/what-is-social-skepticism/.

52 "Steve McIntyre," Wikipedia (Wikimedia Foundation, October 5, 2021), https://en.wikipedia.org/wiki/Steve_McIntyre.

53 Michael E. Mann, Raymond S. Bradley, and Malcolm K. Hughes, "Global-Scale Temperature Patterns and Climate Forcing over the Past Six Centuries," *Nature* 392, no. 6678 (January 1998): 779–787, https://doi.org/10.1038/33859.

54 Michael E. Mann, Raymond S. Bradley, and Malcolm K. Hughes, "Northern Hemisphere Temperatures during the Past Millennium: Inferences, Uncertainties, and Limitations," *Geophysical Research Letters* 26, no. 6 (1999): 759–762, https://doi.org/10.1029/1999gl900070.

55 J. T. Houghton, *Climate Change 2001: The Scientific Basic* (Cambridge, UK: Cambridge University Press, 2001).

56 Marco Evers et al., "Climate Catastrophe: A Superstorm for Global Warming Research," DER SPIEGEL (DER SPIEGEL, April 1, 2010), https://www.spiegel.de/international/world/climate-catastrophe-a-superstorm-for-global-warming-research-a-686697.html.

57 Stephen McIntyre and Ross McKitrick, "Corrections to the Mann Et. Al. (1998) Proxy Data Base and Northern Hemispheric Average Temperature Series," *Energy & Environment* 14, no. 6 (2003): 751–771, https://doi.org/10.1260/095830503322793632.

58 Stephen McIntyre and Ross McKitrick, "The M&M Critique of the MBH98 Northern Hemisphere Climate Index: Update and Implications," *Energy & Environment* 16, no. 1 (2005): 69–100, https://doi.org/10.1260/0958305053516226.

59 Stephen McIntyre and Ross McKitrick, "Hockey Sticks, Principal Components, and Spurious Significance," *Geophysical Research Letters* 32, no. 3 (2005), https://doi.org/10.1029/2004gl021750.

60 Kevin Aylward, "Best Science Blog - the 2007 Weblog Awards," The Internet Archive Wayback Machine, November 1, 2007, https://web.archive.org/web/20071117034042/http://2007.weblogawards.org/polls/best-science-blog-1.php.

61 Warren Pearce, "Are Climate Sceptics the Real Champions of the Scientific Method?," The Guardian (Guardian News and Media, July 30, 2013), https://www.theguardian.com/science/political-science/2013/jul/30/climate-sceptics-scientific-method.

62 Nic Lewis, "A Major Problem with the Resplandy Et Al. Ocean Heat Uptake Paper," Climate etc., November 6, 2018, https://judithcurry.com/2018/11/06/a-major-problem-with-the-resplandy-et-al-ocean-heat-uptake-paper/.

63 L. Resplandy et al., "Quantification of Ocean Heat Uptake from Changes in Atmospheric O2 and CO2 Composition," Nature 563, no. 7729 (2018): 105–108, https://doi.org/10.1038/s41586-018-0651-8.

64 Nic Lewis, "Resplandy et al. Part 5: Final Outcome," Climate etc., September 26, 2019, https://judithcurry.com/2019/09/25/resplandy-et-al-part-5-final-outcome/.

65 Edward Skidelsky, "Words That Think for Us - the Tyranny of Denial" (Prospect Magazine, January 27, 2010), https://www.prospectmagazine.co.uk/magazine/words-that-think-for-us-3.

66 Ben Jervey, "House Science Committee Hearing Pits Three Fringe Climate Deniers against Mainstream Climate Scientist Michael Mann," DeSmog, March 29, 2017, https://www.desmog.com/2017/03/29/house-science-committee-hearing-lamar-smith-michael-mann-climate-consensus-deniers/.

67 Michael D. Lemonick, "Climate Heretic: Judith Curry Turns on Her Colleagues," Scientific American (Scientific American, November 2010), https://www.scientificamerican.com/article/climate-heretic/.

68 "Climate Science Glossary," Skeptical Science, accessed November 10, 2021, https://skepticalscience.com/misinformers.php.

69 Prof Michael E. Mann (@MichaelEMann) "Testimonies at the EPW Senate Hearing: #Science vs #AntiScience #Dessler #Curry." Twitter, January 16, 2014. https://twitter.com/MichaelEMann/status/423914894207877120.

70 E&ENews, February 25, 2015, http://www.eenews.net/climatewire/2015/02/25/stories/1060013991.

71 "Climate Misinformation by Source," Skeptical Science, accessed October 6, 2021, https://skepticalscience.com/misinformers.php.

72 "Climate Disinformation Database," DeSmog, April 21, 2021, https://www.desmog.com/climate-disinformation-database/.

73 "Roger Pielke Jr.," Environmental Studies Program (University of Colorado Boulder, August 17, 2021), https://www.colorado.edu/envs/roger-pielke-jr.

74 "Roger A. Pielke, Sr.," Cooperative Institute for Research in Environmental Sciences, September 29, 2021, https://cires.colorado.edu/research-group/roger-pielke-sr.

75 "Richard Lindzen," Wikipedia (Wikimedia Foundation, October 5, 2021), https://en.wikipedia.org/wiki/Richard_Lindzen.

76 "Richard A. Muller," Wikipedia (Wikimedia Foundation, August 20, 2021), https://en.wikipedia.org/wiki/Richard_A._Muller.

77 "John R. Christy," Faculty and Staff (University of Alabama Huntsville), accessed October 6, 2021, https://www.uah.edu/science/departments/atmospheric-science/faculty-staff/dr-john-christy.

78 John H. Perkins, "Beware Climate Neo-Scepticism," Nature 522, no. 7556 (2015): 287, https://doi.org/10.1038/522287c.

79 Bjøorn Lomborg, *The Sceptical Environmentalist: Measuring the State of the World* (Cambridge, UK: Cambridge University Press, 2001).

80 Naomi Oreskes, "There Is a New Form of Climate Denialism to Look out for – so Don't Celebrate Yet," The Guardian (Guardian News and Media, December 16, 2015), https://www.theguardian.com/commentisfree/2015/dec/16/new-form-climate-denialism-dont-celebrate-yet-cop-21.

81 Ibid.

82 "Dr. James E. Hansen," Earth Institute (Columbia University), accessed October 6, 2021, http://www.columbia.edu/~jeh1/.

83 "Tom Wigley," Wikipedia (Wikimedia Foundation, December 11, 2020), https://en.wikipedia.org/wiki/Tom_Wigley.

84 "Ken Caldeira," Caldeira Lab | Carnegie's Department of Global Ecology (Carnegie Institution for Science), accessed October 7, 2021, https://dge.carnegiescience.edu/labs/caldeira-lab.

85 "Emanuel, Kerry A. | MIT Department of Earth, Atmospheric and Planetary Sciences," MIT EAPS Directory (Massachusetts Institute of Technology), accessed October 7, 2021, https://eapsweb.mit.edu/people/kokey.

86 Philip Shabecoff, "Global Warming Has Begun, Expert Tells Senate," The New York Times (The New York Times, June 24, 1988), https://www.nytimes.com/1988/06/24/us/global-warming-has-begun-expert-tells-senate.html.

87 "Top NASA Scientist Arrested (Again) in White House Protest," Fox News (FOX News Network, December 20, 2014), https://www.foxnews.com/science/top-nasa-scientist-arrested-again-in-white-house-protest.

88 Rebecca Weisser, "Leftists Repent over Lab Leak Farce," The Spectator Australia, June 2, 2021, https://www.spectator.com.au/2021/06/leftists-repent-over-lab-leak-farce/.

89 Charles Calisher et al., "Statement in Support of the Scientists, Public Health Professionals, and Medical Professionals of China Combatting COVID-19," *The Lancet* 395, no. 10226 (2020), https://doi.org/10.1016/s0140-6736(20)30418-9.

90 Kristian G. Andersen et al., "The Proximal Origin of SARS-COV-2," *Nature Medicine* 26, no. 4 (2020): 450–452, https://doi.org/10.1038/s41591-020-0820-9.

91 Charles Calisher et al., "Statement in Support of the Scientists," no. 10226.

92 Matthew Crawford, "Science Has Become a Cartel," UnHerd, May 21, 2021, https://unherd.com/2021/05/how-scientists-sacrificed-scepticism/.

93 Nicholas Wade, "The Origin of COVID: Did People or Nature Open Pandora's Box at Wuhan?," Bulletin of the Atomic Scientists, May 5, 2021, https://thebulletin.org/2021/05/the-origin-of-covid-did-people-or-nature-open-pandoras-box-at-wuhan/.

94 Ibid.

95 Becket Adams, "PolitiFact Retracts Wuhan Lab Theory 'Fact-Check'" (Washington Examiner, May 22, 2021), https://www.washingtonexaminer.com/opinion/politifact-retracts-wuhan-lab-theory-fact-check.

96 Dan Kahan, "Weekend Update: The Distracting, Counterproductive '97% Consensus' Debate Grinds On," The Cultural Cognition Project (The Yale Law School, July 27, 2013), http://www.culturalcognition.net/blog/2013/7/27/weekend-update-the-distracting-counterproductive-97-consensu.html.

97 Michael Oppenheimer et al., "The Limits of Consensus," *Science* 317, no. 5844 (2007): 1505–1506, https://doi.org/10.1126/science.1144831.

98 Arnulf Grübler and Nebojsa Nakicenovic, "Identifying Dangers in an Uncertain Climate," *Nature* 412, no. 6842 (2001): 15, https://doi.org/10.1038/35083752.

99 Michael Oppenheimer et al., *Discerning Experts: The Practices of Scientific Assessment for Environmental Policy* (University of Chicago Press, 2019).

100 Naomi Oreskes et al., "Scientists Have Been Underestimating the Pace of Climate Change," Scientific American Blog Network (Scientific American, August 19, 2019), https://blogs.scientificamerican.com/observations/scientists-have-been-underestimating-the-pace-of-climate-change/?amp.

101 J. A. Curry and P. J. Webster, "Climate Change: No Consensus on Consensus."

102 Silke Beck, "Moving beyond the Linear Model of Expertise? IPCC and the Test of Adaptation," *Regional Environmental Change* 11, no. 2 (2010): 297–306, https://doi.org/10.1007/s10113-010-0136-2.

Chapter Three

THE CLIMATE CHANGE RESPONSE CHALLENGE

"To act on the belief that we possess the knowledge and the power which enable us to shape the processes of society entirely to our liking, knowledge which in fact we do *not* possess, is likely to make us do much harm."

—Nobel laureate economist Friedrich Hayek[1]

The climate change response challenge is often portrayed as simple: increasing CO_2 emissions causes problems, therefore stop fossil fuel emissions. Our failure to stop fossil fuel emissions, after decades of international policies, is often blamed on a lack of political will.[2] Well, political will is lacking at least in part because CO_2 emissions are the byproduct of dependable energy delivered by fossil fuels that has underpinned two centuries of human and economic development.[3]

In a 2018 interim Special Report requested by the UNFCCC,[4] the IPCC mapped out a pathway to limiting the temperature increase in 2100 to 1.5°C above pre-industrial levels. Limiting warming to 1.5°C is assessed to require dramatic emissions reductions by the year 2030 and complete carbon neutrality by around 2050. This would entail unprecedented transformations of energy, transportation, agricultural, urban, and industrial systems. However, no feasible and effective pathway has been identified to achieve any of these objectives in such a short time frame.

Human-caused climate change has been characterized as a "tragedy of the horizon" that imposes a cost on future generations that the current one has no direct incentive to fix.[5,6] The possibility of delayed global rewards is challenged as inadequate justification to take action and incur immediate costs. However, this has led to the concern that we are operating a multigenerational Ponzi scheme by continuing to burn fossil fuels.

Additional background factors that influence the decisions of individual nations to participate in collective action to urgently reduce emissions include: uncertainties and complexity surrounding the underlying climate change science, the priority of economic development, and the unknown impacts of future technologies. There is substantial uncertainty about the magnitude, rate, regionality, and timing of adverse climate impacts, which is confounded by natural climate variability.

Economist Robert Pindyck argues that this uncertainty creates insurance value, pushing us towards early and stronger actions to reduce CO_2 emissions.[7] However, the very uncertainties over climate change that create insurance value prevent determination of exactly how large that insurance value is. How much of a premium we should be willing to pay for such insurance depends in part on society's degree of risk aversion, which is complex and difficult to evaluate.

Towards understanding the broad challenges associated with the climate policy debate, the following topics are examined in this chapter:

- Inconvenient truths associated with emissions reductions
- Disagreements on how to characterize the climate change problem in terms of the policy response
- Framing of solutions in the context of sustainability
- The assumption that global warming is the most important problem facing the world

3.1 Inconvenient Truths

"QTIIPS: Quantitatively Trivial Impact + Intense Political Symbolism"[8] (Keith Hennessey)

Until we come to grips with some inconvenient truths and difficult choices, there is a substantial risk that the issue of emissions mitigation will continue to be dominated by symbolic proclamations by the UNFCCC and strategies that are simply ineffectual. The Paris Agreement has created a wide gap between ambition and obligation by adopting ambitious temperature targets without specifying the means to reach them. What we wish to be true is very different from what is possible based on a rational examination of economics and technology.

The proposed stabilization of CO_2 emissions has revealed and created new problems in terms of energy policy. Energy policy is driven by a complicated mix of economics and economic development, energy security and reliability, environmental quality and health issues, and resource availability.

It is becoming increasingly apparent that we don't know how to address the challenge of rapidly stabilizing atmospheric concentrations of CO_2 at a low level. The green energy revolution has barely begun. Large-scale sequestration of CO_2 emissions is an idea that is far from reality.[9]

Even if the technological issues can be overcome, there are issues of social, economic, and political feasibility that inhibit such rapid change. Further, carbon neutrality strategies by 2050 involve a high risk of ineffectiveness because the strategies need to be adopted and implemented by all countries in order to meet global emissions targets.

The fundamental premises behind stabilization targets are badly matched to the actual problem of the intergenerational management of climate change, both scientifically and politically.[10] The focus on infeasible policy proposals for climate stabilization are impeding more productive policy action on climate change—a topic that is addressed in Part Three of this book.

The UNFCCC emissions reduction policies have brought us to a point between a rock and a hard place, whereby the emissions reduction policy with its extensive costs and questions of feasibility are inadequate for making a meaningful dent in slowing down the expected warming in the twenty-first century. Emissions mitigation policy goals have been set that almost certainly will not be met since they aim beyond the scope of the knowable and doable and what is politically feasible.

3.2 The Sustainability Trap

"There is always a well-known solution to every human problem—neat, plausible, and wrong." (American journalist H.L. Mencken)[11]

Sustainability refers generally to the capacity for the biosphere and human civilization to coexist. According to the UN report *Our Common Future*, "sustainable development" is defined as development that "meets the needs of the present without compromising the ability of future generations to meet their own needs."[12]

Sustainable development is at the heart of the UNFCCC agenda, with established synergies between the UNFCCC 2030 Agenda for Sustainable Development and the Paris Climate Change Agreement.[13,14] Under the combined banner of sustainable development and mitigation of climate change, international diplomacy and resources are being directed toward mitigating climate change by reducing emissions. As a result, resources are being redirected away from the needs of Africa for grid electricity and the needs of South Asia to reduce their vulnerability to extreme weather events and water shortages (see Chapter Thirteen).

Sustainability implies stasis and balance, in the sense of a desire for humanity to achieve a lasting equilibrium with our planet.[15] This concept of stasis and balance also undergirds the ideal of using CO_2 as a control knob to maintain a stable climate. The utility of the sustainability concept is being increasingly challenged precisely because the world and its climate are thought to be continually out of balance and in constant disequilibrium.

3.2.1 Resilience and the Tension with Sustainability

"Hence we are to believe that the most adaptable species that has ever existed, a species so sophisticated that it can survive in outer space, requires an absolutely stable average temperature and sea level in order to survive? This defies common sense." (Environmental finance expert Mario Loyola)[16]

Where sustainability aims to put the world into stationary balance, resilience looks for ways to manage in a continually imbalanced world. Resilience is the ability to bounce back in the face of shocks. It is the capability of a social group to absorb disturbances and reorganize to retain essentially the same functional structure.

Climate resilience is the ability to anticipate, prepare for, and respond to hazardous events or trends related to climate variability and change. Resilience planning is often associated with acute events—floods, hurricanes, or wildfires. Resilience planning also accounts for trends like rising sea levels, worsening air quality, and population migration.

As a practical matter, adaptation has historically been driven by local crises associated with extreme weather and climate events. Early examples of US infrastructure designed to reduce vulnerability to extreme weather events include the levees and floodways built in response to the Great Mississippi Flood of 1927[17] and the construction of the Herbert Hoover Dike in Florida in response to the Lake Okeechobee Hurricane in 1928.[18]

Resilience carries a connotation of returning to the original state as quickly as possible. Advocates of adaptation to climate change are not arguing for simply responding to events and changes after they occur; they are arguing for anticipatory adaptation. However, in adapting to climate change, we need to acknowledge that we don't know how the climate will evolve in the twenty-first century (see Part Two of this book). We are certain to be surprised, and we will make mistakes along the way.

The conflict between sustainability and resilience was demonstrated by New York City in the aftermath of Superstorm Sandy. Superstorm Sandy hit New York City hardest in the location where it was most recently redeveloped: Lower Manhattan, which should have been the least vulnerable part of the island. But it was rebuilt to be sustainable, not resilient, according to urban planner and developer Jonathan Rose.[19] As a result, New York City demonstrated its fragility in the face of flooding and disruptions to the electricity grid.

Possible scenarios of incremental worsening of weather and climate extremes over the course of the twenty-first century do not change this fundamental fact: most regions of the world, including the United States and Europe, are not well adapted to the current range of weather events or to the range that has been experienced over the past century.

Supporters of the sustainability paradigm have often been reluctant to embrace resilience and adaptation with regard to climate change. If we adapt to unwanted change, the reasoning goes, we give a pass to those responsible for causing the problem (producers and emitters of fossil fuels) and we lose the moral authority to pressure them to stop. However, there is an imbalanced tradeoff between climate mitigation and adaptation. Mitigation via emissions reductions helps no one for decades to come, whereas people are helped in the near term as well as in the future by locally designed and implemented adaptation measures to address real vulnerabilities.

3.2.2 *Thrivability and Anti-Fragility*

"Not only can we work to minimize our footprint but we can also create positive handprints." (Jean Russell, founder of the Thrivability movement)[20]

There is a key assumption behind resilience that the *status quo* is good. However, "bounce back" resilience can be viewed as a distraction that delays societal movement forward.[21] Jean Russell, the founder of the Thrivability movement, states that it is not enough to repair the damage our progress has brought. And it is also not enough to manage our risks and be more shock-resistant. Thrivability fosters advances and improvement in what we do. Rather than the bouncing back of resilience, thrivability is associated with bouncing forward.[22]

Russell outlines four modes of engaging with our environment:[23]

- Survival: trying to ensure personal survival and that of the group or nation; responding to basic needs.
- Sustainable: able to be maintained at a given rate or level over time; mitigate damage, austerity; endure in a stable world.
- Resilient: able to withstand or recover quickly from difficult conditions.
- Thrivable: to develop vigorously, to prosper and flourish; anti-fragile, generate, transform.

Each mode includes and also transcends the previous mode: thrivability transcends survival, sustainability, and resilience modes.[24]

Mathematical statistician Nassim Nicholas Taleb's concept of "antifragility" focuses on learning from adversity, and developing approaches that enable us to thrive from high levels of volatility, particularly unexpected extreme events.[25] Anti-fragility goes beyond bouncing back to becoming even better as a result of encountering and overcoming challenges. Anti-fragile systems are dynamic rather than static, thriving and growing in new directions rather than simply maintaining the *status quo*.

Anti-fragile systems actually need extreme, random events to strengthen and grow; they become brittle and atrophy in the absence of these random events. Taleb makes an important point: biological systems in nature are inherently antifragile—they are constantly evolving and growing stronger as a result of random events. In contrast, man-made systems tend to be fragile, having a hard time coping with random events. Taleb highlights a key paradox: our focus in modern times on removing or minimizing randomness has actually had the perverse effect of increasing fragility.[26]

Strategies to increase antifragility include: economic development, reducing the downside from volatility, developing a range of options, tinkering with small experiments, and developing and testing transformative ideas. Antifragility is consistent with decentralized actions including policy innovation that create flexibility and redundancy in the face of volatility. Such strategies provide an "innovation dividend" that is analogous to biodiversity in the natural world, enhancing resilience in the face of future shocks.

In summary, focusing only on sustainability can hamper resilience and thrivability and the overall prosperity of societies.

3.3 Warming Is Not the Only Problem

"Climate change is the greatest threat to global health in the twenty-first century." (World Health Organization 2015)[27]

The climate change movement has called for putting humanity on a war footing to combat global warming.[28] The exceptional status of the perceived threat from climate change driven by fossil fuel emissions obscures the broader ecological problems associated with our impact on the planet. These ecological problems include: land use and deforestation resulting in habitat destruction and loss of species; overfishing; pollution of air, water, and soil from agrochemicals, mineral extraction, burning of wood and dung for energy as well as burning of fossil fuels; and degradation of the Earth's surface from mining, littering, and household and industrial waste.

Stepping back from the details of various policies and targets, we need to remind ourselves that the point of climate policy is to make the world a better place, now and for future generations. Climate change is a problem that deserves consideration, but how should climate change policies be prioritized relative to other problems and threats?

Current climate policy, aimed at net-zero emissions, is very expensive and with little benefits to date; hence there is concern that it will drain potential future economic gains and leave less money for policies that will enhance prosperity and well-being. Over spending on ineffective climate policies means under spending on effective climate policies and on other opportunities we have to improve life for billions of people, now and into the future.

Economist Bjorn Lomborg regards climate change to be a moderate problem in a sea of problems, big and small. Lomborg argues that if we truly want to make the world a better place, we should not let our preoccupation with climate change distract our focus away from other crucial problems. Lomborg argues that we can improve the human condition far more effectively by further opening the world to free trade, ending tuberculosis, and improving access to nutrition, contraception, healthcare, education, and technology.[29]

Climate change is just one of many threats facing our world today, a point made clear by the COVID-19 pandemic. Why should climate change be prioritized over other threats? There is a wide range of threats that we could face in the twenty-first century: solar electromagnetic storms that would take out all space-based electronics including the Global Positioning System (GPS) and electricity transmission lines; pandemics; global financial collapse; a mega volcanic eruption; asteroid strike; cascade of mistakes that triggers a thermonuclear, biological, chemical or cyber war; the rise of terrorism; and huge debt associated with a pandemic or other threats that causes collapse of trade, commerce, manufacture and wealth.[30] How to objectively assess the risk of climate change is the topic of Chapter Ten.

Vast sums spent on climate change mitigation come from the same funds that support human development and effectively insure us against all threats. The focus on climate change has arguably contributed to the world being ill prepared for COVID-19, reflected by the statement from the World Health Organization (WHO) that climate change is the biggest health threat for the twenty-first century.

In 50 years, we may be looking back on UNFCCC climate policies as using chemotherapy to try to cure a head cold, all the while ignoring more serious diseases.

3.4 Tame Problem or Wicked Mess?

"Climate change is an issue that presents great scientific and economic complexity, some very deep uncertainties, profound ethical issues, and even lack of agreement on what the problem is." (Economist Mike Toman of the World Bank)[31]

In addressing a specific problem, frames shape how we conceptualize it. Framing includes what is deemed to be relevant, what is excluded, and even what answers are considered appropriate. A framing bias occurs when a narrow frame pre-ordains the conclusion to a much more complex problem.

The climate change problem has been framed narrowly as being caused by excess carbon dioxide in the atmosphere, which can be solved by eliminating fossil fuel emissions. This framing dominates the United Nations negotiations on climate change. This framing is focused on control of the climate, through restricting CO_2 emissions.

In their Wrong Trousers essay, social scientists Gwythian Prins and Steve Rayner argue that we have made the wrong cognitive choices in our attempts to define the problem of climate change and its solution using a narrow frame, by relying on strategies that worked previously for "tame" problems.[32] A tame problem is well defined, well understood, and the appropriate solutions are agreed upon. Cost-benefit analyses and mitigation techniques are appropriate for tame problems, and the potential harm from miscalculation is bounded.

Complex problems are different from those that are merely complicated. In the presence of feedbacks and circularity, causal mechanisms are not easily elucidated. Problems related to the environment, such as climate change, and most problems related to human health are complex problems. Complex problems require a much larger frame to accommodate uncertainty, ambiguity, chaos, and contradictions. Any framing of a complex problem is provisional, requiring acknowledgment of what is outside the frame and its potential importance.

Prins and Rayner contend that climate change is better characterized as a "wicked" problem. A wicked problem is characterized by multiple problem definitions, contentious methods of understanding, chronic conditions of ignorance, and lack of capacity to imagine future eventualities of both the problem and the proposed solutions.[33] The complex web of causality may result in surprising unintended consequences of attempted solutions that generate new vulnerabilities or exacerbate the original harm. Further, wickedness makes it difficult to identify points of irrefutable failure or success in either the science or the policies.

Climate change has also been categorized as a social "mess."[34] Attributes of social messes include: no unique correct view of the problem; contradictory solutions; missing or uncertain data; ideological, economic, and political constraints; consequences difficult to imagine; great resistance to change; and problem solvers who are not in contact with the problems or potential solutions.

Diagnosis of a problem as "wicked" is tantamount to acknowledging the futility of control. Instead, the focus shifts away from attempting to control towards achieving a better understanding of the problem. Dealing with wicked problems promotes a focus on considering problems from multiple perspectives, designing programs that accommodate complexity and ambiguity, management strategies that account for crises and surprises, improving policy and evaluation capabilities, and strengthening the collaborative capacities of the policy system.[35]

Notes

1 Freidrich von Hayek, "Friedrich Von Hayek Prize Lecture," The Sveriges Riksbank Prize in Economic Sciences in memory of Alfred Nobel 1974 (NobelPrize.org), accessed October 8, 2021, https://www.nobelprize.org/prizes/economic-sciences/1974/hayek/lecture/.
2 Angel Gurría, "Climate Change: A Matter of Political Will," OECD, June 3, 2008, https://www.oecd.org/env/tools-evaluation/climatechangeamatterofpoliticalwill.htm.
3 Alex Epstein, *The Moral Case for Fossil Fuels* (New York, NY: Portfolio/Penguin, 2015).
4 Valerie Masson-Delmotte, ed., "Global Warming of 1.5°C. An IPCC Special Report on the Impacts of Global Warming of 1.5°C above Pre-Industrial Levels and Related Global Greenhouse Gas Emission Pathways, in the Context of Strengthening the Global Response to the Threat of Climate Change, Sustainable Development, and Efforts to Eradicate Poverty" (Geneva, CH: IPCC, 2018).
5 Mark Carney, "Breaking the Tragedy of the Horizon—Climate Change and Financial Stability—Speech by Mark Carney," Bank of England, September 29, 2015, https://www.bankofengland.co.uk/speech/2015/breaking-the-tragedy-of-the-horizon-climate-change-and-financial-stability.
6 Mark Carney, "Resolving the Climate Paradox—Speech by Mark Carney," Bank of England, September 22, 2016, https://www.bankofengland.co.uk/speech/2016/resolving-the-climate-paradox.
7 Robert S. Pindyck, "What We Know and Don't Know about Climate Change, and Implications for Policy," *Environmental and Energy Policy and the Economy* 2 (January 2021): 4–43, https://doi.org/10.1086/711305.
8 Keith Hennessey, "Is the Paris Agreement QTIIPS?," Keith Hennessey, June 2, 2017, https://keithhennessey.com/2017/06/01/is-the-paris-agreement-qtiips/.
9 "Explainer: 10 Ways 'Negative Emissions' Could Slow Climate Change," Carbon Brief, April 11, 2016, https://www.carbonbrief.org/explainer-10-ways-negative-emissions-could-slow-climate-change.
10 Maxwell T. Boykoff et al., "Discursive Stability Meets Climate Instability: A Critical Exploration of the Concept of 'Climate Stabilization' in Contemporary Climate Policy," *Global Environmental Change* 20, no. 1 (2010): 53–64, https://doi.org/10.1016/j.gloenvcha.2009.09.003.
11 H.L. Mencken, *Prejudices: Second Series* (London, UK: Jonathan Cape, 1921).
12 Gro Harlem Brundtland, "Our Common Future: Report of the World Commission on Environment and Development." (Geneva, CH: United Nations-Document A/42/427, 1987).
13 "Transforming Our World: The 2030 Agenda for Sustainable Development," United Nations (United Nations | department of economic and social affairs, 2015), https://sdgs.un.org/2030agenda.

14 "The Paris Agreement," Unfccc.int (United Nations, 2015), https://unfccc.int/process-and-meetings/the-paris-agreement/the-paris-agreement.

15 Charles Kibert et al., *The Ethics of Sustainability*, 2011.

16 Mario Loyola, "Twilight of the Climate Change Movement," The American Interest, August 7, 2017, https://www.the-american-interest.com/2016/03/31/twilight-of-the-climate-change-movement/.

17 Matthew Percy, "After the Flood.," *Journal of the Illinois State Historical Society* 95, no. 2 (2002): 172–201.

18 "Lake Okeechobee / Herbert Hoover Dike," Jacksonville District, U.S. Army Corps of Engineers, accessed October 12, 2021, https://www.saj.usace.army.mil/HHD/.

19 Andrew Zolli, "Learning to Bounce Back," The New York Times (The New York Times, November 3, 2012), https://www.nytimes.com/2012/11/03/opinion/forget-sustainability-its-about-resilience.html.

20 Jean Russell, "Resilience Ain't Enough," The Thrivable Society, February 7, 2013, https://thrivable.net/2013/02/resilience-aint-enough/.

21 John Hagel, "A Contrarian View on Resilience," Edge Perspectives with John Hagel, April 1, 2013, https://edgeperspectives.typepad.com/edge_perspectives/2013/04/a-contrarian-view-on-resilience.html.

22 Jean Russell, "Resilience Ain't Enough."

23 Ibid.

24 Ibid.

25 Nassim Nicholas Taleb, Antifragile: Things That Gain from Disorder (New York, NY: Random House, 2012).

26 Ibid.

27 "WHO Calls for Urgent Action to Protect Health from Climate Change – Sign the Call," World Health Organization (World Health Organization, October 6, 2015), https://www.who.int/news/item/06-10-2015-who-calls-for-urgent-action-to-protect-health-from-climate-change-sign-the-call.

28 Rob Merrick, "UK Must Fight Climate Change on 'War Footing' like Defeat of Nazis, Theresa May Told," The Independent (Independent Digital News and Media, April 29, 2019), https://www.independent.co.uk/news/uk/politics/uk-climate-change-theresa-may-environment-protest-lucas-ed-miliband-second-world-war-nazis-a8891801.html.

29 Bjorn Lomborg, *Cool It: The Skeptical Environmentalist's Guide to Global Warming* (New York, NY: Vintage, 2008).

30 Michael J. Kelly, "Warming Is Not the Only Threat," The Critic Magazine, June 23, 2020, https://thecritic.co.uk/issues/july-august-2020/warming-is-not-the-only-threat/.

31 "A Wicked Problem: Controlling Global Climate Change," World Bank, September 30, 2014, https://www.worldbank.org/en/news/feature/2014/09/30/a-wicked-problem-controlling-global-climate-change.

32 Gwyn Prins and Steve Rayner, "The Wrong Trousers: Radically Rethinking Climate Policy," LSE Research Online (James Martin Institute for Science and Civilization, University of Oxford and the MacKinder Centre for the Study of Long-Wave Events, London School of Economics and Political Science, July 27, 2009), http://eprints.lse.ac.uk/24569/.

33 Horst W. Rittel and Melvin M. Webber, "Dilemmas in a General Theory of Planning," *Policy Sciences* 4, no. 2 (1973): 155–169, https://doi.org/10.1007/bf01405730.

34 Robert E. Horn and Robert P. Weber, "New Tools For Resolving Wicked Problems," Random Musing (MacroVU(r), Inc. and Strategy Kinetics, LLC, 2007), https://www.strategykinetics.com/New_Tools_For_Resolving_Wicked_Problems.pdf.

35 Brian W. Head, "Forty Years of Wicked Problems Literature: Forging Closer Links to Policy Studies," *Policy and Society* 38, no. 2 (September 2018): 180–197, https://doi.org/10.1080/14494035.2018.1488797.

Chapter Four

MIXING SCIENCE AND POLITICS

"When you mix politics and science, you get politics."
—John M. Barry, author of *The Great Influenza*[1]

The relationship between scientific knowledge and political action is far from straightforward. Scientific knowledge invariably has blind spots and is provisional. Policy for complex issues is ever more reliant on knowledge, while science can deliver ever less certainty for society's most complex problems.[2] This tension creates a paradox that is normally resolved through political decisions, and not through the dissemination of "truth" in the sense of uncontested knowledge. Political considerations include public opinion, fiscal priorities, diplomatic considerations, and assessment of political and economic risks and rewards.[3]

Policy decisions related to scientific findings can be fairly straightforward (tame problems), or they can be highly contentious (wicked messes). Good decisions are those that lead to desired outcomes. However, the climate change problem provides much scope for disagreement among reasonable and intelligent people as to what constitutes desired outcomes, let alone how to achieve them.

In politicized arenas, science has become increasingly like litigation, where truth seeking has become secondary to advocacy on behalf of a preferred position or policy solution.[4] Encroachment of politics into socially-relevant science is unavoidable.[5] Problems arise when:[6]

- Driven by external pressures or for their own political purposes, scientists ignore data and research paths that would make their political point weaker or undermine their ideological perspective.
- Politicians interfere with the activities of science.
- Narrow framing of the scientific problem by policy makers, whereby government funding draws the efforts of scientists towards a narrow range of projects that supports preferred policies.
- Politicians, advocacy groups, journalists, and even scientists attempt to intimidate or otherwise silence scientists whose research is judged to interfere with their policy preferences or political agendas.

4.1 Models of the Science-Policy Interface

"Our ignorance is vast at every single step in this process as we go from emissions to concentrations to climate forcing to changes in temperature and other climatic attributes over to the impact assessments and damage estimates." (Environmental and energy scientist Ferenc Toth)[7]

The science-policy interface is comprised of social processes that encompass relations between scientists and other actors in the policy process, with the aim of enriching decision-making.[8]

There are three general models for the relationship between science and policy.[9] In the decisionistic model, scientific knowledge is present but values and beliefs dominate the political choice. In the technocratic model, this dependency between knowledge and politics is reversed, with politics relegated to the role of executive arm of science-driven policy. The pragmatic model recognizes that there is interdependence between values and facts, which creates an intersection between science and policy.

In the *Honest Broker*, policy scientist Roger Pielke Jr. articulated four idealized roles of scientists in policy making.[10] The pure scientist simply provides decision-makers with some fundamental information, allowing them to do what they want with it. The science arbiter answers factual questions decision-makers think are relevant. The issue advocate narrows the scope of choice and presents factual information to support a specific choice. The honest broker provides the decision-maker with as much information as possible in an effort to expand (or at least clarify) the scope of choice for decision-making.

Stealth advocacy occurs when scientists claim to be focusing on science but are really seeking to advance a political agenda. Stealth advocacy is inherent to the IPCC. While the IPCC claims that it is "policy-neutral" and "never policy-prescriptive,"[11] the entire framing of the IPCC Reports is around the mitigation of climate change through emissions reductions. A linguistic dissection of the IPCC reports showed that policy prescription in the IPCC assessment reports includes tacit policy models and other forms of tacit policy argumentation.[12]

4.2 Politicizing Climate Science

"Imperatives like skepticism and disinterestedness are being junked to fuel political warfare that has nothing in common with scientific methodology." (Physician John P. A. Ioannidis)[13]

Politicization of science is the manipulation of science for political gain. In support of a particular political agenda, it occurs when scientific research is influenced or manipulated by the government, business, advocacy groups, the media, or even scientists themselves. Political bias influences the scientific questions that are asked, the interpretation of the findings, what is cited, and what gets canonized. Often this occurs by filtering factual statements by scientists and the media with an eye to downstream political use.

Politicization of climate science became institutionalized in 1988, when the UN founded the IPCC to support a political agenda that became the UNFCCC in 1992. The narrow focus on "dangerous anthropogenic interference" has marginalized research on natural climate variability and also ignores impacts of a warmer climate that would be beneficial.

The UNFCCC climate policy "cart" has been well out in front of the climate science "horse" throughout its history. When the UNFCCC was established, there was as yet no clear signal of human-caused warming in the observations, as per the IPCC First Assessment Report (FAR) in 1990:

> "The size of this [global] warming is broadly consistent with predictions of climate models, but it is also of the same magnitude as natural climate variability."[14]

Scientific support for the 1992 UNFCCC Treaty was not based on observations, but rather on theoretical understanding of the greenhouse effect and simulations from first-generation global climate models. A key element of the UNFCCC treaty was the precautionary principle, as articulated at the Rio Earth Summit in 1992:

> "In order to protect the environment, the precautionary approach shall be widely applied by States according to their capabilities. Where there are threats of serious or irreversible damage, lack of full scientific certainty shall not be used as a reason for postponing cost-effective measures to prevent environmental degradation."[15]

It wasn't until the IPCC's Second Assessment Report in 1995 that:

> "The balance of evidence suggests a *discernible* human influence on global climate."[16]

Apparently, "discernible" was a sufficient threshold for the 1997 Kyoto Protocol, which established legally binding obligations for developed countries to reduce their greenhouse gas emissions in the period 2008–12. At the time, "discernible" was hotly contested because the word entered into the IPCC Second Assessment Report Summary for Policy Makers to appease policy makers that wanted a stronger conclusion.[17]

Experts on the politics of climate change Jaeger and O'Riordan describe early efforts made by Bert Bolin, the first Chairman of the IPCC, to keep the IPCC science separate from the policy process. This was a growing challenge, since politics were increasingly making the science irrelevant. Jaeger and O'Riordan stated that it was frankly impossible for such a panel actually to remain aloof from the political processes that both shape its existence and respond to its propositions.[18]

With the realization that there was an inexorable movement towards a climate treaty, there was an outpouring of skepticism from the scientific community. John Zillman, who served as Vice President and President of the World Meteorological Organization (WMO) during the period 1987–2003, stated that: "the greenhouse debate has now become decoupled from the scientific considerations that had triggered it."[19] Zillman further quoted these statements made by Pierre Morel, Director of World Climate Research Programme (WCRP):

"The increasing direct involvement of the United Nations [...] in the issues of global climate change, environment, and development bears witness to the success of those scientists who have vied for 'political visibility' and 'public recognition' of the problems associated with the earth's climate. The consideration of climate change has now reached the level where it is the concern of professional foreign-affairs negotiators and has therefore escaped the bounds of scientific knowledge (and uncertainty)."[20]

Skepticism about the IPCC was alive and well in the first half of the 1990s. However, with the publication of the IPCC Second Assessment Report in 1995, the IPCC's powerful political supporters acted to quell public criticism by characterizing it as the politicization of science by vested interests. By 2007, climate change had soared to the top of the international political agenda, and the IPCC was awarded the 2007 Nobel Peace Prize jointly with Al Gore.

The authority of the IPCC wasn't seriously challenged until November 2009, with the unauthorized release of emails from the Climatic Research Unit of the University of East Anglia (Climategate).[21] The emails revealed that a number of IPCC authors had violated the norms of science by evading Freedom of Information Act requests for data, manipulating the peer review process, downplaying uncertainty, and otherwise attempting to squash disagreement and dissent from "skeptics." There was a growing realization that in terms of the IPCC's public value, the social practices and quality of knowledge production matter as much as the content of the knowledge itself.

In 2010, the UN InterAcademy Council (IAC) was asked to review the IPCC's processes for developing its reports. The IAC report identified "significant shortcomings in each major step of [the] IPCC's assessment process."[22] Among its recommendations, the IAC said significant improvements were necessary, and criticized the IPCC for claiming to have high confidence in many statements for which there is actually little evidence. The IAC said that the IPCC was too insular, and could benefit from a greater variety of perspectives.[23]

Concerns that the IPCC had lost its scientific objectivity were personified by controversies surrounding IPCC Chairman Rachendra Pachauri (2002–15). The IPCC leadership under Pachauri was heavily criticized for his dismissal of an error in the Fourth Assessment Report that mistakenly said that Himalayan glaciers were in danger of disappearing by 2035. He failed to implement a conflict-of-interest policy for IPCC authors. Further, Pachauri engaged extensively with green lobbyists and advocacy groups and aggressively advocated for a range of policy solutions to climate change.[24]

The IPCC has been characterized as a "knowledge monopoly" by IPCC lead author Richard Tol.[25] Tol raises concerns that the quality of the assessment reports may have declined, and that the IPCC and its defenders may have used its power to hinder competitors. Nevertheless, the IPCC retains its authority in the UNFCCC deliberations, with its Sixth Assessment Reports published in 2021–22.

4.3 Scientizing Climate Policy

"What's the use of having developed a science well enough to make predictions if, in the end, all we're willing to do is stand around and wait for them to come true?" (Nobel Laureate chemist F. Sherwood Rowland)[26]

Scientization of policy is a response to intractable political conflict that attempts to transform the political issues into scientific ones. The rationale for scientization is the belief that science can act as a neutral arbiter of policy—if we could only determine the facts of a matter, the appropriate course of action would become clear. The problem is that science is neither neutral nor capable of answering political questions. The answers that science gives depend on what questions are asked, which inevitably involve value judgments. Science is not designed to answer questions about how the world ought to be, which is the domain of politics.

The phrase "follow the science" has a virtuous ring to it. However, science does not lead anywhere. It can illuminate various courses of action and quantify the risks and tradeoffs. But science cannot make choices for us. By following the science, decision-makers avoid taking responsibility for the choices they make. When science is invoked to legitimize the transfer of responsibility from democratic to technocratic bodies, it insulates the decisions from the realm of political debate.

Climate change is an example where science has been used as a proxy for political debate, which is partly a debate over the economic interests of this generation versus future generations. Science can easily get damaged in such proxy political debates. Both scientists and policy makers are at fault in expecting science to act as a neutral arbiter of different policy choices.

Policy makers are culpable in the misuse of science for policy making by:

- regarding science as a vehicle to avoid "hot potato" policy issues
- expecting black-and-white answers to complex problems
- demanding scientific arguments for their desired policies
- using scientific facts as a substitute for matters of public concern.

Scientists are culpable in the misuse of science for policy making by:

- naivete about expecting scientific evidence to drive policy
- conflating expert judgment with evidence
- playing power politics with their expertise
- combining expert knowledge with values that entangle disputed facts with identity-defining group commitments.

In political debates, "I believe in science"[27] is a statement often made by people who do not understand much about it. They use such statements about science as a way of declaring support for scientific institutions and belief in scientific propositions that are outside their knowledge and understanding.

In the context of climate change, "I believe in science" uses the overall reputation of science to give authority to the climate change consensus, shielding it from questioning and skepticism. "I believe in science" can be a signifier of social group identity that supports massive government legislation to limit or ban fossil fuels. Belief in science makes it appear that disagreement with this solution is equivalent to a rejection of the scientific method and worldview, and therefore the evident benefits that science has delivered. When exposed to science that challenges their political biases, these same "believers" are quick to claim "pseudo-science," without considering (or even understanding) the actual evidence or arguments.[28]

On the other side of the climate debate, calls for "sound science" are made that weaponize uncertainty and rigor to make it more difficult to use science in regulatory decision-making.[29] Individuals promoting sound science work to amplify uncertainty, create doubt, and undermine scientific discoveries that threaten their interests. The "sound science" tactic exploits a fundamental feature of the scientific process: science does not produce absolute certainty but is provisional and subject to change in the face of new evidence.

4.4 Climate Scientists and Power Politics

"the imprimatur of science is being smuggled into deliberations that actually deal with values and politics." (Environmental policy expert Chuck Herrick and environmental philosopher Dale Jamieson)[30]

In 2009, I had a discussion with a scientist who had served as a Coordinating Lead Author for the IPCC. This scientist was actively engaged in garnering signatures from scientists in support of a public statement about taking action on climate change. I argued that he was acting as a policy advocate; he claimed that he was not. He stated that the science demanded that CO_2 emissions be reduced urgently; he argued that he was not a policy advocate because he was not advocating for any *particular* emissions reduction technology or strategy.

As a group, scientists tend to be naive and unrealistic regarding the translation of their scientific evidence into policies, expecting public policy to be scientized. Scientists often assume that important improvements in policy making would ensue if only policy makers would incorporate the knowledge of researchers. To the extent that they want to ground policy-making exclusively in evidence, scientists misconceive policy-making and politics, which always involves making judgements about trade-offs between different goals.[31]

Scientists indirectly accrue political power by determining what evidence is deemed acceptable in political debates, what decisions ought to be made, what should be taught in schools, and what should be reported in the news. Scientists more directly accrue power through appointments to prestigious advisory boards, lucrative consultancies, and direct lines of communication with key policy makers.

The idea that science and scientists deserve special privilege in politics often leads to the temptation to exploit that privilege for political gain. While short-term political gains may be achieved, such exploitation ultimately works against science being afforded special privilege.[32]

Activist climate scientists have been using their privileged position to advance a clear political agenda. Such power politics by activist scientists has inflamed and polarized the climate change debate. Climate science risks losing its privileged position in the policy debates, unless disputed facts and value judgments can be disentangled from the identity-defining group commitments of activist climate scientists.[33]

There are also indirect political pressures that come from within the scientific community and the institutions that support science. Pressure to support the climate consensus and green policies comes from universities and professional societies, scientists themselves who are green activists, journalists and from federal funding agencies in terms of research funding priorities. Because evaluations by one's colleagues are so central to success in academia, it is easy to induce fear of social sanctions for expressing the ideas that, though not necessarily shown to be factually or scientifically wrong, are widely unpopular.

This political activism extends to the professional societies that publish journals and organize conferences. This activism has a gatekeeping effect on what gets published, who gets heard at conferences, and who receives professional recognition. Virtually all professional societies whose membership has any link to climate research have issued policy statements on climate change. These statements reflect the IPCC consensus and include calls to action. Here are conclusions from two of them:

- "Human-induced climate change requires urgent action" (American Geophysical Union)[34]
- "The APS reiterates its 2007 call to support actions that will reduce the emissions, and ultimately the concentration, of greenhouse gases" (American Physical Society)[35]

The biggest problem occurs when such advocacy influences the editorial policies of journals that are published by these professional societies. The journal *Science* is one of the highest impact scientific journals in the world. In 2015, the Chief Editor of *Science*, Marcia McNutt, published an op-ed in the journal entitled *The beyond-two-degree inferno*, stating "The time for debate has ended. Action is urgently needed."[36] Here is the concern: *Science* sends out for review only a small fraction of the submitted papers. Apart from the role the Chief Editor may have in selecting which papers go out for review or eventually get published, McNutt's essay sends a message to the other editors, reviewers, and scientists that papers challenging aspects of the climate consensus are not to be published in *Science*. This further leads to a perception of favored status for papers by activist authors that sound the "alarm."

The subsequent Chief Editor of *Science*, Jeremy Berg, articulated a different perspective. Berg stated that he is concerned about the loss of public trust in science, and how the behavior of scientists has contributed to that by straying into policy commentator roles, notably in the case of climate change.[37] Activism and advocacy by editors of scientific journals reduces the credibility of the journals, introduces biases into the science, and interferes with policy processes that are informed by science.

How many studies providing quality data and analyses relevant to climate controversies have gone unpublished because the researcher feared repercussions, did not see the value of reporting it, or did not want the results to become widely known? How many skeptical papers were not published by activist editorial boards? How many published papers have buried results in order to avoid highlighting findings that conflict with preferred narratives? I am aware of anecdotal examples of each of these actions, but the total number is unknowable.

4.5 Institutional Politics of Climate Science

"If we were driving in the wrong direction—in the direction where no new ideas can be accepted—then even if scientific work goes on, the progress would be stifled" (Astrophysicist Thomas Gold)[38]

Today, nearly all of scientific research is funded by governments and therefore by taxpayer dollars. In the United States, the current social contract between scientists and society dates back to Vannevar Bush and the conclusion of World War II. To address the emerging Soviet threat, US leaders demanded a broad, sustained program of research and development. The National Science Foundation (NSF) was established in 1950. Science in the United States has been very well funded, and scientists have delivered societal benefits in agriculture, energy, human health, information technology, and so on.[39]

However, as a result of this support science has strayed from the ideal of independent, self-motivated individuals, and has become increasingly bureaucratized. For research areas where the political stakes are high such as climate change, any dissent from the official consensus may be hazardous to a researcher's career.

Meteorologist and science policy analyst Bill Hooke describes how stresses in the twenty-first century, including funding levels, have frayed the fabric of the social contract between scientists and society.[40] Hooke is concerned that many scientists have responded to these stresses by resorting to advocacy. While Hooke rightfully raises concerns about the behaviors and motives of scientists, the other side of the social contract is at least equally problematical. Politicians have used climate science to support a political agenda.

Here is how research funding motivates what is going on with regard to climate science. Success for individual researchers, particularly at the large research universities, is driven by the research dollars that they attract—big lab spaces, high salaries, institutional prestige, and career advancement. At the Program Manager level within a funding agency, success is reflected in growing the size of their program (e.g., more funding) and having some high-profile results (e.g., press releases). At higher levels, Divisional administrators are competing for budget dollars against the other Divisions – tying their research programs to a national/international policy priority such as climate change helps in this competition. At the agency level, success is reflected in growing, or at least preserving, the agency's budget. Aligning yourself, your program, your agency with the current political imperatives is a key to success.

When a general attitude or a viewpoint has become established in a subject, either naturally or to be consistent with political values, then it is very easy to obtain funds to do work in that subject on the basis of "shoehorn science." If a proposal conforms to the consensus framework, it is relatively easy to get funded. After a sufficient amount of diligent shoehorn science, facts have become forced into a pattern that has become preordained.

It is very difficult to obtain federal research funding for climate science research that challenges the existing paradigm. Difficulty in the peer review process is only part of the problem. One problem is reflected in an email I received from a scientist employed at NASA (US):

> "I was at a small meeting of NASA-affiliated scientists and was told by our top manager that he was told by his NASA boss that we should not try to publish papers contrary to the current global warming claims, because he (the NASA boss) would then have a headache countering the undesirable publicity."[41]

The biggest problem is that calls for proposals from the federal funding agencies make an implicit assumption of the dominance of human caused global warming in the topics for which they are requesting research proposals (shoehorning). Hence scientists have little motivation to work on anything else. In the area of climate science, government funding provides little motivation for research that is needed to improve fundamental understanding and modeling of climate dynamics (see Part Two of this book). The end result is "climate taxonomy"—research that analyzes the results of climate model simulations to infer dire societal consequences.

Scientists with perspectives that are not consistent with the consensus are at best marginalized, in that they find it difficult to obtain funding and get papers published by gatekeeping journal editors. At worst, they are ostracized by labels of "denier" or "heretic" (see Section 2.4.2). Independent scientists (of independent, or no financial means) are increasingly asking important and fundamental questions that aren't relevant to government research funding priorities. In a recent interview, I was describing the "climate auditors" and the important role that they play in quality control of climate science (Section 2.4.1). The interviewer said that "it is good to see the system working." Well, if the auditors are essential for the system to work, they are few in number and receive no government funding.

Policymakers bear the responsibility of the mandate that they give to panels of scientific experts. In the case of climate change, the UNFCCC demanded of the IPCC too much precision in support of a pre-ordained policy solution where complexity, chaos, and our current understanding does not support such precision. Asking scientists to provide simple policy-ready answers for complex matters results in an impossible situation for scientists and misleading outcomes for policy makers.

These concerns raise the issue as to whether the social contract between climate science, the institutions that support climate science, and policy makers is operating in a manner that is healthy for either the science or the policy process.

Notes

1 John Kopp and Bob McGovern, "100 Years Ago, 'Spanish Flu' Shut down Philadelphia—and Wiped out Thousands," PhillyVoice, September 27, 2018, https://www.phillyvoice.com/100-years-ago-spanish-flu-philadelphia-killed-thousands-influenza-epidemic-libery-loan-parade/.

2 Nico Stehr and Reiner Grundmann, "How Does Knowledge Relate to Political Action?," *Innovation: The European Journal of Social Science Research* 25, no. 1 (2012): 29–44, https://doi.org/10.1080/13511610.2012.655572.

3 Peter Gluckman, "Policy: The Art of Science Advice to Government," *Nature* 507, no. 7491 (2014): 163–165, https://doi.org/10.1038/507163a.

4 Geoff Costeloe, "Arguing in America," Quillette (Quillette, July 6, 2020), https://quillette.com/2020/07/06/arguing-in-america/.

5 Lucas Bergkamp, "Politics and the Changing Norms of Science," Climate Etc., October 24, 2016, https://judithcurry.com/2016/10/25/politics-and-the-changing-norms-of-science/.

6 Heather Douglas, "Scientific Integrity in a Politicized World," in Logic, Methodology and Philosophy of Science: Proceedings of the 14th International Congress (Nancy): Logic and Science Facing the New Technologies (London, UK: College Publications, 2014), 253–268.

7 Ferenc L Toth, "Practice and Progress in Integrated Assessments of Climate Change," *Energy Policy* 23, no. 4–5 (1995): 253–267, https://doi.org/10.1016/0301-4215(95)90152-w.

8 Sybille van den Hove, "A Rationale for Science–Policy Interfaces," Futures 39, no. 7 (2007): 807–826, https://doi.org/10.1016/j.futures.2006.12.004.

9 Habermas Jürgen and Jeremy J. Shapiro, *Toward a Rational Society Student Protest, Science, and Politics* (Boston, MA: Beacon Press, 1971).

10 Roger A. Pielke, The Honest Broker: Making Sense of Science in Policy and Politics (Cambridge, UK: Cambridge University Press, 2007).

11 "Intergovernmental Panel on Climate Change," IPCC, accessed November 15, 2021, https://archive.ipcc.ch/organization/organization.shtml.

12 Kjersti Fløttum et al., "Synthesizing a Policy-Relevant Perspective from the Three IPCC 'Worlds'—a Comparison of Topics and Frames in the SPMs of the Fifth Assessment Report," *Global Environmental Change* 38 (2016): 118–129, https://doi.org/10.1016/j.gloenvcha.2016.03.007.

13 John P.A. Ioannidis, "How the Pandemic Is Changing Scientific Norms," Tablet Magazine, September 9, 2021, https://www.tabletmag.com/sections/science/articles/pandemic-science.

14 Ulrich Cubasch et al., "Introduction," *Climate Change 2013—The Physical Science Basis Contribution of Working Group I to the Fifth Assessment Report of the Intergovernmental Panel on Climate Change*, 2013, 119–158, https://doi.org/10.1017/cbo9781107415324.007.

15 United Nations, *Report of the United Nations Conference on Environment and Development: Rio De Janeiro, 3–14 June 1992* (New York, NY: United Nations, 1993).

16 Ed Houghton et al., *Climate Change 1995: The Science of Climate Change (IPCC)* (Cambridge, UK: Cambridge University Press, 1998), 4.

17 Bernie Lewin, *Searching for the Catastrophe Signal: The Origins of the Intergovernmental Panel on Climate Change* (London, UK: GWPF, 2017).

18 Jill Jaeger and Timothy O'Riordan, eds., *Politics of Climate Change: A European Perspective* (London, UK: Routledge, 1996).

19 Bernie Lewin, *Searching for the Catastrophe Signal: The Origins of the Intergovernmental Panel on Climate Change* (London, UK: GWPF, 2017).

20 Ibid.

21 Steven Mosher and Thomas Fuller, *Climategate: The Crutape Letters* (Lexington, KY: Createspace, 2010).

22 InterAcademy Council, *Climate Change Assessments: Review of the Processes and Procedures of the IPCC* (Amsterdam: InterAcademy Council, 2010), 13.

23 Ibid.

24 Donna Laframboise, *Into the Dustbin: Rajendra Pachauri, the Climate Report & the Nobel Peace Prize* (Port Dover, CA: Ivy Avenue Press, 2013).

25 Richard S. Tol, "Regulating Knowledge Monopolies: The Case of the IPCC," *Climatic Change* 108, no. 4 (2011): 827–839, https://doi.org/10.1007/s10584-011-0214-6.

26 Paul Brodeur, "In the Face of Doubt," The New Yorker (The New Yorker, June 2, 1986), https://www.newyorker.com/magazine/1986/06/09/in-the-face-of-doubt.

27 Evan Lehmann, "Hillary Clinton Declares, 'I Believe in Science,'" Scientific American (Scientific American, July 29, 2016), https://www.scientificamerican.com/article/hillary-clinton-declares-i-believe-in-science/.

28 Robert Tracinski, "Why I Don't 'Believe' in 'Science,'" The Bulwark, March 26, 2019, https://www.thebulwark.com/why-i-dont-believe-in-science/.

29 Christie Aschwanden, "There's No Such Thing as 'Sound Science'," FiveThirtyEight (ABC News, December 6, 2017), https://fivethirtyeight.com/features/the-easiest-way-to-dismiss-good-science-demand-sound-science/.

30 Charles N Herrick and Dale Jamieson, "Junk Science and Environmental Policy: Obscuring Public Debate with Misleading Discourse," *Philosophy and Public Policy Quarterly* 21, no. 2–3 (2001): 11–16.

31 Roger Pielke, "Should Scientists Rule?," The Breakthrough Institute, January 7, 2013, https://thebreakthrough.org/articles/should-scientists-rule.

32 Judith Curry, "Is Much of Our Effort to Combat Global Warming Actually Making Things Worse?," Climate Etc., May 26, 2016, https://judithcurry.com/2016/05/23/is-much-of-our-effort-to-combat-global-warming-actually-making-things-worse/.

33 Peter Tangney, "Between Conflation and Denial – the Politics of Climate Expertise in Australia," *Australian Journal of Political Science* 54, no. 1 (2018): 131–149, https://doi.org/10.1080/10361146.2018.1551482.

34 AGU, "Position Statement on Climate Change," AGU, August 2013, https://www.agu.org/Share%20and%20Advocate/Share/Policymakers/Position%20Statements/Human-induced%20climate%20change%20requires%20urgent%20action.

35 APS, "15.3 Statement on Earth's Changing Climate," American Physical Society, November 14, 2015, https://www.aps.org/policy/statements/15_3.cfm.

36 Marcia McNutt, "The beyond-Two-Degree Inferno," *Science* 349, no. 6243 (March 2015): 7–7, https://doi.org/10.1126/science.aac8698.

37 David Matthews, "Science Editor-in-Chief Sounds Alarm over Falling Public Trust," Times Higher Education (THE), February 16, 2017, https://www.timeshighereducation.com/news/science-editor-chief-sounds-alarm-over-falling-public-trust.

38 Thomas Gold, "New Ideas in Science," *Journal of Scientific Exploration* 3, no. 2 (1989): 103–112.

39 "A Timeline of NSF History," National Science Foundation—Where Discoveries Begin, accessed August 3, 2022, https://www.nsf.gov/about/history/overview-50.jsp.

40 William Hooke, "Reaffirming the Social Contract between Science and Society," Eos, March 17, 2015, https://eos.org/opinions/reaffirming-the-social-contract-between-science-and-society.

41 Judith Curry, email message, October 8, 2015.

Part Two

UNCERTAINTY OF TWENTY-FIRST CENTURY CLIMATE CHANGE

"We need both new wine and new bottles—new ideas as well as new institutions to make them vibrant."

—Philosopher Robert Frodeman[1]

The information available on climate change for policy-making purposes is plagued by large inherent uncertainties. Apart from uncertainties about future emissions, this includes uncertainties in projections from global climate models as well as climate change impacts, economic costs, and policy responses. Additional uncertainties are associated with the technological, social, and political contexts surrounding the policy response options.

Part Two describes a new framework for thinking about the climate change problem. This framework does not attempt to resolve the plethora of problems identified in Part One. Instead, it seeks to bypass most of the existing problems that have contributed to the acrimonious public debate and policy gridlock surrounding climate change.

At the heart of this reframing is a better understanding and accommodation of all aspects of uncertainty surrounding climate change. This new framework moves away from producing consensus science that supports the linear model of "predict-then-act," to a scientific process that supports a scenario-rich robust decision-making framework. This new framework opens up space for disagreement among scientists and broadens participation to include individuals with a wide range of expertise.

Part Two focuses on plausible outcomes of climate change in the twenty-first century that are relevant to policy making, including natural climate variability plausible worst-case scenarios. This formulation bypasses debates over the historical and paleoclimate data records. It starts the climate change clock in the year 2000, which characterizes the current climate to which humanity has more-or-less adapted. A focus on regional climate variability in the context of local vulnerabilities is regarded to be more important for decision-making than changes in global mean temperature.

This framework for the climate change problem sets the stage for a response framework (Part Three) based on robust decision-making aimed at improving human well-being in the twenty-first century.

Note

1   Robert Frodeman, "Socrates Untenured: New Wine and New Bottles," *Issues in Science and Technology* 36, no. 4 (Summer 2020): 28–29.

Chapter Five

THE CLIMATE CHANGE "UNCERTAINTY MONSTER"

"Science [...] can never solve one problem without raising ten more problems."
—Irish playwright George Bernard Shaw[1]

In the linear model of expertise and decision-making (Section 4.1.1), uncertainty and doubt are enemies of action. Attempts to hide or simplify uncertainty are at the heart of some of the most acrimonious debates over climate change.

Uncertainty is a state of incomplete knowledge arising from a lack of information or from disagreement about what is known or even knowable. The fundamental mischaracterization of the climate change problem as a tame rather than a wicked problem (Section 3.4) has resulted in institutionalized efforts to ignore, simplify, reduce, and control uncertainty.

In frontier research, the objective is to extend knowledge in ways that change how we think about a particular topic. Doubt and uncertainty are inherent at the knowledge frontier. Researchers using the same data and hypotheses can come to different conclusions in the context of a vast universe of different possible research designs.[2] While extending the knowledge frontier often reduces uncertainty in some dimensions, inevitably it leads to greater uncertainty in other dimensions as unanticipated complexities are discovered. Careful consideration of uncertainties is not a central element of frontier research.

However, uncertainty assessment and characterization are paramount for science that is targeted at policy-making. The greatest challenges are presented by frontier research that is policy relevant, such as climate change.

This chapter presents a consilience of diverse ideas about uncertainty, which provides the foundation for a strategy to accommodate uncertainty at the climate science-policy interface and in decision-making.

5.1 The Uncertainty Monster

"He who fights with monsters might take care lest he thereby become a monster." (Nineteenth-century German philosopher Friedrich Nietzsche)[3]

With a heritage in monster theory,[4] Dutch social scientist Jeroen van der Sluijs introduced the concept of the "uncertainty monster" in context of the different ways that the scientific community responds to the monstrous uncertainties associated with environmental problems.[5] The "monster" arises from the confusion and ambiguity associated with knowledge versus ignorance, objectivity versus subjectivity, facts versus values, prediction versus speculation, and science versus policy. The uncertainty monster gives rise to discomfort and fear, particularly with regard to our reactions to things or situations we cannot understand or control, including the presentiment of radical unknown dangers. Specifically, in the context of decision-making, the uncertainty monster is perceived as something to be feared and avoided.

In 2011, I coauthored a paper with Peter Webster entitled "Climate Science and the Uncertainty Monster,"[6] which adapted van der Sluijs' concept of the uncertainty monster to climate science. This paper was motivated by the Royal Society's Workshop on Handling Uncertainty in Science, which was held in London in March 2010.[7,8] This was a highly multidisciplinary meeting that discussed how scientists from a range of disciplines handle the issue of uncertainty. Speakers included luminaries from the fields of theoretical physics, statistics, philosophy, mathematics, cosmology, health, economics and finance, and climate.

Apart from one talk on communicating uncertainty about climate change, there was no mention of consensus. At the time the Workshop was held, the world was in the grip of Climategate. This was a period when I was questioning the consensus-seeking process of the IPCC, its inadequate treatment of uncertainty, and overconfidence in its conclusions. The Royal Society Workshop was seminal in providing a new anchor for my thinking, and I decided to start a new research project on uncertainty in climate science. Once I spotted Jeroen van der Sluijs' uncertainty monster paper, I knew that I had found my theme, and the "climate uncertainty monster" was born.

In our uncertainty monster paper, we adapted van der Sluijs' strategies of coping with the uncertainty monster at the climate science-policy interface:

Monster hiding. Uncertainty hiding or the "never admit error" strategy can be motivated by a political agenda or because of fear that uncertain science will be judged as poor science by the outside world. However, the monster may be too big to hide and hiding enrages the monster. Strategies that attempt to silence dissent are examples of monster hiding; ample examples were provided by the Climategate emails.[9]

Monster exorcism. The uncertainty monster exorcist focuses on reducing uncertainty through advocating for more research. However, monster theory predicts that reducing uncertainty will prove to be in vain in the long run for complex problems. For each head of the uncertainty monster that science chops off, several new monster heads will pop up due to unforeseen complexities, analogous to the Hydra beast of Greek mythology. In climate science, there was hope in the 1990s that the uncertainty in climate model predictions could be reduced with more research. However, increased understanding continues to add additional dimensions to the problem, increasing uncertainty.

Monster adaptation. Monster adapters attempt to transform the monster by subjectively quantifying and simplifying the assessment of uncertainty. Monster adaptation was formalized by the IPCC in the context of characterizing uncertainty in a consensus approach (e.g., Moss and Schneider 2000).[10] This strategy has bred overconfidence in the IPCC's conclusions.

Monster detection. As described in Section 2.4 on skeptics, the first type of uncertainty detective is a scientist challenging existing theses and working to extend knowledge frontiers. A second type is a watchdog auditor, whose main concern is accountability, quality control, and transparency of the science. A third type is a merchant of doubt, who distorts and magnifies uncertainties as an excuse for inaction.

Monster assimilation. Monster assimilation is about learning to live with the monster and giving uncertainty an explicit place in the contemplation and management of environmental risks. Assessment and communication of uncertainty and ignorance, along with extended peer communities, are essential in monster assimilation. The challenge to monster assimilation is the ever-changing nature of the monster and the birth of new monsters.

5.2 Uncertainty Typologies

"Real knowledge is to know the extent of one's ignorance." (Chinese philosopher Confucius)[11]

Notions of uncertainty range from everyday usage in common parlance to specific definitions appearing in the philosophical and scientific literature. There have been several attempts at systematic taxonomies of uncertainty.[12,13,14]

The nature of uncertainty is expressed by the distinction between epistemic uncertainty and ontic uncertainty. Epistemic uncertainty is associated with imperfections of knowledge, which may be reduced by further research and empirical investigation. Ontic (often referred to as aleatory) uncertainty is associated with inherent variability or randomness; as such, ontic uncertainties are not reducible.

Epistemic uncertainty of the state of the climate system includes uncertainty due to limitations of measurement devices, insufficient data, extrapolations and interpolations, and variability over time or space. Epistemic uncertainties in global climate models include missing or inadequately treated physical processes, uncertainty in the numerical value of physical parameters, errors associated with the discretization and algorithmic approximations, and errors in the specification of external forcing. Natural internal variability of the climate system contributes to ontic uncertainty in climate simulations.

The location of uncertainty relates to what we are uncertain about. The location may relate to a measurement, a data set, the nature of an event, or aspects related to a model such as the model parameters or structure.

5.2.1 Level of Uncertainty

"Not only does God definitely play dice, but He sometimes confuses us by throwing them where they can't be seen." (Theoretical physicist Stephen Hawking)[15]

The level of uncertainty relates to the spectrum of knowledge and ignorance, and as such is closely related to the levels of incertitude described above. A complete logical structure of the level of uncertainty is characterized as a progression between deterministic understanding and total ignorance: statistical uncertainty, scenario uncertainty, and recognized ignorance.

Complete certainty implies deterministic knowledge, with no uncertainty.

Statistical uncertainty can be described adequately in statistical terms. Outcomes can never be known precisely, but precise, decision-relevant probability statements can be provided for each outcome.

Scenario uncertainty implies that a range of plausible outcomes (scenarios) are enumerated, but with a weak basis for ranking them in terms of likelihood. A scenario is a plausible but unverifiable description of how the system and/or its driving forces may develop in the future. Scenarios may be regarded as a range of discrete possibilities with no a priori allocation of likelihood. An example of scenario uncertainty of relevance to climate change is associated with future greenhouse gas emission scenarios used to force global climate models.

Deep uncertainty (recognized ignorance) refers to fundamental uncertainty in the mechanisms being studied and a weak scientific basis for developing scenarios; future outcomes lie outside of the realm of regular or quantifiable expectations. Reducible ignorance may be resolved by conducting further research, whereas irreducible ignorance implies that research cannot improve knowledge. *Border with ignorance* denotes knowledge of the presence or possibility of ignorance.[16]

Total ignorance: is the deepest level of uncertainty, and refers to potential issues to the extent that we do not even know that we do not know. *Conscious ignorance*, where we know what we don't know, is contrasted with unacknowledged or *meta-ignorance*, where we don't even consider the possibility of error.[17]

An alternative taxonomy for levels of uncertainty uses the terms known knowns, known unknowns, unknown unknowns, and unknowable unknowns.[18] With respect to the classification above, a known known encompasses both deterministic certainty and statistical uncertainty. The known unknowns encompass scenario uncertainty. The unknown unknown corresponds to ignorance, including border with ignorance where the unknowns are perhaps suspected. Unknown unknowns may be the targets of frontier research or philosophical speculations, or may correspond to future circumstances or outcomes that are impossible to predict or to know when or where to look for them. Unknown knowns refer to that which is known, but ignored for some reason.

5.3 Uncertainty and the IPCC

"Some things are believed because they are demonstrably true. But many other things are believed simply because they have been asserted repeatedly – and repetition has been accepted as a substitute for evidence." (Economist Thomas Sowell)[19]

Understanding uncertainty associated with the complex, nonlinear, and chaotic climate system, let alone managing it, is a very challenging endeavor. Hence it is tempting for scientists and policy makers to simplify uncertainty to make it appear that the appropriate considerations have been undertaken.

How has the IPCC dealt with the challenge of uncertainty and assigning confidence levels to its conclusions? Prior to 2000, uncertainty was dealt with by the IPCC in an ad hoc manner. The *Uncertainty Guidance* paper published in 2000 recommended steps for assessing uncertainty in the IPCC Assessment Reports.[20] It recommended a common vocabulary to express quantitative levels of confidence based on the amount of evidence and the degree of agreement (consensus) among experts.

The actual implementation of this guidance in the IPCC Reports has focused more on communicating uncertainty rather than on actually characterizing it[21] or acknowledging areas of incomplete knowledge or ignorance. Defenders of the IPCC uncertainty characterization argue that subjective consensus expressed using simple terms is more easily understood by policy makers. While adhering (mostly) to the principles set forth in the "Uncertainty Guidance" paper, the IPCC oversimplifies the characterization of uncertainty by substituting expert judgment for a thorough understanding of uncertainty. They look at "evidence for" and "evidence against" (but somehow neglect a lot of the evidence against), and completely neglect to acknowledge ignorance.

In 2011, the journal *Climatic Change* published *Special Issue: Guidance for Characterizing and Communicating Uncertainty and Confidence in the Intergovernmental Panel on Climate Change.*[22] A total of 15 papers were invited for this *Special Issue*, including one that I authored. The point was made in several papers (including mine) that the greatest need was to acknowledge areas of ignorance and disagreement, particularly with regard to potentially high impact outcomes.[23] Overconfidence is an inevitable result of neglecting ignorance. There were also several substantive critiques of the consensus-seeking expert judgment process used by the IPCC.[24]

The IPCC faces a daunting challenge in characterizing uncertainty because unquantifiable uncertainties and ignorance dominate the quantifiable uncertainties. However, this challenge does not justify the IPCC's oversimplification of characterizing uncertainty in the climate system (monster simplification), which has led to misleading overconfidence and has fueled public distrust. More significantly, overconfidence can lead to poor policy decisions and outcomes.

5.4 Taming the Uncertainty Monster

"I used to be scared of uncertainty; now I get a high out of it." (Actor Jensen Ackles)[25]

Symptoms of an enraged uncertainty monster include increased levels of confusion, ambiguity, discomfort, and doubt. Evidence that the monster is currently enraged includes the acrimony of the public debate on climate change, militant consensus enforcement, and the doubt that is being expressed regarding the feasibility and impacts of rapid emissions reductions.

The climate change uncertainty monster is too big to hide, exorcise or adapt. Increasing concern that scientific and political dissent is underexposed by the IPCC's consensus approach argues for ascendancy of the monster detection and assimilation approaches.

Objectives of taming the monster at the institutional level are to improve the environment for dissent in scientific arguments, make climate science less political, clarify the political values and visions in play, expand political debate, and encourage a broader range of experts to participate in the evaluation of climate science and its institutions.

"Physicist Richard Feynman's famous address on Cargo Cult Science clearly articulates the scientist's responsibility:

"Details that could throw doubt on your interpretation must be given, if you know them. If you make a theory [...] then you must also put down all the facts that disagree with it, as well as those that agree with it. In summary, the idea is to try to give all of the information to help others to judge the value of your contribution; not just the information that leads to judgment in one particular direction or another."[26]

Individual scientists can tame the uncertainty monster by clarifying the confusion and ambiguity associated with knowledge versus ignorance and objectivity versus subjectivity. At the 2010 Royal Society Workshop, statistician David Spiegelhalter provided the following advice:[27]

• We should try to quantify uncertainty where possible.
• All useful uncertainty statements require judgment and are contingent.
• We need clear language to honestly communicate deeper uncertainties with due humility and without fear.
• For public confidence, trust is more important than certainty.

The "hopeful monster" is a colloquial term used in evolutionary biology to describe the production of new major evolutionary groups.[28] New information technology and the open knowledge movement are enabling the emergence of the hopeful monster. Social computing has unrealized potential to facilitate understanding of complex issues, drive public policy innovation and provide transparency. These new technologies facilitate the rapid diffusion of information and sharing of expertise, giving hitherto unrealized power to peer communities. Climategate illustrated the importance of the blogosphere (and increasingly, Twitter) as an empowerment of the extended peer community whereby, "criticism and a sense of probity needed to be injected into the system by the extended peer community from the (mainly) external blogosphere."[29]

While the uncertainty monster will undoubtedly evolve and even grow, it can be tamed through acknowledgment, and understanding. Most importantly, we can learn to live with the monster by adapting our policies to explicitly accommodate uncertainty and disagreement.

Notes

1 George Bernard Shaw, "Speech at the Einstein Dinner," *New York Times*, October 29, 1930, 12.
2 Nate Breznau et al., "Observing Many Researchers Using the Same Data and Hypothesis Reveals a Hidden Universe of Uncertainty," *MetaArXiv* (March 24, 2021), https://doi:10.31222/osf.io/cd5j9.
3 Friedrich Wilhelm Nietzsche, *Beyond Good and Evil* (Mineola, NY: Dover Publications, 1997).

4 Martijntje Smits, "Taming Monsters: The Cultural Domestication of New Technology," *Technology in Society* 28, no. 4 (November 2006): 489–504, https://doi.org/10.1016/j.techsoc.2006.09.008.

5 Jeroen van der Sluijs, "Uncertainty as a Monster in the Science–Policy Interface: Four Coping Strategies," *Water Science and Technology* 52, no. 6 (January 2005): 87–92, https://doi.org/10.2166/wst.2005.0155.

6 Judith Curry and Peter Webster, "Climate Science and the Uncertainty Monster," *Bulletin of the American Meteorological Society* 92, no. 12 (2011): 1667–1682, https://doi.org/10.1175/2011bams3139.1.

7 "Handling Uncertainty in Science," Royal Society, 2010, https://royalsociety.org/science-events-and-lectures/2010/uncertainty-science/.

8 T. N. Palmer and P. J. Hardaker, "Handling Uncertainty in Science," *Philosophical Transactions of the Royal Society A: Mathematical, Physical and Engineering Sciences* 369, no. 1956 (2011): 4681–4684, https://doi.org/10.1098/rsta.2011.0280.

9 Steven W. Mosher and Thomas Fuller, *Climategate: The Crutape Letters* (Lexington, KY: CreateSpace, 2010).

10 R H Moss and S H Schneider, "Uncertainties in the IPCC TAR: Recommendations to Lead Authors for More Consistent Assessment and Reporting," in *Guidance Papers on the Cross Cutting Issues of the 3rd Assessment Report of the IPCC*, ed. Pachauri R K (Rajendra K.), T. (Tomihiro) Taniguchi, and K. (Kanako) Tanaka (Geneva, CH: Intergovernmental Panel on Climate Change, 2000), 33–51.

11 Disciples of Confucius, *The Analects of Confucius* (translated), (475–221 BC).

12 W.E. Walker et al., "Defining Uncertainty: A Conceptual Basis for Uncertainty Management in Model-Based Decision Support," *Integrated Assessment* 4, no. 1 (2003): 5–17, https://doi.org/10.1076/iaij.4.1.5.16466.

13 Arthur Caesar Petersen, *Simulating Nature: A Philosophical Study of Computer-Simulation Uncertainties and Their Role in Climate Science and Policy Advice* (Boca Raton, FL: CRC Press, 2012).

14 David J. Spiegelhalter and Hauke Riesch, "Don't Know, Can't Know: Embracing Deeper Uncertainties When Analysing Risks," *Philosophical Transactions of the Royal Society A: Mathematical, Physical and Engineering Sciences* 369, no. 1956 (2011): 4730–4750, https://doi.org/10.1098/rsta.2011.0163.

15 Stephen Hawking, "Does God Play Dice," Stephen Hawking, 1999, https://www.hawking.org.uk/in-words/lectures/does-god-play-dice.

16 Silvio O. Funtowicz and Jerome R. Ravetz, "Science for the Post-Normal Age," *Futures* 25, no. 7 (1993): 739–755, https://doi.org/10.1016/0016-3287(93)90022-l.

17 Gabriele Bammer and Michael Smithson, *Uncertainty and Risk: Multidisciplinary Perspectives* (London, UK: Earthscan, 2009).

18 Donald Rumsfeld, "Defense Department Briefing" (C-SPAN, February 12, 2002), https://www.c-span.org/video/?168646-1%2Fdefense-department-briefing.

19 Thomas Sowell, *Economic Facts and Fallacies* (New York, NY: Basic Books, 2011).

20 R H Moss and S H Schneider, "Uncertainties in the IPCC TAR."

21 Arthur Caesar Petersen, *Simulating Nature*.

22 Gary Yohe and Michael Oppenheimer, eds., "Special Issue: Guidance for Characterizing and Communicating Uncertainty and Confidence in the Intergovernmental Panel on Climate Change," *Climatic Change* 108, no. 4 (2011).

23 Judith Curry, "Reasoning about Climate Uncertainty," *Climatic Change* 108, no. 4 (September 2011): 723–732, https://doi.org/10.1007/s10584-011-0180-z.

24 Gary Yohe and Michael Oppenheimer, eds., "Special Issue."

25 Armin W. Doerry, "Noise and Noise Figure for Radar Receivers.," October 2016, https://doi.org/10.2172/1562649.

26 Richard Feynman, "Cargo Cult Science," Caltech's 1974 Commencement Address (Caltech, 1974), https://calteches.library.caltech.edu/51/2/CargoCult.htm.

27 David Spiegelhalter, "Quantifying Uncertainty," Royal Society Downloads, 2010, https://downloads.royalsociety.org//audio/DM/DM2010_03/Speigelhalter.mp3.

28 Stephen Jay Gould, "The Return of Hopeful Monsters," *Natural History* 86, no. June/July (1977): 22–30.

29 Jerome Ravetz, "Climategate: Plausibility and the Blogosphere in the Post-Normal Age.," Watts Up With That?, February 9, 2010, http://wattsupwiththat.com/2010/02/09/climategate-plausibility-and-the-blogosphere-in-the-post-normal-age/.

Chapter Six

CLIMATE MODELS

"Models try to squeeze the blooming, buzzing, confusion into a miniature Joseph Cornell box, and then, if it more or less fits, assume that the box is the world itself."
                    —South African writer and financial engineer Emanuel Derman[1]

A wide range of models are used in climate science to study the climate system and its behavior across multiple temporal and spatial scales. Climate models range from simple energy balance models to exceedingly complex Earth system models. Global climate models (GCMs) create a coarse-grained simulation of the Earth's climate system using computers. GCM development has followed a pathway mostly driven by scientific curiosity, although efforts have been constrained by computational limitations. GCMs were originally designed as a tool to help understand how the climate system works, integrating community knowledge so that the implications of collective knowledge can be explored. GCMs are used by researchers to represent aspects of climate that are difficult to observe, experiment with theories by enabling otherwise infeasible calculations, understand a complex system of equations that would otherwise be impenetrable, and explore the climate system to identify unexpected outcomes. As such, GCMs have emerged as an important tool in climate research.

While GCMs continue to be used by scientists to increase understanding about how the climate system works, their outputs play a central role in developing international, national, and local policies. In the context of the IPCC and the UNFCCC, climate models are used to assess the causes of recent climate change, predict future climate states, provide guidance for emissions reduction policies, support local adaptation policies, and to provide inputs for Integrated Assessment Models to assess the social cost of carbon.

There is considerable uncertainty and disagreement about the extent to which climate model simulations provide accurate information about the world. Scientists lack the data they need to comprehensively test a model's performance, and scientists disagree as to which metrics should be used for climate model evaluation. Conclusions about model reliability are made more difficult insofar as the climate system is nonlinear, complex, and not well understood theoretically.

Initially, climate research programs aimed at the reduction of the uncertainties in climate models.[2,3] However, since the mid-1990s, it has been increasingly recognized that more research does not necessarily reduce the overall uncertainties regarding future climate change.[4] Ongoing research continues to reveal unforeseen complexities in the climate system that add to the perceived uncertainty of climate models; examples include ice sheet dynamics, under ocean and under ice geothermal heat flux, atmospheric chemistry processes that impact the aerosol indirect radiative effect, and solar indirect effects. Furthermore, there are unresolvable limits to the reduction of scientific uncertainties owing to limits to our capacity to handle complexity, computer limitations, and the inherent unpredictability of the climate system.

Because of these distinctive features of GCMs, their epistemology has been of particular interest to philosophers of science.[5,6,7,8] Relevant topics in the epistemology of computer simulation include how and when trust in simulation results can be justified and how uncertainties associated with simulation results should be probed and characterized.

6.1 Global Climate Models

"People build the most complicated model they can think of, include everything, then run it once on the largest computer with the highest resolution they can afford, then wonder how to interpret the results." (Climate modeler Reto Knutti)[9]

GCMs have modules that simulate the atmosphere, ocean, land surface, sea ice, and glaciers. GCMs use complex mathematical equations that can only be approximately solved on computers. The atmospheric module simulates radiative transfer through the atmosphere and the circulations that produce the evolution and variations of winds, temperature, humidity, and pressure. The ocean module simulates the oceanic circulation, how it transports heat and exchanges heat and moisture with the atmosphere. The land surface module simulates how vegetation, soil, and snow cover exchange energy and moisture with the atmosphere. GCMs also include modules for sea ice and glacier ice. Chemical processes are increasingly being included in Earth System Models, to simulate the global carbon cycle and to improve the simulation of atmospheric composition. Some of the equations in climate models are based on the laws of physics such as Newton's laws of motion and the laws of thermodynamics. However, there are key processes in GCMs that are approximated and not based on physical laws.

To solve these complex equations on a computer, GCMs divide the atmosphere, oceans, and land into a three-dimensional grid system. Discretized versions of equations are then calculated for each cell in the grid repeatedly for each time step that makes up the simulation period. The number of cells in the grid system determines the model resolution, whereby each grid cell effectively has a uniform temperature, wind field, and so on. Common resolutions for GCMs are about 100–200 kilometers in the horizontal direction, one kilometer vertically, and a time-stepping resolution of about 30 minutes. While GCMs could represent processes somewhat more realistically at higher resolutions, the computing time required to do the calculations increases substantially. A doubling of resolution requires about 10 times more computing power.

Owing to computer limitations, tradeoffs must be made between model resolution, model complexity, and the length and number of simulations to be conducted. Because of the relatively coarse spatial and temporal resolutions of the models, there are many important processes that occur on scales that are smaller than the model resolution, such as clouds and rainfall. These subgrid-scale processes are represented using parameterizations, which are simple formulas that attempt to approximate the actual processes based on the grid-scale information.

Parameterizations are based on observations or more detailed physical process models. Parameterizations related to clouds and precipitation remain the most challenging and are responsible for the biggest differences between the outputs of different GCMs.[10]

When simulating the past or future climate using a GCM, the objective is to simulate correctly the spatial variation of climate conditions in some average sense—typically decades or more—in such a way that the statistics of the simulated climate match the statistics of the actual climate.

More than 20 international climate modeling groups contribute climate model simulations to the IPCC assessment reports. Many of the individual climate modeling groups contribute simulations from multiple different models. Why are there so many different climate models? There are literally thousands of different choices made in the construction of a climate model—these include model resolution, complexity of the submodels, and the choice of parameterizations for subgrid-scale processes. Each different set of choices produces a different model with different sensitivities. Further, different modeling groups have different focal interests, such as long paleoclimate simulations, details of ocean circulations, nuances of the interactions between aerosol particles and clouds, or the carbon cycle. These different interests focus limited computational resources on a particular aspect of simulating the climate system, at the expense of others.

6.1.1 Complexity and Chaos

"[S]ince the climate system is complex, occasionally chaotic, dominated by abrupt changes and driven by competing feedbacks with largely unknown thresholds, climate prediction is difficult, if not impracticable." (Computer modelers Rial et al.)[11]

Variations in climate can be caused by external forcing such as solar variations, volcanic eruptions, or changes in atmospheric composition such as an increase in carbon dioxide. Climate can also change owing to internal processes within the climate system. The best-known example of internal climate variability is El Niño/La Niña (ENSO). Modes of decadal to centennial to millennial internal variability arises from the slow circulations in the oceans and their interactions with the atmosphere. As such, the ocean serves as a flywheel on the climate system, storing and releasing heat on decadal to millennial timescales and thus acting to stabilize the climate. As a result of the time lags and storage of heat in the ocean, the climate system is never in equilibrium.

Many processes in the atmosphere and oceans are nonlinear, which means that there is no simple proportional relation between cause and effect. The nonlinear dynamics of the atmosphere and oceans are described by the Navier–Stokes equations (based on Newton's laws of motion), which form the basis of prediction of winds and circulation in the atmosphere and oceans. The solution of Navier–Stokes equations is one of the most vexing problems in all of mathematics: the Clay Mathematics Institute has declared this to be one of the top seven problems in all of mathematics and is offering a one million prize for its solution.[12]

Climate model complexity arises from nonlinearity of the equations, high dimensionality (millions of degrees of freedom), and the linking of multiple subsystems. Weather has been characterized as being in a state of deterministic chaos, owing to the sensitivity of weather forecast models to small perturbations in initial conditions of the atmosphere. The source of the chaos is nonlinearities in the Navier–Stokes equations. A consequence of sensitivity to initial conditions is that beyond a certain time, the system will no longer be predictable in a deterministic sense; for weather, this predictability timescale is a matter of weeks.

Coupling a nonlinear, chaotic atmospheric model to a nonlinear, chaotic ocean model gives rise to additional modes of instability and chaos. How to characterize this is virtually impossible using current theories of nonlinear dynamical systems, particularly under conditions of externally forced changes (e.g., increasing concentrations of CO_2 carbon dioxide). This situation has been referred to as "pandemonium."[13]

6.1.2 Model Calibration and Tuning

"Indeed, whether climate scientists like to admit it or not, nearly every model has been calibrated precisely to the twentieth-century climate records – otherwise it would have ended up in the trash. 'It's fair to say all models have tuned it,' says Isaac Held, a scientist at the Geophysical Fluid Dynamics Laboratory."[14]

There is an extremely large number of weakly constrained choices for selecting model parameters and parameterizations. Calibration is necessary to address parameters that are unknown or inapplicable at the model resolution, and also in the linking of submodels. As the complexity of a model grows, model calibration becomes an increasingly important issue. Model calibration is accomplished by kludging (or tuning) of parameters. A kludge required in one model may not be required in another model that has greater structural adequacy or higher resolution.[15]

Reducing prediction error is a fundamental objective of model calibration. A serious challenge to improving complex nonlinear models through calibration is that model complexity and analytic impenetrability precludes the precise evaluation of the location of parameter(s) that are producing the prediction error.[16] For example, if a model is producing shortwave (solar) surface radiation fluxes that are substantially biased relative to observations, it is exceedingly difficult to determine whether the error arises from the radiative transfer model, the magnitude of incoming solar radiation at the top of the atmosphere, concentrations of the gases that absorb shortwave radiation, physical and chemical properties of the aerosols in the model that reflect shortwave radiation, cloud properties, convective parameterization that influences the distribution of water vapor and clouds, and/or characterization of surface reflectivity.

Whether a new or revised parameterization adds to or subtracts from the overall reliability of the model may have more to do with some entrenched features of model calibration than it does with the new parameterization's fidelity to reality when considered in isolation.

Agreement between model and observational data does not imply that the model gets the correct answer for the right reasons. Continual ad hoc model calibration can mask underlying deficiencies in model structural form. A remarkable article was published in *Science* entitled "Climate scientists open up their black boxes to scrutiny."[17] Nearly every model has been calibrated to produce observed variations in the twentieth-century global mean surface temperature. For such models, matching the twentieth-century historic temperatures is no longer a useful metric for determining whether the models are good or bad.

For example, all of the GCMs used in the IPCC Fourth Assessment Report (AR4) reproduce the time series for the twentieth century of globally averaged surface temperature anomalies; yet they have different feedbacks and sensitivities and produce markedly different simulations of the twenty-first century climate.[18] Further, tuning climate models to observations during the period 1975–2000 tunes the model to a warming phase of natural internal variability, resulting in oversensitivity of the models to CO_2 as the warming from internal variability is aliased with CO_2 driven warming. The CMIP5 and CMIP6 models (used for the IPCC Fifth and Sixth Assessment Reports) have generally employed less tuning to the global twentieth-century climate, resulting in larger discrepancies between the climate model simulations and the historical observations.[19]

6.1.3 Ensemble Modeling Techniques

"Studying complexity for me has led to the opposite conclusion: prediction is possible only in the most mundane cases. Everything else points to structuring your activity based on uncertainty and unpredictability." (Complexity scientist Joe Norman)[20]

A key approach in climate science is the comparison of results from multiple model simulations with each other and against observations. These simulations have typically been performed by separate models with consistent boundary conditions and prescribed emissions or radiative forcings, as in the Coupled Model Intercomparison Project (CMIP) phases.[21]

Multi-model ensembles (MMEs) are useful in sampling and quantifying model uncertainty, within and between generations of climate models. The primary value of MMEs is to provide a well-quantified model range.

Increased computing power has made it possible to investigate simulated internal variability and to provide robust estimates of forced model responses, using Large Initial Condition Ensembles (ICEs), also referred to as Single Model Initial Condition Large Ensembles (SMILEs).[22,23,24] Such ensembles employ a single GCM in a fixed configuration but starting from a variety of slightly different initial states. In some experiments, these initial states differ only slightly. As the climate system is chaotic, such tiny changes in initial conditions lead to different evolutions for the individual realizations of the climate system as a whole.

A third common modelling technique is the perturbed parameter ensemble (PPE) methods. These methods are used to assess uncertainty based on a single model, with individual parameters perturbed to reflect the full range of their uncertainty.[25,26] Statistical methods can then be used to detect which parameters are the main causes of uncertainty across the ensemble. PPEs have been used frequently in simpler models and are being applied to more complex models. A caveat of PPEs is that the estimated uncertainty will depend on the specific parameterizations of the underlying model and may well be an underestimation of the true uncertainty.

Together, the three ensemble methods (MMEs, SMILEs, PPEs) allow investigation of climate model uncertainty arising from internal variability, initial and internal boundary conditions, model formulations and parameterizations.[27]

6.2 Climate Model Inadequacies and Uncertainties

"Rather than reducing biases stemming from an inadequate representation of basic processes, additional complexity has multiplied the ways in which these biases introduce uncertainties in climate simulations." (Climate modelers Bjorn Stevens and Sandrine Bony)[28]

"Model imperfection" is a general term that describes our limited ability to simulate a system and is categorized here in terms of model inadequacy and model uncertainty. Model inadequacy reflects our limited understanding of the climate system and inadequacies of numerical solutions employed in computer models. Model uncertainty is associated with uncertainty in model parameters and sub-grid parameterizations, and also uncertainty in initial conditions.

The key challenge is how to understand, represent, and reason about climate model imperfections in order to build confidence in the conclusions that scientists reach using a climate model. Climate model imperfections can be categorized as follows.

Framing and context inadequacy addresses errors and uncertainties introduced by the exclusion of portions of the real world that leave an invisible range of other uncertainties. This includes "unknown knowns" or "known neglecteds" that are ignored for some reason. As an example, most GCMs omit processes that relate to solar indirect effects on the climate.

Model structural uncertainty refers to uncertainty in model functional form, which is the conceptual modeling of the physical system. This includes the model equations and which subsystems to include.

Model technical uncertainty arises from the implementation of the model solution on a computer, including solution approximations and numerical errors related to the translation from laws representing continuous space to discrete representation at grid points. Developing a tractable solution on the computer typically involves compromising the original formulation of model structural form by making simplifying assumptions. The coarse resolution of climate models necessitates many approximations.

Parameter uncertainty includes uncertain constants and other parameters that are largely contained in sub-grid scale parameterizations. The greatest uncertainties are associated with parameterizations of clouds and precipitation.

Input uncertainty relates to uncertainty in model inputs that describe the system and the external forces that drive system changes. In historical climate simulations, uncertainties associated with atmospheric aerosol particles produces substantial uncertainty in the mid-twentieth century, before satellite observations were available.

Initial condition uncertainty arises in simulations of nonlinear and chaotic dynamical systems.[29,30] If the initial conditions are not known exactly for a dynamical model based on the Navier–Stokes equations, the forecast trajectory will diverge from the actual trajectory, and it cannot be assumed that small perturbations have small effects. Initial condition uncertainty is typically addressed by multiple simulations with slightly altered initial conditions, to produce an ensemble of simulations.

Model outcome uncertainty, sometimes referred to as prediction error, arises from the propagation of the aforementioned uncertainties through the model simulation. Climate model prediction error is evaluated by comparisons with observations. The challenges of evaluating simulations of a complex system with millions of degrees of freedom is daunting, particularly in the face of inadequate observations.

Uncertainty quantification error arises from the error quantification procedure itself. Model uncertainty is typically characterized by conducting an ensemble of simulations that vary model parameters and initial conditions systematically.[31,32]

New model forms with increasing complexity are generally regarded as better by the climate modeling community. Structural forms of GCMs have undergone significant change in the twenty-first century, largely by adding more atmospheric chemistry, an interactive carbon cycle, additional prognostic equations for cloud microphysical processes, increasingly complex land surface models, ocean geochemistry, and ice sheet models. Several models have undergone significant structural changes to their dynamical core for the atmosphere and oceans.

The term "Butterfly Effect" was coined to describe the sensitive dependence of chaotic systems to initial condition uncertainty.[33] The Butterfly Effect revealed that in deterministic nonlinear dynamical systems, slight differences in an initial condition can yield wildly different outputs, raising concerns about the impact of observational errors in the initial conditions for a model simulation. By analogy to the Butterfly Effect, the "Hawkmoth Effect" implies that if the model structure is only slightly wrong, then the results may not be close to the correct solution.[34,35] Due to the Hawkmoth Effect, even a model with good approximation to the equations of the climate system may not produce output that accurately reflects the future climate. The nonlinear compound effects of even a small tweak to the model structure can be so great that the marginal performance benefits of additional subroutines or processes may be zero or even negative. In short, adding detail to the model can make it less accurate, and complex models may be less informative than simple models.[36]

6.3 Sociology and Epistemology of Climate Modeling

"It is an unfortunate fact that when you raise the question of the reliability of many simulations you are often told about how much manpower went into it, how large and fast the computer is, how important the problem is, and such things, which are completely irrelevant to the question that was asked." (Mathematician Richard Hamming)[37]

I spent much of my career in the 1990s attempting to exorcise the climate model uncertainty monster. Like others,[38] I thought the answer to improving climate models lay in improving parameterizations of physical processes such as clouds and sea ice (the focus of my personal research at the time), combined with increasing model resolution.

Circa 2002, I was becoming increasingly aware of the complexity and fundamental inadequacies of GCMs.[39] I began thinking about climate model uncertainty and how it was (or rather, wasn't) characterized and accounted for in assessments such as the IPCC assessment reports. A seminal event in the evolution of my thinking on this subject was a challenge that I received at the blog *Climate Audit* to host a discussion on climate models.[40] This exchange increased my understanding of why some scientists and engineers from other fields find climate models unconvincing.

The 2010 Royal Society *Workshop on Handling Uncertainty in Science* motivated me to become a serious monster detective on the subject of climate models.[41] As an outcome of my climate model uncertainty detection, I began to question the following:

- whether the ever-growing complexity of GCMs was helping our understanding of the climate system;
- whether GCMs were fit-for-purpose for policy relevant objectives such as attributing the causes of warming and variations in weather and climate extremes;

- whether GCMs were useful for predicting future climate change; and
- whether GCMs were of use for understanding or predicting regional climate variability.

In spite of the alleged climate consensus, I came to realize that I was not alone in having these concerns. GCM outputs are used by scientists from a variety of fields, and also by economists, regulatory agencies, and policy makers. Hence, GCMs have received considerable scrutiny from a broad community of scientists, engineers, software experts, and philosophers of science. Scientists from inside the climate modeling community are increasingly voicing similar concerns.[42]

Given these concerns, how did GCMs become the dominant tool for both climate research and policy? The policy-driven imperative of climate prediction has resulted in the accumulation of power and authority around GCMs, based on the promise of using GCMs to set emissions reduction targets and for regional predictions of climate change. Complexity of model representation has become a central normative principle in evaluating climate models, good science, and policy utility. However, not only are GCMs resource-intensive and intractable, they are also characterized by over parameterization and inadequate attention to uncertainty. The hope for useful regional predictions of climate change is unlikely to be realized based on the current path of model development. The advancement of climate science has arguably been slowed by the focus of resources on this one path of climate modeling.[43]

6.3.1 Assessing Confidence in Climate Models

"The purpose of models is not to fit the data but to sharpen the questions." (Mathematician Samuel Karlin)[44]

The authority of GCMs in both climate research and policy issues raises the following questions:[45]

- Why haven't extensive disputes about uncertainties compromised the authority of GCM simulations?
- How did scientists manage to reach a conceptual consensus about GCMs in spite of persisting scientific gaps, myriad uncertainties, and limited means of model validation?
- Why did scientists develop trust in their delicate model constructions?

There are two basic approaches for building confidence in models:

- Formal approach: explicit analysis of model errors, including a detailed analysis of sub-model interactions[46]
- Informal approach: modelers' personal judgment as to the complexity and adequacy of the models

For GCMs, the informal approach has dominated, because model complexity limits the extent to which model processes, interactions, and uncertainties can be understood and evaluated.

Scientists who develop climate models and use their results are convinced (at least to some degree) of their usefulness for research. They are convinced because of the model's relation to the physical understanding of the processes involved, consistency of the simulated responses among different models and different model versions, and the ability of the model to reproduce the observed mean state of the climate and some elements of variability. Climate models have inherited some measure of confidence from the successes of numerical weather prediction.[47]

"Comfort" relates to the sense that the model developers themselves have about their model. This comfort includes the history of model development, the reputations of the various modeling groups and the individuals that contributed, and consistency of the simulated responses among different model modeling groups and different model versions. From the perspective of a user of the outputs of climate model simulations, the sanctioning of these models by the IPCC is an important factor. However, such factors can be regarded as a form of truthiness that appeals to the authority of the modelers.

Confidence in a forecast model's outputs depends critically on the evaluation of the forecasts against real-world observations, both using historical data (hindcasts) and actual forecasts. For model confirmation against observations, there is no straightforward definition of model performance. A critical issue in such comparisons is having reliable global observational datasets with well-characterized error statistics, which represents a separate challenge in itself.[48]

Attempts to assess climate model similarity to the observed climate are fraught with challenges: inadequacy of data in terms of coverage and quality; selection of variables to confirm and on which time and space scales; a vast and multi-dimensional parameter space to be explored; model initialization and internal variability; and concerns about circularity with regards to data used in both model tuning and confirmation. Owing to the long timescales present, particularly in the ocean component, there is no possibility of a true cycle of model forecast confirmation and improvement, since the life cycle of an individual model version (order of a few years) is substantially less than the simulation period (order of centuries). Philosopher Wendy Parker argues that known climate model error is too pervasive to allow climate model confirmation to be of use.[49]

6.3.2 Fitness for Purpose

"[A]ll models are wrong, but some are useful." (Statistician George Box)[50]

Because of the complexity of GCMs, the notion of a correct or incorrect model is not well defined. The relevant issue is how well the model reproduces reality and whether the model is fit for its intended purpose.[51]

All models are imperfect; we don't need a perfect model, just one that serves its purpose. Airplanes are designed using models that are inadequate in their ability to simulate turbulent flow. Financial models based upon crude assumptions about human behavior are used to manage risk. Since GCM simulations are being used as the basis for international climate and energy policy, it is important to assess the adequacy of climate models for this purpose, as well as for basic climate research.

Some of the purposes that GCMs are used for include:

1. Hypothesis testing via numerical experiments, to understand how the climate system works, including its sensitivity to altered forcing.
2. Attribution of the causes of past climate variability and change.
3. Attribution of the causes of extreme weather events.
4. Simulation of future states of the climate, on timescales ranging from years to decades to centuries.
5. Projections of future regional climate variation for use in adaptation decision-making.
6. Projections of future statistics of extreme weather events.

The same climate model configurations are used for this plurality of applications, ranging from basic research to policy applications. There is continual tension among climate modeling groups about allocation of computer resources to higher model resolutions versus more complex physical parameterizations versus large ensembles of simulations. Different applications would be optimized by different choices.

The primary inadequacies of GCMs are related to their limited ability to predict natural fluctuations in the climate system. It is well known among climate modelers that model bias (when compared with observations) is often much greater than the signals that the models attempt to predict.[52]

Large biases seriously challenge the internal consistency of the projected change, and consequently they challenge the plausibility of the projected climate change. Both basic physics and past observations confirm that our ability to predict natural fluctuations of the climate system is limited by such biases. These limitations confound all six of the above applications of GCMs, but arguably make GCMs not fit-for-purpose for #3 through #6.

Assessing the adequacy of climate models for the purpose of predicting future climate (#5 and #6) is particularly difficult and arguably impossible.[53] It is often assumed that if climate models reproduce current and past climates reasonably well, then we can have confidence in future predictions. However, empirical accuracy, to a substantial degree, may be due to tuning rather than to the model structural form. Further, the model may lack representations of processes and feedbacks that would significantly influence future climate change. Therefore, reliably reproducing past and present climate is a necessary but not a sufficient condition for a model to be adequate for long-term projections, particularly for high-forcing scenarios that are well outside those previously observed in the instrumental record.

The fundamental inadequacies of GCMs are not widely realized in part because of a desire by climate scientists to communicate the aspects of climate science that are well settled. By deemphasizing the failures of GCMs, climate modelers Tim Palmer and Bjorn Stevens argue that scientists inadvertently contribute to complacency with the state of climate modeling, and fail to communicate the need and importance of developing better climate models.[54] Owing to political sensitivities, there has been reluctance among climate scientists to be too openly critical of the GCMs. Unfortunately, this has led to faulty impressions about the source of our confidence and about our ability to meet the scientific challenges posed by climate change.

6.4 Are GCMs the Best Tools?

"Mathematical models are a great way to explore questions. They are also a dangerous way to assert answers." (Mathematician Andrea Saltelli)[55]

In the 1990s, the perceived policy urgency required a quick confirmation of dangerous human-caused climate change. GCMs were invested with this authority by policy makers desiring a technocratic basis for their proposed policies.[56] However, both the scientific and policy challenges of climate change are much more complex than was envisioned in the 1990s. Complexity of model representation has become a central normative principle in evaluating climate models and their policy utility. However, not only are GCMs resource-intensive and intractable to interpret, they are also pervaded by over parameterization and inadequate attention to uncertainty.

Whereas GCMs are fit for the purpose of testing the basic tenets of our understanding of global climate change, climate modelers Tim Palmer and Bjorn Stevens argue they are inadequate for addressing the needs of society struggling to anticipate the impact of pending changes to weather and climate. In particular, for applications that require regional climate model output, Palmer and Stevens argue that the current generation of models is not fit for purpose.[57]

The end result is that the climate modeling enterprise has attempted a broad range of applications driven by needs of policy makers, using models that are not fit for this purpose. Further impediments to the utility of GCMs for policy purposes are: the requirement of massive computing and personnel resources, slowness to incorporate new processes from new insights, infeasibility of performing extensive sensitivity and uncertainty analyses, and unresponsiveness for rapidly exploring the implications of different model assumptions and policy scenarios.

With regards to fitness for purpose of global/regional climate models for climate adaptation decision-making, an excellent summary is provided by a team of scientists from the Earth Institute and the Red Cross Climate Center of Columbia University:

> "Climate model projections are able to capture many aspects of the climate system and so can be relied upon to guide mitigation plans and broad adaptation strategies, but the use of these models to guide local, practical adaptation actions is unwarranted. Climate models are unable to represent future conditions at the degree of spatial, temporal, and probabilistic precision with which projections are often provided which gives a false impression of confidence to users of climate change information."[58]

Complex computer simulations have come to dominate the field of climate science and its related fields. Traditional knowledge sources of theoretical analysis and challenging theory with observations have been underutilized. In an article aptly titled "The perils of computing too much and thinking too little," atmospheric scientist Kerry Emanuel raises the concern that inattention to theory is producing climate researchers who use these vast resources ineffectively, and that the opportunity for true breakthroughs in understanding and prediction is being diminished.[59]

GCMs clearly have an important role to play, particularly in scientific research. However, driven by the urgent needs of policy makers, the advancement of climate science is arguably being slowed by the focus of resources on this one path of climate modeling. A major challenge is that nearly all of the resources are being spent on GCMs and IPCC production runs, with little time and funds left over for theoretical and model innovations. We need a plurality of climate models that are developed and utilized in different ways for different purposes. For many issues of decision support, the GCM centric approach may not be the best approach. The numerous problems with GCMs, and concerns that these problems will not be addressed in the near future given the current development path, suggest that alternative frameworks should be explored—this is the topic of Chapter Eight. New approaches are particularly needed for the science-policy interface.

Notes

1 Emanuel Derman, *Models Behaving Badly: Why Confusing Illusion with Reality Can Lead to Disaster, on Wall Street and in Life* (New York, NY: Free Press, 2012).
2 "WCRP History," WCRP—World Climate Research Programme, February 13, 2013, https://www.wcrp-climate.org/about-wcrp/about-history.
3 "History," International Geosphere-Biosphere Programme, accessed May 2, 2022, http://www.igbp.net/about/history.4.1b8ae20512db692f2a680001291. html.
4 Michael Oppenheimer et al., "Negative Learning," *Climatic Change* 89, no. 1–2 (June 2008): 155–172, https://doi.org/10.1007/s10584-008-9405-1.
5 Eric Winsberg, "Sanctioning Models: The Epistemology of Simulation," *Science in Context* 12, no. 2 (1999): 275–292, https://doi.org/10.1017/s0269889700003422.
6 Eric B. Winsberg, *Science in the Age of Computer Simulation* (Chicago, IL: University of Chicago Press, 2010).
7 Wendy S. Parker, "Confirmation and Adequacy-for-Purpose in Climate Modelling," *Aristotelian Society Supplementary Volume* 83, no. 1 (January 2009): 233–249, https://doi.org/10.1111/j.1467-8349.2009.00180.x.
8 Arthur Caesar Petersen, *Simulating Nature: A Philosophical Study of Computer-Simulation Uncertainties and Their Role in Climate Science and Policy Advice* (Boca Raton, FL: CRC Press, 2012).
9 Jon Turney, "All Scientific Models Are Wrong but Some at Least Are Useful: Aeon Essays," Aeon (Aeon Magazine, December 16, 2013), https://aeon.co/ essays/all-scientific-models-are-wrong-but-some-at-least-are-useful.
10 Mark D. Zelinka et al., "Clearing Clouds of Uncertainty," *Nature Climate Change* 7, no. 10 (October 29, 2017): 674–678, https://doi.org/10.1038/nclimate3402.
11 José A. Rial et al., "Nonlinearities, Feedbacks and Critical Thresholds within the Earth's Climate System," *Climatic Change* 65, no. 1/2 (2004): 11–38, https:// doi.org/10.1023/b:clim.0000037493.89489.3f.
12 "The Millennium Prize Problems," Clay Mathematics Institute, accessed May 2, 2022, https://www.claymath.org/millennium-problems/millennium-prize-problems.
13 D.A Stainforth et al., "Confidence, Uncertainty and Decision-Support Relevance in Climate Predictions," *Philosophical Transactions of the Royal Society*

A: Mathematical, Physical and Engineering Sciences 365, no. 1857 (2007): 2145–2161, https://doi.org/10.1098/rsta.2007.2074.

14 Paul Voosen, "Climate Scientists Open up Their Black Boxes to Scrutiny," *Science* 354, no. 6311 (2016): 401–402, https://doi.org/10.1126/science.354.6311.401.

15 Thorsten Mauritsen et al., "Tuning the Climate of a Global Model," *Journal of Advances in Modeling Earth Systems* 4, no. 3 (August 7, 2012), https://doi.org/10.1029/2012ms000154.

16 Johannes Lenhard and Eric Winsberg, "Holism, Entrenchment, and the Future of Climate Model Pluralism," *Studies in History and Philosophy of Science Part B: Studies in History and Philosophy of Modern Physics* 41, no. 3 (September 2010): 253–262, https://doi.org/10.1016/j.shpsb.2010.07.001.

17 Paul Voosen, "Climate scientists open up."

18 Stephen E. Schwartz, "Uncertainty Requirements in Radiative Forcing of Climate Change," *Journal of the Air & Waste Management Association* 54, no. 11 (November 2004): 1351–1359, https://doi.org/10.1080/10473289.2004.1047 1006.

19 Femke J. Nijsse, Peter M. Cox, and Mark S. Williamson, "Emergent Constraints on Transient Climate Response (TCR) and Equilibrium Climate Sensitivity (ECS) from Historical Warming in CMIP5 and CMIP6 Models," *Earth System Dynamics* 11, no. 3 (2020): 737–750, https://doi.org/10.5194/esd-11-737-2020.

20 Joe Norman (@normonics), "Studying Complexity for Me Has Led to the Opposite Conclusion: Prediction Is Possible Only in the Most Mundane Cases. Everything Else Points to Structuring Your Activity Based on Uncertainty and Unpredictability. Https://T.co/T9ompQHZEV," Twitter, November 14, 2021, https://twitter.com/normonics/status/1459937922245963776.

21 Veronika Eyring et al., "Overview of the Coupled Model Intercomparison Project Phase 6 (CMIP6) Experimental Design and Organization," *Geoscientific Model Development* 9, no. 5 (May 26, 2016): 1937–1958, https://doi.org/10.5194/gmd-9-1937-2016.

22 J. E. Kay et al., "The Community Earth System Model (CESM) Large Ensemble Project: A Community Resource for Studying Climate Change in the Presence of Internal Climate Variability," *Bulletin of the American Meteorological Society* 96, no. 8 (August 1, 2015): 1333–1349, https://doi.org/10.1175/bams-d-13-00255.1.

23 Nicola Maher et al., "The Max Planck Institute Grand Ensemble: Enabling the Exploration of Climate System Variability," *Journal of Advances in Modeling Earth Systems* 11, no. 7 (June 4, 2019): 2050–2069, https://doi.org/10.1029/2019ms001639.

24 Megan C. Kirchmeier-Young, Francis W. Zwiers, and Nathan P. Gillett, "Attribution of Extreme Events in Arctic Sea Ice Extent," *Journal of Climate* 30, no. 2 (2017): 553–571, https://doi.org/10.1175/jcli-d-16-0412.1.

25 Reto Knutti et al., "Challenges in Combining Projections from Multiple Climate Models," *Journal of Climate* 23, no. 10 (May 15, 2010): 2739–2758, https://doi.org/10.1175/2009jcli3361.1.

26 Hideo Shiogama et al., "Multi-Parameter Multi-Physics Ensemble (MPMPE): A New Approach Exploring the Uncertainties of Climate Sensitivity," *Atmospheric Science Letters* 15, no. 2 (November 4, 2013): 97–102, https://doi.org/10.1002/asl2.472.

27 Wendy S. Parker, "Ensemble Modeling, Uncertainty and Robust Predictions," *WIREs Climate Change* 4, no. 3 (2013): 213–223, https://doi.org/10.1002/wcc.220.

28 Bjorn Stevens and Sandrine Bony, "What Are Climate Models Missing?," *Science* 340, no. 6136 (May 31, 2013): 1053–1054, https://doi.org/10.1126/science.1237554.

29 Leonard A. Smith, "What Might We Learn from Climate Forecasts?," *Proceedings of the National Academy of Sciences* 99, no. suppl_1 (February 19, 2002): 2487–2492, https://doi.org/10.1073/pnas.012580599.

30 T.N. Palmer et al., "Representing Model Uncertainty in Weather and Climate Prediction," *Annual Review of Earth and Planetary Sciences* 33, no. 1 (May 2005): 163–193, https://doi.org/10.1146/annurev.earth.33.092203.122552.

31 M. Leutbecher and T.N. Palmer, "Ensemble Forecasting," *Journal of Computational Physics* 227, no. 7 (March 20, 2008): 3515–3539, https://doi.org/10.1016/j.jcp.2007.02.014.

32 Leonard A. Smith, "Predictability Past, Predictability Present," *Predictability of Weather and Climate*, 2006, 217–250, https://doi.org/10.1017/cbo9780511617652.010.

33 Edward N. Lorenz, *The Essence of Chaos* (Seattle, WA: Univ. of Washington Press, 2008).

34 Erica Lucy Thompson, "Modelling North Atlantic Storms in a Changing Climate," Imperial College London, March 1, 2013, https://doi.org/10.25560/14730.

35 Leonard A. Smith, "What Might We Learn from Climate Forecasts?," *Proceedings of the National Academy of Sciences* 99, no. suppl_1 (February 19, 2002): 2487–2492, https://doi.org/10.1073/pnas.012580599.

36 Erica L. Thompson and Leonard A. Smith, "Escape from Model-Land," *Economics* 13, no. 1 (March 8, 2019), https://doi.org/10.5018/economics-ejournal.ja.2019-40.

37 Richard R. Hamming, in *Art of Doing Science and Engineering Learning to Learn* (Hoboken, NJ: CRC Press, 2014), 135.

38 David A. Randall and Bruce A. Wielicki, "Measurements, Models, and Hypotheses in the Atmospheric Sciences," *Bulletin of the American Meteorological Society* 78, no. 3 (March 1, 1997): 399–406, https://doi.org/10.1175/1520-0477(1997)0782.0.co;2.

39 Leonard A. Smith, "What Might We Learn from Climate Forecasts."

40 Judith Curry, "Curry Reviews Jablonowski and Williamson," Climate Audit, February 4, 2008, https://climateaudit.org/2008/02/03/curry-reviews-jablonski-and-williamson/.

41 "Handling Uncertainty in Science," Royal Society, 2010, https://royalsociety.org/science-events-and-lectures/2010/uncertainty-science/.

42 Tim Palmer and Bjorn Stevens, "The Scientific Challenge of Understanding and Estimating Climate Change," Proceedings of the National Academy of Sciences 116, no. 49 (December 2, 2019): 24390–24395, https://doi.org/10.1073/pnas.1906691116.

43 Simon Shackley et al., "Uncertainty, Complexity and Concepts of Good Science in Climate Change Modelling: Are GCMs the Best Tools?," *Climatic Change* 38, no. 2 (February 1998): 159–205, https://doi.org/10.1023/a:1005310109968.

44 Samuel Karlin, "The Eleventh R. A. Fisher Memorial Lecture—Kin Selection and Altruism," *Proceedings of the Royal Society of London. Series B. Biological Sciences* 219, no. 1216 (1983): 327–353, https://doi.org/10.1098/rspb.1983.0077.

45 Matthias Heymann, "Understanding and Misunderstanding Computer Simulation: The Case of Atmospheric and Climate Science—an Introduction," *Studies in History and Philosophy of Science Part B: Studies in History and Philosophy of Modern Physics* 41, no. 3 (2010): 193–200, https://doi.org/10.1016/j.shpsb.2010.08.001.

46 Patrick J. Roache, *Fundamentals of Verification and Validation* (Socorro, NM: hermosa publ, 2009).

47 Reto Knutti, "Should We Believe Model Predictions."

48 John J. Bates and Jeffrey L. Privette, "A Maturity Model for Assessing the Completeness of Climate Data Records," *Eos, Transactions American Geophysical Union* 93, no. 44 (October 30, 2012): 441–441, https://doi.org/10.1029/2012eo440006.

49 Wendy S. Parker, "Confirmation and Adequacy-for-Purpose."

50 Box George E P. and Norman Richard Draper, *Empirical Model-Building and Response Surfaces* (Chichester, UK: Wiley, 1987).

51 Wendy S. Parker, "Confirmation and Adequacy-for-Purpose."

52 Thorsten Mauritsen et al., "Tuning the Climate."

53 Christoph Baumberger, Reto Knutti, and Gertrude Hirsch Hadorn, "Building Confidence in Climate Model Projections: An Analysis of Inferences from Fit," *WIREs Climate Change* 8, no. 3 (January 12, 2017), https://doi.org/10.1002/wcc.454.

54 Tim Palmer and Bjorn Stevens, "The Scientific Challenge."

55 Andrea Saltelli et al., "Five Ways to Ensure That Models Serve Society: A Manifesto," *Nature* 582, no. 7813 (June 24, 2020): 482–484, https://doi.org/10.1038/d41586-020-01812-9.

56 Simon Shackley et al., "Uncertainty, Complexity and Concepts."

57 Tim Palmer and Bjorn Stevens, "The Scientific Challenge."

58 Hannah Nissan et al., "On the Use and Misuse of Climate Change Projections in International Development," *WIREs Climate Change* 10, no. 3 (March 14, 2019), https://doi.org/10.1002/wcc.579.

59 B. Stevens, "The Perils of Computing Too Much and Thinking Too Little," Eos, June 25, 2020, https://eos.org/editor-highlights/the-perils-of-computing-too-much-and-thinking-too-little.

Chapter Seven

IPCC SCENARIOS OF TWENTY-FIRST CENTURY CLIMATE CHANGE

"[A]ll of our knowledge is about the past, and all our decisions are about the future."
　　　　—Ian Wilson, author of From Scenario Thinking to Strategic Action[1]

Climate policy discussions are framed by the IPCC Assessment Reports. At the center of the IPCC approach to climate policy analysis are scenarios of the future climate. The IPCC uses global climate models, driven by scenarios of future emissions, as the basis for generating climate futures.

Scenarios refer to "a plausible, comprehensive, integrated and consistent description of how the future might unfold while refraining from a concrete statement on probability."[2] Scenarios of climate futures play a fundamental role in characterizing societal risks and policy response options.[3]

This chapter examines the IPCC's scenarios of future climate change.

7.1 Emissions Scenarios

"There isn't, you know, like a Mad Max scenario among the SSPs [emissions scenarios], we're generally in the climate-change field not talking about futures that are worse than today." (Brian O'Neill, one of the lead architects of the Shared Socioeconomic Pathways developed for the IPCC Sixth Assessment Report)[4]

The IPCC projections of future climate change are driven by changes to radiative forcing arising from changes to concentrations of greenhouse gases and aerosols associated with human activity—primarily from emissions associated with fossil fuels. One approach to generating scenarios is simply to specify different levels of radiative forcing, which are referred to as "pathways." Another approach is to develop socioeconomic and emission scenarios to provide plausible descriptions of how future emissions and their radiative forcing may evolve. To capture a range of possible future emissions scenarios, energy system modelers use Integrated Assessment Models (IAMs) that simulate both future energy technologies and emissions, and also incorporate assumptions about population, land use, socioeconomic development and policy assumptions.

The most recent set of scenarios used by the IPCC are the Representative Concentration Pathways (RCP) and scenarios developed from the Shared Socioeconomic Pathways (SSP).[5,6]

The RCPs are a set of four climate scenarios for the end of the twenty-first century. The RCPs were formulated for use in the IPCC Fifth Assessment Report, to reflect different potential climate outcomes (RCP2.6, RCP4.5, RCP6.0, and RCP8.5).[7] The number (e.g., 8.5) reflects the additional radiative forcing (in Watts per square meter, W/m^2, which is a measure of energy per unit time and per unit area) in 2100 at the top of the atmosphere from greenhouse gas emissions and other factors, relative to pre-industrial times. To date (in 2020), radiative forcing relative to pre-industrial levels is about 2.5 W/m^2.

For the IPCC Sixth Assessment Report (AR6), the SSPs were formulated from five socioeconomic and technological trajectories that reflect pathways that the world might follow in the twenty-first century.[8] Each pathway has a baseline in which no climate policies are enacted after 2010. Additional SSP scenarios are linked to different climate policies to generate different outcomes for the end of the twenty-first century. A subset of SSP scenarios has been selected for the IPCC AR6, with a radiative forcing of 1.9, 2.6, 3.4, 4.5, 6.0, 7.0, or 8.5 W/m^2 in 2100.[9]

Table 7.1 compares the SSP scenarios used for the Coupled Model Intercomparison Project (CMIP) Phase 6 (IPCC AR6) climate model simulations in terms of gigatons of CO_2 emitted per year, for the year 2050. For reference, emissions in 2021 are about 36 gigatons of carbon dioxide ($GtCO_2$) per year.[10] The UNFCCC objective is net zero emissions by 2050.

The IPCC warns against treating emissions and SSP scenarios as predictions because they reach far into the future.[11] We should rightly approach projections

Table 7.1 $GtCO_2/yr$ emissions by 2050 under different SSP scenarios.[13,14]

Scenario	$GtCO_2/yr$
SSP5–8.5	82
SSP4–6.0	48
SSP2–4.5	42
SSP4–3.4	20
SSP1–2.6	18
IEA STEPS	36
IEA APC	22

far into the future with humility and acknowledge that there is a great deal of uncertainty. However, for 30-year projections to 2050, the range of plausible emissions scenarios can be narrowed.

The International Energy Agency (IEA) takes a different approach from the IPCC, presenting two emissions scenarios to 2050 (Table 7.1).[12] Policies that have actually been implemented are described in a scenario referred to as STEPS, which projects continued emissions through 2050 at the rate of about 36 $GtCO_2$ per year. The trajectory that would be achieved if all countries met their current commitments under the Paris Agreement is referred to as APC, which projects emissions declining to about 22 $GtCO_2$ per year by 2050.[15] The implication of the IEA STEPS scenario is that maintaining the policies already implemented would result in global CO_2 emissions out to 2050 that are similar to what they are today (in 2021).

The IEA analysis suggests that the world is entering an extended plateauing of emissions. For climate change to 2050, SSP2–4.5 and SSP4–3.4 are the most likely of the IPCC scenarios and should be the focus of impact assessments and policy planning.[16]

The most striking aspect of the comparison between the IPCC and IEA scenarios to 2050 is the strong divergence of the extreme emissions scenarios RCP8.5 and SSP5–8.5 from the IEA scenarios, with the 8.5 emissions values more than twice as high as the IEA STEP scenario at 2050.

In evaluating these scenarios, it is important to recognize that predicting future emissions is inherently uncertain, particularly as the time horizon increases. Poorly understood carbon feedbacks (such as methane emissions from thawing permafrost) could lead to higher forcing levels. However, such speculative feedbacks are unlikely to arise from the relatively modest warming expected between now and 2050.

7.1.1 Extreme Emissions Scenario

"Anyone, including me, who has built their understanding on what level of warming is likely this century on that RCP8.5 scenario should probably revise that understanding in a less alarmist direction." (David Wallace-Wells, author of *The Uninhabitable Earth*)[17]

The extreme emissions scenario RCP8.5 has commonly been referred to as the "business as usual" (BAU) scenario.[18] Referring to RCP8.5 as BAU implies that it is probable in the absence of stringent climate mitigation. Positioning the extreme RCP8.5 scenario as the only clearly defined baseline has made this scenario central to assessments of climate change impacts.

RCP8.5 paints a dystopian future that is fossil-fuel intensive and excludes any climate mitigation policies. RCP8.5 drives climate model projections of nearly 5°C of warming by the end of the century, relative to pre-industrial temperatures. Via RCP8.5, this outlook of the future was subsequently adopted for thousands of academic studies that project future climate impacts on people and the environment, to evaluate the costs and benefits of adaptation and mitigation policies, and to estimate the cost effectiveness of policies designed to meet mitigation targets.

Over the past several years, there has been substantial debate over the 8.5 scenarios – whether they are plausible and whether or not they should be used for policy-making purposes. The IPCC AR6 (Chapter One) makes the following statement:

"In the scenario literature, the plausibility of the high emissions levels underlying scenarios such as RCP8.5 or SSP5–8.5 has been debated in light of recent developments in the energy sector."[19]

The 8.5 scenarios can only emerge under a very narrow range of circumstances, comprising a severe course change from recent energy use. Both the RCP8.5 and the SSP5–8.5 scenarios have drawn criticism owing to the assumptions around future coal use, requiring up to 6.5 times more coal use in 2100 than today—an amount larger than some estimates of economically-recoverable coal reserves. A recent elicitation of energy experts gives SSP5–8.5 only a 5 percent chance of occurring among all of the possible no-policy baseline scenarios; the likelihood of SSP5–8.5 becomes much lower when recent and future commitments for policy actions are considered.[20]

The IEA analysis (Table 7.1) indicates that the worst-case scenarios, RCP8.5/SSP5–8.5, are off the table for the next several decades, and will stay there unless we actively choose to follow them. In many ways, the world has moved beyond the no-policy baseline scenarios through a combination of technological innovation, falling costs of clean energy sources and climate policies already enacted. These changes are unlikely to be reversed, even in the absence of new policies and technologies.

It is difficult to overstate the importance of the shift in expectations for future emissions that is represented by the difference in the new IEA scenarios versus RCP8.5. The IPCC, the US National Climate Assessment, and a majority of published papers have centered their analyses on RCP8.5 as a reference scenario against which climate impacts and policies are evaluated.[21] Further, RCP8.5 is being used by the insurance sector for projecting climate change impacts and also by state and local governments for regional adaptation planning.[22,23]

The new IEA scenarios have rendered obsolete much of the climate impacts literature of the past decade, which have focused on RCP8.5, and the accompanying media reporting. It is now clear that climate impact assessments have been biased in an alarming direction by continued inclusion, and even sole reliance, on RCP8.5. The end result is that the scientific literature has become imbalanced in an apocalyptic direction, which is amplified in the media to stoke alarm about climate change.

When I first started writing this subsection, regarding RCP8.5 as an implausible scenario was a minority perspective. In late 2021, the Nationally Determined Contributions (NDC) Synthesis Report written in preparation for the UNFCCC 26th Conference of the Parties (COP26) shows no sign of RCP8.5; this scenario has been quietly lost from the international climate policy makers agenda.[24]

7.2 Climate Sensitivity to CO_2 Emissions

"The truth is rarely pure and never simple." (Poet Oscar Wilde)[25]

Atmospheric CO_2 concentration has increased from its pre-industrial level of 280 parts per million (ppm) to around 412 ppm in 2021.[26] Human-caused warming depends not only on increasing greenhouse gases, but also on how sensitive the climate is to this increase.

Climate sensitivity is defined as the global surface warming that occurs when the concentration of CO_2 in the atmosphere is doubled, relative to pre-industrial levels.[27] The magnitude of the sensitivity of Earth's climate to increasing concentrations of CO_2 is at the heart of the scientific debate on anthropogenic climate change, and also the public debate on the appropriate policy response to increasing CO_2 in the atmosphere.

The magnitude of the climate sensitivity to a doubling of CO_2 concentration is associated with substantial uncertainties, which are highly consequential for policy making. If climate sensitivity to CO_2 concentration is low, then the impacts of increased emissions are expected to be relatively benign, and the benefits of emissions reductions much reduced. If the sensitivity is high, then even moderate increases of CO_2 could produce significant impacts. Climate sensitivity and estimates of its uncertainty are key inputs into the economic models that drive cost-benefit analyses and estimates of the social cost of carbon.

Since the 1979 Charney Report,[28] estimates in international assessment reports have put equilibrium climate sensitivity (ECS) between 1.5 and 4.5°C of warming for a doubling of CO_2 relative to pre-industrial levels. The ECS range between 1.5 and 4.5°C is characterized by the IPCC AR5 as the *likely* range

(66 percent probability); the AR5 stated that ECS is *extremely unlikely* (95 percent probability) to be below 1.0°C and *very unlikely* (90 percent probability) to exceed 6°C.[29]

This range has remained stubbornly wide, despite many individual studies claiming to narrow it. Most recently, the IPCC AR6 narrowed the *likely* range of ECS to 2.5–4.0°C and the *very likely* range to be between 2.0 and 5.0°C.[30] This narrowing of the range on the low end is disputed.[31]

Understanding of the uncertainties is illuminated by examining the different methods of estimating climate sensitivity and the confounding effects of feedback processes.

7.2.1 Different Ways of Estimating Sensitivity

"[W]e were under the necessity of collecting together a variety of facts, and of entering into long trains of reasoning." (Geologist Charles Lyell)[32]

Climate scientists use multiple lines of evidence to assess climate sensitivity:

- Process understanding
- Climate model simulations
- Historical observations
- Paleoclimatic observations

By itself, the equilibrium warming effect of a doubling of atmospheric CO_2 is slightly more than 1°C.[33] The large values of ECS arise from positive feedbacks that amplify the warming effect of CO_2. Water vapor feedback is positive: a warmer atmosphere may have more water vapor, which itself is a powerful greenhouse gas. Water vapor feedback is partially canceled by the fact that atmospheric temperature decreases more slowly with height above the Earth's surface when atmospheric water vapor increases. Warmer temperatures also result in less snow and sea ice cover, allowing the Earth to absorb more of the sun's radiation. Some simple estimates of these three feedbacks increase the ECS to around 2°C.[34]

Higher values of ECS arise primarily from positive cloud feedbacks. The magnitude, and even the sign, of cloud feedbacks is the greatest source of uncertainty. Elements of cloud feedback include changes in the latitudinal distribution of clouds, changes in the distribution of cloud height (changes in low versus high clouds), changes to the phase of clouds (ice versus liquid), changes in cloud particle size (associated with changes in concentration and/or composition of aerosol particles), changes in the precipitation efficiency of clouds, and even changes in how clouds are distributed over the daily solar cycle.[35] It is difficult for climate models to simulate any of these cloud processes correctly owing to their small scale, let alone predict how they will change in the future. Further, cloud processes modulate the magnitudes of the water vapor, lapse rate, and the surface albedo feedbacks.[36]

ECS can be determined from climate model simulations by doubling the concentration of CO_2, allowing several centuries for the warming to equilibrate. To avoid the necessity of long simulations, effective climate sensitivity is commonly evaluated from a 150-year simulation in response to a sudden quadrupling of CO_2. The spread of ECS values from the CMIP5 climate models (used in the AR5) was 2.0–4.7°C; the spread has increased for the CMIP6 models (used in the AR6) to between 1.8 and 5.5°C.[37,38] Far from the science being settled in relation to the response of the climate system to increasing greenhouse gas concentrations, insofar as it is represented by climate models, uncertainties appear to be growing. The main cause of the overall upward shift in ECS in CMIP6 relative to CMIP5 is a larger positive cloud feedback, driven by changes to the cloud parameterizations in many CMIP6 models.[39]

Climate sensitivity represents an emergent property of climate models— that is, it is not directly parameterized or tuned. However, collections of parameterizations are often chosen to produce an expected value of climate sensitivity, particularly with regards to sub-grid-scale cloud processes. Otherwise plausible models and parameter selections have been discarded because of perceived conflict with this warming rate, or aversion to a model's climate sensitivity being outside an accepted range.

The IPCC AR4 (2007) relied almost exclusively on climate models in its assessment of ECS. The IPCC AR5 (2013) relied on both climate model determinations and determinations from the historical record, noting substantial disagreements between these two methods. The IPCC AR6 (2021) did not rely on climate model simulations in their assessment of climate sensitivity, relying on the other three methods in its assessment.

Climate sensitivity is also estimated from instrumental records of surface temperatures and ocean heat content, combined with estimates of how climate forcings (e.g., greenhouse gases, solar, volcanoes, aerosols) have changed in the past.[40] Using this information, an energy budget method is used to deduce the ECS. The greatest source of uncertainty in the energy budget method using historical data is uncertainty in the amount and composition of aerosol particles and their interactions with cloud radiative properties (the so-called aerosol indirect effect). Assumptions are also needed about ocean heat storage.

Paleoclimate proxies are used to evaluate the sensitivity of past climates, by comparing paleoclimate changes in the Earth's temperatures to estimates of changes in forcings. The two most informative periods are the last glacial maximum (around 20,000 years ago), which was about 3–7°C colder than today, and a mid-Pliocene period (roughly three million years ago), which was 1–3°C warmer than today.[41] The limits on cooling during the last glacial maximum give the best single evidence that high values of climate sensitivity are unlikely. However, paleoclimate estimates are associated with very large uncertainties in the estimated temperatures and forcings. Further, estimates of climate sensitivity based on past climate states may not be representative of the current state of the climate system.

Based on a recent review article by Sherwood et al., the IPCC AR6 integrated all of the different streams of inputs using a Bayesian analysis method.[42] The IPCC AR6 narrowed the likely range of ECS to 2.5–4.0°C.[43] Subsequent to publication of the AR6, independent climate scientist Nic Lewis (see also Section 2.4.1) published in the journal *Climate Dynamics* an extensive critique of the methodology used in the Sherwood et al. paper to determine the ECS.[44] Key concerns raised by Lewis include errors, outdated input values, and objective versus subjective Bayesian priors in the analysis. Lewis' analysis found that climate sensitivity is estimated to be much lower and better constrained than in the Sherwood et al. analysis – median 2.16°C (1.75–2.7°C in the 17–83 percent *likely* range), and 1.55–3.2°C in the 5–95 percent *very likely* range. This analysis implies that climate sensitivity is considerably more likely to be below 2°C than above 2.5°C.[45]

7.2.2 Transient Climate Response

"One aim of the physical sciences has been to give an exact picture of the material world. One achievement of physics in the twentieth century has been to prove that that aim is unattainable." (Mathematician and philosopher Jacob Bronowski)[46]

The Transient Climate Response (TCR) is of greater utility in providing an observational constraint on climate sensitivity. TCR is the amount of warming that would occur at the time when CO_2 doubles, having increased gradually by 1 percent each year over a period of 70 years. Relative to the ECS, observationally-determined values of TCR avoid the problems of uncertainties in ocean heat uptake and the fuzzy boundary in defining equilibrium arising from a range of timescales for the longer-term feedback processes (e.g., ice sheets). TCR is more generally related to peak warming and better constrained by historical warming, than ECS. The IPCC AR6 judged the *very likely* range of TCR to be 1.2–2.4°C.[47] In contrast to ECS, the upper bound of the TCR is more tightly constrained.

A related but more flexible parameter is the Transient Response to cumulative carbon Emissions, or TCRE, which relates global temperature change to the cumulative emissions. TCRE is based upon a linear relationship between global average temperature and cumulative emissions, which holds for cumulative emissions of up to 2 teratons of carbon. For a given value of TCRE, we can calculate the amount of warming expected over a future period in response to scenarios of cumulative carbon emission. The IPCC AR6 provides a *very likely* range for TCRE of 1.0–2.3°C, with a median value of 1.6°C.[48]

7.3 IPCC Projections of Manmade Global Warming for the Twenty-First Century

"The commanding general is well aware that the forecasts are no good. However, he needs them for planning purposes." (Anecdote provided by economist and Nobel Laureate Kenneth Arrow, who worked as a long-range weather forecaster in the United States military during World War II).[49]

The IPCC AR6 provides the following projections for twenty-first century global warming, relative to a baseline period of 1850–1900, for five different emissions scenarios (Table 7.2).[50] For the near-term period 2021–40, there is little dependence on emissions scenario. For the end-of-century period 2081–2100, there is a factor of 3 increase in the warming for the extreme emissions scenario (SSP5–8.5) versus the lowest scenario (SSP1–1.9). Within each emissions scenario, there is a substantial range of about 30–40 percent of the average amount of warming, which is associated primarily with uncertainty in climate sensitivity.

Table 7.2 Projected changes in global surface temperature for three 20-year time periods for five emissions scenarios. Temperature differences are relative to the baseline period 1850–1900. Changes relative to the recent reference period 1995–2014 may be calculated approximately by subtracting 0.85°C, the best estimate of the observed warming from 1850–1900 to 1995–2014 (IPCC AR6 Summary for Policy Makers (SPM) Table SPM.1).[53]

Scenario	Near-term, 2021–40		Mid-term, 2041–60		Long term, 2081–2100	
	Best estimate (°C)	Very likely range (°C)	Best estimate (°C)	Very likely range (°C)	Best estimate (°C)	Very likely range (°C)
SSP1–1.9	1.5	1.2–1.7	1.6	1.2–2.0	1.4	1.0–1.8
SSP1–2.6	1.5	1.2–1.8	1.7	1.3–2.2	1.8	1.3–2.4
SSP2–4.5	1.5	1.2–1.8	2.0	1.6–2.5	2.7	2.1–3.5
SSP3–7.0	1.5	1.2–1.8	2.1	1.7–2.6	3.6	2.8–4.6
SSP5–8.5	1.6	1.3–1.9	2.4	1.9–3.0	4.4	3.3–5.7

The IPCC AR6 procedure for providing projected estimates of global average temperature change over the 21st century diverged markedly from the AR4 and AR5, which relied totally on simulations from global climate and Earth-system models, equally weighting all climate model simulations that were submitted. The CMIP6 models used by the AR6 overall provided substantially more warming than the CMIP5 models used in the AR5, including some models with values of climate sensitivity that were judged to be unrealistically high.[51] Hence, the AR6 combined the CMIP6 projections with observational constraints on simulated past warming and best estimates of the climate sensitivity to increased CO_2 (ECS, TCR). The reduction of the constrained projections relative to the original CMIP6 simulations reported by the AR6 (AR6 Figure 4.11) is substantial, up to 20 percent for the higher emissions scenarios.[52]

The implications of the new constrained procedure used by the AR6 in making projections of global air surface temperature change are profound. It breaks the hegemony of the global climate models in dominating the IPCC's conclusions about twenty-first century climate change.

7.4 Climate Impact-Drivers

"The future ain't what it used to be." (Yogi Berra)[54]

The AR6 rationale for constraining the projections of average global surface temperature using observations arguably produces more plausible projections. However, for other impacts of warming (e.g., precipitation, sea ice), the CMIP6 projections are used directly, which arguably inflates the impacts for a given emissions scenario. To address this issue to some extent, the AR6 uses "global warming levels," targeting impacts relative to 1.5, 2, 3, and 4°C of warming.[55] This bypasses the problem of uncertainty in climate sensitivity, but ignores the timing of the warming level which is a key element for policy making.

The AR6 defines climatic impact-drivers as physical climate system conditions that affect an element of society or ecosystems.[56] Depending on system tolerance, the impact drivers and their changes can be beneficial, detrimental, neutral, or a mixture across interacting system elements and regions. While recognizing that climate impact drivers can lead to beneficial outcomes, the IPCC AR6 focuses on adverse impacts connected to hazards and risk.

This section addresses climate impact-drivers associated with extreme climate and weather events and sea level rise. Projections of climate impact-drivers rely on understanding the causes of historical changes, and their relationship with a warming climate.

The IPCC uses a two-part framework to identify changes in climate: *detection*, which identifies a change in the climate; and *attribution*, which explains why the identified change may have occurred.[57] Without detection of a change over the historical record, and then attribution to human-caused warming, there is little basis for any future projections of change in the particular impact driver.

7.4.1 Detection of Changes in Extreme Weather and Climate Events

"Hindsight is notably cleverer than foresight." (Chester W. Nimitiz, Fleet Admiral of the US Navy and Commander in Chief of the US Pacific Fleet during World War II)[58]

The IPCC AR6 Summary for Policy Makers highlights the following impacts that have adverse consequences:[59]

- It is *virtually certain* that heatwaves have become more frequent and more intense across most land regions since the 1950s, while cold waves have become less frequent and less severe, with *high confidence* that human-induced climate change is the main driver of these changes.
- The frequency and intensity of heavy precipitation events have increased since the 1950s over most land areas (*high confidence*), and human-induced climate change is *likely* the main driver.
- Human-induced climate change has contributed to increases in agricultural and ecological droughts in some regions due to increased land evapotranspiration (*medium confidence*).
- It is *likely* that the global proportion of major (Category 3–5) tropical cyclone occurrence has increased over the last four decades. Event attribution studies and physical understanding indicate that human-induced climate change increases heavy precipitation associated with tropical cyclones (*high confidence*), but data limitations inhibit clear detection of past trends on the global scale.

It is significant what is *not* mentioned in the Summary for Policy Makers. Chapters Eleven and Twelve in the IPCC AR6 identify the following event types for which there is either no change or low confidence in any change. There is no observed significant trend in meteorological or hydrological droughts, no trends in extratropical storms, no trends in the total number of tropical cyclones, and no trends in tornadoes, hail, or lightning associated with severe convective storms.[60,61] The AR6 is equivocal on floods: global hydrological models project a larger fraction of land areas to be affected by an increase in river floods than by a decrease in river floods (*medium confidence*).[62]

Every extreme weather event now gets associated in the media with manmade global warming. Any change in the intensity or frequency of extreme weather events is incremental at most; even if a change in the event type has been detected over the past 50+ years, attributing any change to emissions-driven warming is not at all straightforward, and quantitative forward projection of any changes is highly uncertain.

7.4.2 Sea Level Rise

"The illusion that we understand the past fosters overconfidence in our ability to predict the future." (Economist Daniel Kahneman, author of *Thinking, Fast, and Slow*)[63]

Arguably the most important climate impact driver that is unambiguously associated with increasing temperatures is global sea level rise. Global warming increases sea level rise through thermal expansion of the volume of sea water and melting of glaciers and ice sheets. Another factor that influences sea level variations is land water storage. Regional sea level change is influenced by large-scale ocean circulation patterns and geologic processes and deformation from the redistribution of ice and water. Local sea level rise is further influenced by vertical land motion from geologic process, ground water withdrawal, and fossil fuel extraction.

The IPCC AR6 estimates that global mean sea level increased by 0.20 (0.15–0.25) meters (0.2 meters = 7.9 inches) between 1901 and 2018. The rate of sea level rise has accelerated in recent decades. At the ocean basin scale, sea levels have risen fastest in the Western Pacific and slowest in the Eastern Pacific over the period 1993–2018.[64]

The observing systems for global sea level rise have advanced significantly in the satellite era, particularly with the advent of satellite altimeters since 1993. Local tide gauges have provided useful data for the past century, and even longer for a few locations. Following the end of the Little Ice Age in the mid-nineteenth century, tide gauges show that the global mean sea level began rising during the period 1820–1860.

Paleoclimate proxies are useful for putting recent sea level rise into geological context. The IPCC AR6 concludes that the rate of global mean sea level rise since the twentieth century is faster than over any preceding century in at least the last three millennia. Further back in time, there is *medium confidence* that the global mean sea level was 5–10 meters higher during the Last Interglacial (132–119 thousand years ago), and 5–25 meters higher during the Mid-Pliocene Warm Period (3.3–3.0 million years ago). At the Last Glacial Maximum

(21–19 thousand years ago), the global mean sea level was 125–134 meters below present.[65]

The IPCC AR6 finds that there is *high agreement* across published global mean sea level projections for 2050 and there is little sensitivity to emissions scenarios. Considering only projections incorporating ice-sheet processes in whose quantification there is at least *medium confidence*, the global sea level projections for 2050, across all emissions scenarios, fall between 0.1 and 0.4 meters (5th–95th percentile *very likely* range) relative to the 1995–2014 baseline period.[66]

Conversely, there is *low agreement* across published global mean sea level projections for 2100, particularly for higher emissions scenarios. Considering only projections representing processes in whose quantification there is at least *medium confidence*, the AR6 global mean sea level projections for 2100 lie between 0.2 and 1.0 meters (5th–95th percentile *very likely* range) under the medium emissions scenario SSP2–4.5.[67] There is deep uncertainty surrounding projections of sea level rise, particularly for the higher emissions scenarios, which is addressed further in Section 9.5

7.5 Climate Predictions or Possible Futures?

"The sooner we depart from the present strategy, which overstates an ability to both extract useful information from and incrementally improve a class of models that are structurally ill suited to the challenge, the sooner we will be on the way to anticipating surprises, quantifying risks, and addressing the very real challenge that climate change poses." (Climate modelers Bjorn Stevens and Tim Palmer)[68]

With regard to projections of future climate change, the IPCC Fourth Assessment Report (2007) provided the following conclusion:

"There is considerable confidence that climate models provide credible quantitative estimates of future climate change, particularly at continental scales and above."[69]

Is this level of confidence in climate model projections justified? It is often assumed that if climate models reproduce current and past climates reasonably well, then we can have confidence in future predictions. However, empirical accuracy, to a substantial degree, may be due to tuning rather than to the model structural form. Further, the model may lack representations of processes and feedbacks that would significantly influence future climate change. Therefore, reliably reproducing past and present climate is not a sufficient condition for a model to be adequate for long-term projections, particularly for high-forcing scenarios that are well outside those previously observed in the instrumental record.

The CMIP5 twenty-first century climate simulations used in the IPCC AR5 are not actual predictions (or projections) of climate change, but rather show the sensitivity of climate change to different emissions scenarios. The CMIP5 simulations neglect solar variations (beyond the 11-year cycle), future volcanic eruptions, and the correct phasing and amplitude of multi-decadal climate variability associated with ocean circulations. The argument for dismissing future natural climate variability has been that it has a much smaller impact on the climate than emissions forcing. However, cumulatively and on decadal to multi-decadal timescales, this is not necessarily true. The CMIP5 simulations are at best possible climate futures, contingent on the model forcing assumptions used.

The CMIP6 climate simulations take several steps to deal with natural climate variability that bring the simulations closer to actual predictions. The CMIP6 twenty-first century simulations are forced by a single scenario of background volcanic forcing, which produces a cooling of about 0.2°C by the end of the twenty-first century relative to simulations with no volcanic forcing.[70] CMIP6 uses a more realistic scenario for future solar activity that exhibits variability at all timescales, providing a plausible course of solar activity until 2300.[71] CMIP6 has attempted to address the challenges associated with natural internal climate variability by using Single Model Initial Condition Large Ensembles (SMILEs; Section 6.1.3).

While the advances made under CMIP6 are a step in the right direction, further attention is needed with respect to natural climate variability—not only to improve projections of twenty-first century climate variability and change, but also to better understand the causes of climate change over the past century. The collection of CMIP5 and CMIP6 simulations provides an insufficient number of scenarios for volcanic and solar activity, and also inadequate treatments of multi-decadal to centennial internal climate variability as well as solar indirect effects. This is addressed in Chapter Eight.

Notes

1 Ian Wilson, "From Scenario Thinking to Strategic Action," *Technological Forecasting and Social Change* 65, no. 1 (September 2000): 23–29, https://doi.org/10.1016/s0040-1625(99)00122-5.

2 Detlef P. van Vuuren et al., "A New Scenario Framework for Climate Change Research: Scenario Matrix Architecture," *Climatic Change* 122, no. 3 (March 2013): 373–386, https://doi.org/10.1007/s10584-013-0906-1.

3 Brian C. O'Neill et al., "The Scenario Model Intercomparison Project (Scenariomip) for CMIP6," *Geoscientific Model Development* 9, no. 9 (September 28, 2016): 3461–3482, https://doi.org/10.5194/gmd-9-3461-2016.

4 Emma Marris, "We're Heading Straight for a Demi-Armageddon," *The Atlantic*, November 3, 2021.

5 Detlef P. van Vuuren et al., "The Representative Concentration Pathways: An Overview," *Climatic Change* 109, no. 1–2 (August 5, 2011): 5–31, https://doi.org/10.1007/s10584-011-0148-z.

6 Keywan Riahi et al., "The Shared Socioeconomic Pathways and Their Energy, Land Use, and Greenhouse Gas Emissions Implications: An Overview," *Global Environmental Change* 42 (January 2017): 153–168, https://doi.org/10.1016/j.gloenvcha.2016.05.009.

7 Detlef P. van Vuuren et al., "The Representative Concentration Pathways."

8 Keywan Riahi et al., "The Shared Socioeconomic Pathways."

9 Brian C. O'Neill et al., "The Scenario Model Intercomparison Project (Scenariomip) for CMIP6," *Geoscientific Model Development* 9, no. 9 (September 28, 2016): 3461–3482, https://doi.org/10.5194/gmd-9-3461-2016.

10 IEA, "Global Energy Review: CO_2 Emissions in 2021," IEA, March 2022, https://www.iea.org/reports/global-energy-review-co2-emissions-in-2021-2.

11 D. Chen et al., "Framing, Context, and Methods Supplementary Material," in *Climate Change 2021: The Physical Science Basis. Contribution of Working Group I to the Sixth Assessment Report of the Intergovernmental Panel on Climate Change* (Geneva, CH: Intergovernmental Panel on Climate Change, 2021), 147–286.

12 IEA, "Net Zero by 2050."

13 Keywan Riahi et al., "The Shared Socioeconomic Pathways."

14 IEA, "Net Zero by 2050," IEA, May 2021, https://www.iea.org/reports/net-zero-by-2050.

15 Ibid.

16 Roger Pielke Jr et al., "Plausible 2005–2050 Emissions Scenarios Project between 2 °C and 3 °C of Warming by 2100," *Environmental Research Letters* 17, no. 2 (February 11, 2022): 024027, https://doi.org/10.1088/1748-9326/ac4ebf.

17 David Wallace-Wells, "Here's Some Good News on Climate Change: Worst-Case Scenario Looks Unrealistic," Intelligencer (Intelligencer, December 20, 2019), https://nymag.com/intelligencer/2019/12/climate-change-worst-case-scenario-now-looks-unrealistic.html.

18 Roger Pielke and Justin Ritchie, "Systemic Misuse of Scenarios in Climate Research and Assessment," *SSRN Electronic Journal*, April 21, 2020, https://doi.org/10.2139/ssrn.3581777.

19 D. Chen et al., "Framing, Context, and Methods."

20 Zeke Hausfather, "Explainer: The High-Emissions 'RCP8.5' Global Warming Scenario," Carbon Brief, August 21, 2021, https://www.carbonbrief.org/explainer-the-high-emissions-rcp8-5-global-warming-scenario.

21 D. Reidmiller et al., eds., "Impacts, Risks, and Adaptation in the United States: The Fourth National Climate Assessment, Volume II," 2018, https://doi.org/10.7930/nca4.2018.

22 Roger Grenier et al., "Quantifying the Impact from Climate Change on U.S. Hurricane Risk," Verisk, 2020, https://www.air-worldwide.com/siteassets/Publications/White-Papers/documents/air_climatechange_us_hurricane_whitepaper.pdf.

23 "State and Local Progress on Adaptation Plans—Georgetown Climate Center," georgetownclimatecenter.org, accessed May 5, 2022, https://www.georgetownclimate.org/adaptation/index.html.

24 "COP26: Update to the NDC Synthesis Report," UNFCCC, November 4, 2021, https://unfccc.int/news/cop26-update-to-the-ndc-synthesis-report.

25 Oscar Wilde, "The Importance of Being Earnest," (February 14, 1895).

26 Zeke Hausfather, "Explainer: How Scientists Estimate Climate Sensitivity," Carbon Brief, June 19, 2019, https://www.carbonbrief.org/explainer-how-scientists-estimate-climate-sensitivity.

27 Ibid.

28 Jule Charney et al., "Carbon Dioxide and Climate: A Scientific Assessment," 1979, https://doi.org/10.17226/12181.

29 Rajendra K. Pachauri and Leo Meyer, *Climate Change 2014: Synthesis Report* (Geneva, CH: Intergovernmental panel on climate change, 2015).

30 "In-Depth Q&A: The IPCC's Sixth Assessment Report on Climate Science," Carbon Brief, August 9, 2021, https://www.carbonbrief.org/in-depth-qa-the-ipccs-sixth-assessment-report-on-climate-science.

31 Nicola Scafetta, "Testing the CMIP6 GCM Simulations versus Surface Temperature Records from 1980–1990 to 2011–2021: High ECS Is Not Supported," Climate 9, no. 11 (2021): 161, https://doi.org/10.3390/cli9110161.

32 Charles Lyell, *Principles of Geology*, vol. III (London, UK: J. Murray, 1830).

33 Brian J. Soden and Isaac M. Held, "An Assessment of Climate Feedbacks in Coupled Ocean–Atmosphere Models," *Journal of Climate* 19, no. 14 (July 15, 2006): 3354–3360, https://doi.org/10.1175/jcli3799.1.

34 S. C. Sherwood et al., "An Assessment of Earth's Climate Sensitivity Using Multiple Lines of Evidence," *Reviews of Geophysics* 58, no. 4 (July 22, 2020), https://doi.org/10.1029/2019rg000678.

35 Judith A. Curry and Peter J. Webster, "Thermodynamic Feedbacks in the Climate System," in *Thermodynamics of Atmospheres and Oceans* (New York, NY: Academic Press, 1999), 351–385.

36 Ibid.

37 Rajendra K. Pachauri and Leo Meyer, *Climate Change*.

38 D. Chen et al., "Framing, Context, and Methods."

39 Mark D. Zelinka et al., "Causes of Higher Climate Sensitivity in CMIP6 Models," *Geophysical Research Letters* 47, no. 1 (January 3, 2020), https://doi.org/10.1029/2019gl085782.

40 Alexander Otto et al., "Energy Budget Constraints on Climate Response," *Nature Geoscience* 6, no. 6 (May 19, 2013): 415–416, https://doi.org/10.1038/ngeo1836.

41 "Why Low-End 'Climate Sensitivity' Can Now Be Ruled Out," Carbon Brief, July 22, 2020, https://www.carbonbrief.org/guest-post-why-low-end-climate-sensitivity-can-now-be-ruled-out.

42 S. C. Sherwood et al., "An Assessment of Earth's Climate Sensitivity."

43 D. Chen et al., "Framing, Context, and Methods."

44 Nicholas Lewis, "Objectively Combining Climate Sensitivity Evidence," Climate Dynamics, 2022 (in press).

45 Ibid.

46 Jacob Bronowski, *The Ascent of Man* (Boston, MA: Little, Brown, 1973).

47 D. Chen et al., "Framing, Context, and Methods."

48 Rajendra K. Pachauri and Leo Meyer, *Climate Change*.

49 Michael Szenberg and Kenneth Arrow, "I Know a Hawk from a Handsaw," in *Eminent Economists: Their Life Philosophies* (Cambridge, UK: Cambridge University Press, 1993), 47.
50 V. Masson-Delmotte et al., eds., "Summary for Policymakers," in *Climate Change 2021: The Physical Science Basis. Contribution of Working Group I to the Sixth Assessment Report of the Intergovernmental Panel on Climate Change* (Geneva, CH: Intergovernmental Panel on Climate Change, 2021), 3–32.
51 Paul Voosen, "UN Climate Panel Confronts Implausibly Hot Forecasts of Future Warming," RealClearEnergy, July 29, 2021, https://www.realclearenergy. org/2021/07/29/un_climate_panel_confronts_implausibly_hot_forecasts_of_ future_warming_787527.html.
52 V. Eyring et al., "Human Influence on the Climate System," in *Climate Change 2021: The Physical Science Basis. Contribution of Working Group I to the Sixth Assessment Report of the Intergovernmental Panel on Climate Change* (Geneva, CH: Intergovernmental Panel on Climate Change, 2021), 3–32.
53 Ibid.
54 Yogi Berra, *The Yogi Book: "I Really Didn't Say Everything I Said!"* (New York City, NY: Workman Publishing Company, 1998).
55 V. Masson-Delmotte et al., eds., "Summary for Policymakers."
56 Ibid.
57 Ibid.
58 Theodore Taylor, *The Magnificent Mitscher* (Annapolis, MD: Naval Inst. Press, 1991).
59 V. Masson-Delmotte et al., eds., "Summary for Policymakers."
60 S. I. Seneviratne et al., "Weather and Climate Extreme Events in a Changing Climate," in *Climate Change 2021: The Physical Science Basis. Contribution of Working Group I to the Sixth Assessment Report of the Intergovernmental Panel on Climate Change* (Geneva, CH: Intergovernmental Panel on Climate Change, 2021), 1513–1766.
61 R. Ranasinghe et al., "Climate Change Information for Regional Impact and for Risk Assessment," in *Climate Change 2021: The Physical Science Basis. Contribution of Working Group I to the Sixth Assessment Report of the Intergovernmental Panel on Climate Change* (Geneva, CH: Intergovernmental Panel on Climate Change, 2021), 1767–1926.
62 H. Douville et al., "Water Cycle Changes," in *Climate Change 2021: The Physical Science Basis. Contribution of Working Group I to the Sixth Assessment Report of the Intergovernmental Panel on Climate Change* (Geneva, CH: Intergovernmental Panel on Climate Change, 2021), 1055–1210.
63 Daniel Kahneman, Thinking, Fast and Slow (New York, NY: Farrar, Straus and Giroux, 2013).
64 B. Fox-Kemper et al., "Ocean, Cryosphere and Sea Level Change," in *Climate Change 2021: The Physical Science Basis. Contribution of Working Group I to the Sixth Assessment Report of the Intergovernmental Panel on Climate Change* (Geneva, CH: Intergovernmental Panel on Climate Change, 2021), 1211–1362.
65 Ibid.
66 Ibid.
67 Ibid.
68 Tim Palmer and Bjorn Stevens, "The Scientific Challenge of Understanding and Estimating Climate Change," *Proceedings of the National Academy of Sciences* 116, no. 49 (December 3, 2019): 24390–24395, https://doi.org/10.1073/ pnas.1906691116.

69 S Solomon et al., eds., *Climate Change 2007: The Physical Science Basis* (Cambridge, UK: Cambridge University Press, 2007).

70 John C. Fyfe et al., "Significant Impact of Forcing Uncertainty in a Large Ensemble of Climate Model Simulations," *Proceedings of the National Academy of Sciences* 118, no. 23 (May 28, 2021), https://doi.org/10.1073/pnas.2016549118.

71 Katja Matthes et al., "Solar Forcing for CMIP6 (v3.2)," *Geoscientific Model Development* 10, no. 6 (June 22, 2017): 2247–2302, https://doi.org/10.5194/gmd-10-2247-2017.

Chapter Eight

ALTERNATIVE METHODS FOR GENERATING CLIMATE CHANGE SCENARIOS

"When it comes to thinking about the impacts of climate change, we must guard against a failure of imagination."
—Center for Naval Analyses Military Advisory Board[1]

There is a growing realization that global climate models are not fit for the purpose of predicting twenty-first century climate variations and change, particularly on decadal and regional scales (Sections 6.3.4, 6.4). Climate model simulations under the auspices of the CMIP program and the IPCC are not providing the full range of scenarios of plausible climate outcomes. The CMIP simulations include very limited scenarios of volcanic eruptions and solar variability. Further, the climate models have inadequate representations of solar indirect effects and multi-decadal to century scale variations in the large-scale ocean circulations. Natural internal climate variability is of particular importance for scenarios of regional climate variations.

There is a need for alternative science-based frameworks for developing scenarios of climate futures, particularly with regard to regional extreme weather and climate events. The framework presented here has an indirect role for climate model simulations in developing scenarios of climate futures, but utilizes a broader range of methods for generating scenarios.[2] Statistical and network-based models, data-driven scenarios, and speculative but physically-based what-if scenarios are also used in scenario generation. This alternative scenario framework reflects a shift away from median and likely values to a possibility framework for scenarios that accepts all outcomes that are not judged to be implausible by our current background knowledge.

The rationale for the scenario generation methods described in this chapter is provided by the risk governance framework that is discussed in Part Three of this book.[3] A broad range of justified scenarios, along with expression of uncertainties, provide inputs for robust decision-making. Robust decision-making incorporates scientific uncertainty into the decision-theoretic framework as knowledge, not ignorance.[4]

8.1 Escape from Model-Land

"Letting go of the phantastic mathematical objects and achievables of model-land can lead to more relevant information on the real world and thus better-informed decision-making." (Computer modeler Erica Thompson, author of Escape from Model-Land)[5]

"Model-land" is a world in which mathematical simulations are evaluated against other mathematical simulations.[6] Decision support in model-land implies taking the output of model simulations at face value, and then interpreting the frequencies from model-land to represent probabilities in the real world.

Climate scientists and other practitioners are using simulations from global climate models to provide precise, quantified, probabilistic projections at high resolution.[7] Climate models that were originally used to guide policy-makers regarding greenhouse gas mitigation strategies are now in demand to inform highly localized and detailed adaptation decisions. However, global climate models are inadequate for predicting regional climate change and the frequency and intensity of extreme weather events.[8] The direct use of climate model output in regional adaptation decision-making is inappropriate and produces an inflated impression of confidence in estimates of future changes.[9] Regional downscaling (statistical or dynamical) of the global climate models does not usually help, since the downscaled outputs inherit the deficiencies of the global climate models.

In escaping from model-land, climate models are not discarded completely; rather, the aim is to use them more effectively. Climate models should be considered not simply as prediction machines, but as scenario generators, sources of insight into complex system behavior, and aids to critical thinking within robust decision frameworks. Such a shift would have implications for how users perceive and use information from climate models and the types of simulations that will have the most value for informing decision-making.

If the scenarios of future climates from global climate model simulations are inadequate, what other strategies can be used to generate scenarios of future climate change that are relevant to decision-making? Simple climate models, process models, and data-driven models can also provide the basis for generating scenarios of future climate. Historical and paleoclimate records provide a rich source of information for developing future scenarios. These alternative methods for generating future climate scenarios are particularly relevant for developing regional scenarios and scenarios of impact variables that are not directly simulated by global climate models. Creatively driven what-if scenarios (often referred to as "storylines") can also be generated, based on our understanding of climate dynamics.[10]

8.2 Emissions and Temperature Targets

"The first rule of climate chess is this: The board is bigger than we think, and includes more than fossil fuels." (Environmental scientist Jon Foley)[11]

The strategy to limit global warming is tied directly to limiting the amount of CO_2 emitted into the atmosphere. Emissions targets are a centerpiece of the UNFCCC Paris Agreement. The goal of the emissions targets is to limit global warming to well below 2.0, preferably 1.5°C, compared to pre-industrial temperatures (typically the baseline period 1851–1900).[12] For reference, the climate has warmed in 2020 by about 1.1°C above this baseline.[13]

How meaningful are these temperature and emissions targets? These targets are usefully interpreted analogously to highway speed limits. A speed limit of 100 km/hr does not imply danger at higher speeds or safety at lower speeds; a speed of 110 km/hr probably isn't associated with a noticeable difference in danger, but a speed of 150 km/hr would be much less safe. On the other hand, even at a speed of 50 km/hr a bad accident can occur, caused by something unforeseen or random.

Using the medium emissions scenario (SSP2–4.5), the IPCC AR6 constrained global mean temperature projections (see Table 7.1) indicate that there is a 50 percent chance that the 1.5°C threshold would be crossed around 2030 and the 2°C threshold would be crossed around 2052.[14] There is uncertainty in the year for which the thresholds would be crossed (2026–42 for the 1.5°C threshold and 2038–72 for the 2°C threshold), mostly owing to the range of values of climate sensitivity to CO_2 among different models.[15]

We now apply the alternative scenario approach to illustrate how natural climate variability could influence the global mean surface temperature change through 2050, and hence influence the time of crossing the 1.5 and 2.0°C thresholds. Specifically, alternative scenarios of volcanic eruptions, solar variability, and internal climate variability are considered. The risk from not accounting for scenarios of natural climate variability is that critical possible future climate outcomes are being ignored, potentially causing maladaptation.

8.2.1 Natural Internal Variability

"Finally, we show that even out to thirty years large parts of the globe (or most of the globe) could still experience no-warming due to internal variability." (Climate modelers Maher et al.)[16]

Variations in global mean surface temperature are linked to recurrent multi-decadal variations in large-scale ocean circulations. Not taking multi-decadal internal variability into account in predictions of future warming runs the risk of over-estimating the warming for the next two to three decades, when the Atlantic Multi-decadal Oscillation is expected to shift into its cold phase.[17,18] The Atlantic Multidecadal Oscillation is a coherent mode of natural variability of sea surface temperatures occurring in the North Atlantic Ocean, with an estimated period of 60–80 years.

While climate models simulate the large-scale ocean circulations and internal climate variability, the simulated magnitude in the multi-decadal band is too low in most models relative to observations, and the phasing of the variability in long-term climate simulations is not synced with the actual, observed climate variations. Averaging multiple simulations from climate models effectively averages out the internal variations, leaving only the forced climate variability (e.g., CO_2 forcing).

A study that used six single model initial condition large ensembles (SMILEs) for twenty-first-century simulations found that on a 15-year timescale, surface temperature trend projections are dominated by internal variability, with little influence of structural model differences or the emissions scenario. On a 30-year timescale, structural model differences and emissions scenario uncertainties play a larger role in controlling surface temperature trend projections. However, even for projections out to thirty years, most of the globe could still experience no warming due to internal variability even with continued CO_2 forcing.[19]

So, for the period between 2020 and 2050, will natural internal variability contribute to warming or cooling, relative to the underlying warming trend from emissions? Most analyses have identified the Atlantic Multidecadal Oscillation (AMO) as having the dominant multi-decadal imprint on global temperatures. It has been estimated that there is a peak-to-trough impact of the AMO on global mean surface temperatures of 0.3–0.4°C.[20,21] The climate has been in the warm phase of the AMO since 1995; hence in 2021, it has been 26 years since the previous shift. Analysis of historical and paleoclimatic records suggests that a shift to the cold phase of the AMO is expected occur within the next 12 years (by 2032), with a 50 percent probability of the shift occurring in the next five years (by 2026).[22]

While we are currently still in the warm phase of the AMO, we are past the peak of the warm phase. Hence, we consider the following three scenarios for the contribution of multi-decadal internal variability to global mean surface temperature change for the period 2020–50:

• The None scenario (0°C) assumes no net impact of multidecadal internal variability on the global mean surface temperature, which is implicit in the constrained projections of the IPCC AR6.

• The Moderate scenario (−0.2°C) assumes a shift to the cool phase of the AMO in the 2020s with a moderate impact;

• The Strong scenario (−0.3°C) assumes a shift to the cool phase of the AMO in the 2030s with a stronger impact.

The Moderate and Strong scenarios are contingent on the assumptions that the AMO is the primary driver of multi-decadal internal climate variability and that a shift to the cold phase of the AMO is expected in the next decade.[23] Other facets of multi-decadal and decadal scale internal variability could come into play over the next three decades, but these scenarios illustrate the magnitude of plausible outcomes over the next three decades.

While the scenarios presented here focus on cooling over the next three decades, it is noted that the same reasoning leads to the expectation that internal variability will contribute to warming during decades in the second half of the twenty-first century.

8.2.2 Volcanoes

"Now is the winter of our discontent." (Shakespeare's King Richard III, in response to a string of very cold winters during the Little Ice Age)[24]

The instrumental period covering the past 150 years has been relatively quiet with regards to volcanic eruptions, and thus it is tempting to ascribe potential volcanism a minor role in future climate projections. However, over the past two millennia, there have been periods with considerably stronger volcanic activity. Clusters of strong tropical eruptions have contributed to sustained cold periods such as the Little Ice Age.

Explosive volcanoes are omitted from scenarios used for future climate projections (CMIP5/CMIP6) because they are unpredictable. Due to the direct radiative effect of volcanic aerosol particles that reach the stratosphere, large volcanic eruptions lead to an overall decrease of global mean surface temperature, which can extend to multi-decadal or even century timescales in the case of clusters of large volcanic eruptions.[25]

Explosive volcanic eruptions of the magnitude of the 1991 Pinatubo eruption or larger have occurred on average twice per century throughout the past 2,500 years.[26] About eight extremely explosive volcanic eruptions (more than five times stronger than Pinatubo) occurred during this period. The largest of these are Samalas in 1257 and Tambora in 1815, the latter resulting in "the year without a summer" with harvest failures across the Northern Hemisphere.[27] It has been estimated that a Samalas-type eruption may occur 1–2 times per millennium on average.[28]

Given the unpredictability of individual eruptions, the CMIP5/CMIP6 climate model simulations either specify future volcanic forcing as zero or as a constant background value.[29] The background value used in the CMIP6 simulations has been estimated from the historical record since 1850. Background estimates of volcanic cooling determined from climate models range from 0.1 to 0.27°C, the differences arising from model structural differences.[30,31]

The IPCC AR6 states that it is *likely* that at least one large eruption will occur during the twenty-first century. The AR6 further acknowledges that a low likelihood, high impact outcome of several large eruptions would greatly alter the twenty-first-century climate trajectory compared to emissions-based projections.[32] How much cooling could happen from explosive volcanic eruptions in the twenty-first-century? A cluster of explosive eruptions such as happened in the first half of the nineteenth century is estimated to have caused 0.5°C cooling averaged over several decades.[33]

The analysis here considers three scenarios of volcanic cooling for the twenty-first century:

- Low Baseline scenario, equivalent to a weak response to the average volcanic forcing over the historical record since 1850, estimated at −0.1°C;
- High Baseline scenario, estimated at −0.27°C;
- Extreme Cluster of volcanoes, analogous to the explosive eruptions that occurred during 1810–40, estimated to have caused a decadal-averaged cooling of −0.5°C.

8.2.3 Solar Variations

"The field of Sun-climate relations [...] in recent years has been corrupted by unwelcome political and financial influence as climate change sceptics have seized upon putative solar effects as an excuse for inaction on anthropogenic warming." (solar physicist and IPCC author Mike Lockwood)[34]

"We argue that the Sun/climate debate is one of these issues where the IPCC's 'consensus' statements were prematurely achieved through the suppression of dissenting scientific opinions." (Climate and solar scientists Ronan Connolly et al.)[35]

The impact of solar variations on the climate is uncertain and subject to substantial debate. However, you would not infer from the IPCC assessment reports that there is debate or substantial uncertainty surrounding this issue.

The Sun goes through cycles of approximately 11 years (the Schwabe Cycle) in which solar activity waxes and wanes. Above the Earth's atmosphere, the difference in Total Solar Irradiance (TSI, measured in watts per square meter W/m^2) between the 11-year maxima and minima is small, on the order of 0.1 percent of the total TSI, or about 1 W/m^2. A multidecadal increase in TSI should cause global warming (all else being equal); similarly, a multidecadal decrease in TSI should cause global cooling. Researchers have speculated that multi-decadal and longer changes in solar activity could be a major driver of climate change.

Exactly how TSI has changed over time has been a challenging problem to resolve. Since 1978, there have been direct measurements of TSI from satellites. However, interpreting any multi-decadal trends in TSI requires comparisons of observations from overlapping satellites. Substantial uncertainty exists in the TSI composites during the period from 1978 to 1992. This is mostly due to the fact that the Active Cavity Radiometer Irradiance Monitor (ACRIM) Satellite Two solar satellite mission was delayed because of the Space Shuttle Challenger disaster in 1986 (ACRIM2 was eventually launched in late 1991). This delay prevented the ACRIM2 record from overlapping with the ACRIM1 record that ended in July 1989. The ACRIM-gap prevents a direct cross-calibration between the two high-quality ACRIM1 and ACRIM2 TSI records.[36]

This rather arcane issue of cross-calibration of two satellite records has profound implications. There are a number of rival composite TSI datasets, disagreeing as to whether TSI increased or decreased during the period 1986–96. Further, the satellite record of TSI is used for calibrating proxy models, whereby past solar variations are inferred from sunspots and cosmogenic isotope measurements.[37] As a result, some of the datasets for past values of TSI (since 1750) have low variability, implying a very low impact of solar variations on global mean surface temperature, whereas datasets with high TSI variability can explain 50–98 percent of the temperature variability since preindustrial times.[38,39]

The IPCC AR5 adopted the low variability solar reconstructions, without discussing this controversy. The AR5 concluded that the best estimate of radiative forcing due to TSI changes for the period 1750–2011 was 0.05 W/m^2 (*medium confidence*). For reference, the forcing from atmospheric greenhouse gases over the same period was 2.29 W/m^2.[40] Thus, the IPCC AR5 message was that changes in solar activity are nearly negligible compared to anthropogenic ones for forcing climate change.

The IPCC AR6 acknowledges a much larger range of estimates of changes in TSI over the last several centuries, stating that the TSI between the

Maunder Minimum (1645–1715) and the second half of the twentieth century increased by 0.7–2.7 W/m², a range that includes both low and high variability TSI data sets.[41] However, the recommended forcing dataset for the CMIP6 climate model simulations used in the AR6 averages two data sets with low solar variability.[42]

The uncertainties and debate surrounding solar variations and their impact on climate was the topic of the *Climate Dialogue*, a remarkable blogospheric experiment.[43] *Climate Dialogue* was the result of a request by the Dutch parliament to facilitate the scientific discussions between climate experts representing the full range of views on the subject. The Dialogue on solar variations included five distinguished scientists with extensive publication records on the topic.[44] One participant was in line with the IPCC AR5, thinking that solar variations are only a minor player in the Earth's climate. Two participants argued for a larger and even dominant role for solar variability, and the other two emphasized uncertainties in our current understanding.

More recently, a review article was published in the journal *Research in Astronomy and Astrophysics*. The article had 23 co-authors with a range of perspectives, but who were united by their agreement not to take the consensus approach of the IPCC. Rather, the paper emphasized where dissenting scientific opinions exist as well as identifying areas where there is scientific agreement. The authors found that the Sun/climate debate is an issue where the IPCC's consensus statements were prematurely achieved through the suppression of dissenting scientific opinions.[45]

Of direct relevance to projections of twenty-first century climate is whether we might expect a substantial change in solar activity relative to the twentieth century. On multidecadal timescales, proxy reconstructions of solar activity reveal occasional phases of unusually high or low solar activity, which are respectively called grand solar minima and maxima.[46] Grand solar maxima occur when solar cycles exhibit greater than average activity for decades or centuries.

Solar activity reached unusually high levels in the second half of the twentieth century, estimated to be the highest in the last 8,000 years, although there is disagreement among reconstructions as to whether this maximum peaked in the 1950s or continued into the 1990s.[47] It has been estimated that about 20 grand maxima have occurred over the last 11 millennia, averaging one per 500 years.[48] During the last 11 millennia, there have been 11 grand solar minima, with intervals between them ranging from a hundred years to a few thousand years. The most recent grand minimum was the Maunder Minimum, during the 1645–1715 period.[49]

There are several reasons to expect lower solar activity during the twenty-first century, relative to the twentieth century. The recently completed solar cycle 24 was the smallest sunspot cycle in 100 years and the third in a trend of diminishing sunspot cycles.[50] Further, a grand maximum is more likely to be followed by a grand minimum than by another grand maximum.[51] Empirically-based projections imply a new solar minimum starting in 2002–04 and ending in 2063–75.[52] It has been estimated that there is an 8 percent chance of the Sun falling into a Grand Minimum during the next 40 years.[53] However, the depth and length of a phase of low solar activity in the twenty-first century is largely uncertain.

If the Sun did fall into a minimum during mid twenty-first century of the magnitude of the Maunder Minimum, how much cooling could we expect? Estimates from climate models and other analytical models expect the cooling to be small, ranging from 0.09 to 0.3°C.[54] These models assume that solar-climate interaction is limited to TSI forcing alone.

However, there is growing evidence that other aspects of solar variability, which are referred to as solar indirect effects, amplify TSI forcing or are independent of TSI forcing.[55] Candidate processes include: solar ultraviolet changes; energetic particle precipitation; atmospheric-electric-field effect on cloud cover; cloud changes produced by solar-modulated galactic cosmic rays; large relative changes in the magnetic field; and the strength of the solar wind. Solar indirect effects can be classified as known unknowns. While these indirect effects are not included in the CMIP6 twenty-first-century projections, we can make some inferences based upon recent publications. Recent research suggests that solar indirect effects could amplify an anomaly in solar insolation by a factor of up to 3–7.[56,57,58] If such an amplification factor is included, then a surface temperature decrease of up to 1°C (or even more) from a Maunder Minimum could occur.

In light of these considerations, three scenarios for solar variability in the mid-twenty-first century are considered here:

- CMIP6 Reference scenario: approximately –0.1°C[59]
- Intermediate: –0.3°C, corresponds to high Maunder minimum estimate without amplification effects, or a weaker minimum with amplification effects[60]
- High: –0.5°C, a low solar scenario (which is not a Maunder Minimum) with amplification by solar indirect effects[61]

The next 20–30 years of observations should reveal a lot about the role of the Sun in climate.

8.2.4 Global Surface Temperature Scenarios to 2050

"Unless we take action on climate change, future generations will be roasted, toasted, fried, and grilled." (Christine LaGarde, Managing Director of the International Monetary Fund)[62]

Synthetic scenarios building upon historical and paleo data, climate model outputs, targeted physical process models and storylines based on physical reasoning (the "storyline" approach) provide a broader range of outcome scenarios than global climate model simulations, particularly with regard to natural climate variability.

The scenarios presented in the previous subsections are integrated here to assess how natural climate variability could change our expectations for the amount of warming expected by 2050, in particular the years in which the 1.5 and 2.0°C thresholds will be crossed. All of the scenarios of natural variability considered here point in the direction of cooling through 2050, for reasons discussed in the preceding subsections.

The final integral temperature change is the sum of temperature changes driven by:

- SSP2–4.5 emissions—three scenarios that span the AR6 *very likely* range (+1.6, +2.0, +2.5°C), referenced to the baseline period 1851–1900
- Volcanoes—three scenarios relative to a nominal baseline of –0.1°C in the CMIP6 simulations (0, –0.17, –0.4°C), referenced to a baseline of 2020
- Solar—three scenarios relative to a nominal baseline of –0.1°C in the CMIP6 simulations (0, –0.2, –0.5°C), referenced to a baseline of 2020
- Natural internal variability—three scenarios (0, –0.2, –0.3°C), referenced to a baseline of 2020

With four variables and three scenarios for each variable, we can produce a total of 81 scenarios by adding combinations of scenario inputs for the individual variables. Three of these outcome scenarios correspond directly to the AR6 values associated with SSP2–4.5, while the others include some combination of the scenarios of natural climate variability.

Figure 8.1 shows a histogram of the 81 different scenario outcomes. The outcome frequencies are indicated on the y-axis. For reference, the temperature in 2020 is 1.1°C above the 1851–1900 baseline, which is indicated by the vertical line.[63] The scenario with the most warming is 2.5°C, which corresponds to the upper bound of the *likely* range from the AR6 (with no additional impacts from natural variability). The scenario with the lowest amount of warming is 0.4°C, which corresponds to the lower bound of the AR6 *very likely* range with the most

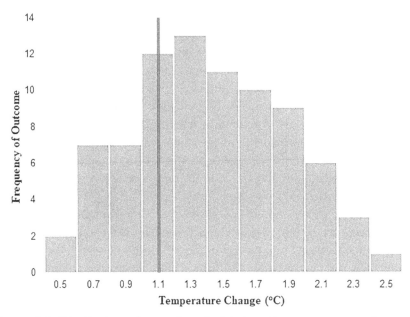

Figure 8.1 Distribution of scenarios of global surface temperature change for 2050, referenced to a baseline of 1851–1900. The vertical line corresponds to 1.1°C, representing the warming to 2020. The y-axis is the frequency of outcomes, based on 81 scenarios.

extreme scenario for each of the components of natural variability—this extreme outcome scenario for 2050 is 0.8°C cooler than the temperature in 2020.

There is weak justification for providing likelihoods of the individual outcomes, which is referred to as scenario uncertainty. While each of these scenario outcomes is arguably plausible, the distribution of outcomes in Figure 8.1 does not in any way reflect the probability of the outcomes. Are some of these scenarios more likely than others? Selecting the intermediate scenario for each variable produces an outcome scenario of +1.43°C, indicating that we would not cross the 1.5°C threshold before 2050 (compared to an expected crossing circa 2030 using the AR6 best estimate for SSP2–4.5). For the intermediate outcomes in Figure 8.1, there are multiple pathways to the same temperature outcomes, supporting a greater likelihood for the intermediate outcomes. However, judgment about the likelihood of individual outcomes rests on the analyst's assessment of the likelihood of the individual input scenarios. Note that the largest uncertainty is associated with the range of temperature increases in the *likely* range from the SSP2–4.5 emissions scenarios.

What do I personally consider to be the most likely scenario outcome for 2050? I would select the lowest emissions-based scenario (+1.6°C), the baseline volcanic scenario (0°C), the intermediate solar scenario (–0.2°C), and the moderate scenario for internal variability (−0.2°C); this would produce a net total warming relative to the 1850–1900 baseline of 1.2°C by 2050—a weak rising temperature trend between 2020 and 2050. However, focusing on a best estimate under conditions of deep uncertainty is misleading, whether it is the IPCC's best estimate or my own. A full exploration of scenario outcome uncertainty is much more meaningful for risk governance and decision-making.

All of the components of natural variability point to cooling during the period 2020 to 2050. Individually these terms are not expected to be large in the moderate scenarios. However when summed, their magnitude approaches, or could even exceed, the magnitude of the emissions-driven warming for the next three decades.

Studies using global climate models to assess the probability of decades in the twenty-first century being characterized by net cooling have mostly focused only on natural internal variability,[64,65] with one study that considers volcanic eruptions plus internal variability.[66] Volcanic-induced cooling becomes increasingly important in facilitating neutral or negative temperature trends on longer timescales, in conjunction with natural internal variability effects. Several studies have addressed the combination of internal and solar variability.[67,68,69] Apart from the wild-card of volcanic eruptions, the big uncertainty is solar indirect effects. The growing likelihood of a solar minimum of some magnitude during the mid-twenty-first century emphasizes the need for a resolution to the debate over low versus high variability solar reconstructions, and an improved understanding of solar indirect effects.

The bottom line is that uncertainty in global temperature projections to 2050 is skewed towards lower values, as uncertainty in near-term scenarios of emissions is decreasing. The confluence of cooling contributions from solar, volcanoes, and natural internal variability during the period 2020–50 could extend by decades the time horizon for keeping the global mean surface temperature below the thresholds of 1.5 and 2.0°C. While solar and volcanic cooling could extend throughout the twenty-first century, natural internal variability is expected to transition to warming effect later in the twenty-first century. This extension has important implications for the urgency of emissions reductions and planning for geoengineering interventions, which is addressed in Chapter Fourteen.

The targets of 1.5 and 2°C are easy to measure and communicate and have been effective at galvanizing political will and public support. However, these targets are vague approximations to some of the dangers of climate change and mis-represent the nature of the scientific knowledge upon which these numbers are claimed to rest.[70] These targets have arguably become a "fetish" that exerts excessive power over our imagination of the climatic future, narrowing our policy options and directing our policy making.[71] Further, these targets encourage goal displacement,[72] which occurs when attention becomes focused on hitting the target while obscuring the real reasons why we are concerned about climate change in the first place—the wellbeing of humans and ecosystems.

8.3 Regional Scenarios of Extreme Events

"We are all now stuck in a science-fiction novel that we're writing together." (Science fiction novelist Kim Stanley Robinson)[73]

Decision makers are demanding regional climate projections for adaptation decision-making and have high expectations regarding accuracy and resolution.[74] The IPCC AR6 recognizes this importance, devoting Chapters Ten to Twelve to regional climate variability and change.

At the regional scale, the importance of internal variability is greater and uncertainties are larger than at the global scale. Further, there are processes and feedbacks of a regional nature that influence the local climate and extreme weather events. Hence projections of regional climate change pose additional challenges.

Statistical and dynamical downscaling are postprocessing methods used to derive high-resolution climate information from global climate model simulations. The IPCC AR5 concluded with *high confidence* that downscaling adds value in regions with highly variable topography and for various small-scale phenomena;[75] this conclusion motivated widespread downscaling efforts that provided very precise regional projections.[76] However, the AR6 provides a more realistic perspective.[77] Regional models necessarily inherit biases from the global models used to provide boundary conditions at the region's borders. Bias adjustment cannot overcome all consequences of unresolved or strongly misrepresented physical processes, such as large-scale circulation biases or local feedbacks, and may instead introduce other biases and implausible climate change signals.

The IPCC AR6 acknowledges that there is no simple recipe for making regional climate projections, and the IPCC AR5 Working Group II (WG II) regarded such projections to be a matter of basic research. The challenge for developing useful regional scenarios of climate change requires integrating carefully culled climate model simulations with historical data and process-based understanding. The IPCC AR6 refers to this as "the distillation of regional climate change information from multiple lines of evidence."[78]

This section illustrates some approaches for developing regional climate scenarios and their applications.

8.3.1 Extreme Weather and Climate Events

"If we do not put a brake on climate change, it will have devastating consequences for all of us—there will be more storms, there will be more heat and catastrophes, more droughts, and there will be rising sea levels and increasing floods." (Former Chancellor of Germany Angela Merkel)[79]

Alarm over global warming is commonly driven by the expectation of greater frequency and/or increased severity of extreme weather and climate events (hereafter referred to as "extreme events"). Extreme events include floods, droughts, hurricanes, temperature extremes, storm surges. Extreme events are climate impact drivers for the region in which they occur.

There are numerous challenges to predicting statistics of future extreme weather and climate events. The relatively low resolution of climate models precludes accurate simulation of extreme weather events such as heat/cold waves, extreme precipitation events, and hurricanes. While the longer timescales associated with drought would seem to make them more amenable to simulation by climate models, the land surface and hydrological modules add another dimension of uncertainty on top of the driving climate projections. Further, the signal of a warming climate occurs against a background of natural internal variability. Therefore, it is difficult to separate the effect of global warming from internal variability in diagnosing from observations the impact of global warming on extreme events.

Various approaches are used to define extreme events. These are generally based on the determination of relative (e.g., 90th percentile) or absolute (e.g., 35°C for a hot day) thresholds above which conditions are considered extreme. They can also be characterized in terms of duration, which is of particular relevance for droughts.

Extreme weather and climate events are commonly characterized by return periods, which is an average time or an estimated average time between events to occur. For example, the return period of a flood might be 100 years, equivalent to a probability of one percent of occurring in any one year. In any given 100-year period, a 100-year event may occur once, twice, more, or not at all. The return period is a statistic, computed from a set of data (observations or climate model output). The IPCC focuses on extreme events since 1950, associated with the most reliable global data sets.[80] Clearly, one cannot determine the size of a 1,000-year event or even a 100-year event based on such records alone; therefore, statistical models are used to predict the magnitude of such an (unobserved) event.

There are several challenges here. Statistical models used to calculate return periods assume stationarity of the time series, whose variability does not change with time. The assumption of stationarity is clearly violated in a changing climate and is arguably violated even in the presence of coherent multi-decadal climate variability. Further, genuinely extreme events are rare and may occur in clusters, and hence are not easily captured by statistical models.

Here is an example from one of the clients of my company who was concerned about the risk from a major hurricane storm surge during the next 50 years. The specific concern related to plans to build a facility on the coast of Tampa Bay in Florida (US), in a location with an elevation of 8.3 feet above sea level. The client hired a well-known risk management company to assess the risk of a hurricane storm surge at the location for the 50, 100, 250, and 500-year return periods. The 100-year storm surge was estimated using a hazard model to be 6.2 feet, and the 500-year storm surge was estimated at 13 feet.

I was asked to assess this analysis. I did a simple investigation into the historical records of hurricanes striking Tampa Bay. In 1921, a major hurricane struck Tampa Bay, bringing an estimated storm surge of 10–12 feet.[81] The worst recorded hurricane to strike Tamp Bay was in 1848, which was at least a Category 4 hurricane with an estimated storm surge of 15 feet—substantially exceeding even the estimated 500-year storm surge.[82]

So far, the analysis has not considered projections of global warming on the risk of storm surge. Simply from warming, little to no change in the frequency of major hurricanes striking Tampa Bay is expected (see Section 13.2.1), although a change to the cold phase of the Atlantic Multidecadal Oscillation in coming decades would actually reduce the hurricane risk to Florida.[83] However, warming under the medium emissions scenario (SSP2–4.5) is expected to cause the local sea level to increase over the next 50 years by 8–15 inches (AR6), a rate greater than simply extrapolating the expected increase from the local tide gauge record of 5.5 inches.[84]

The best analyses of future extreme events are derived from a comprehensive analysis of historical events, not just from the meteorological data record but also from historical ship reports, newspaper articles, and even diaries. Where available, paleoclimatic records should be used. The bottom line is that if it has happened before, it can happen again. The challenge is to understand the causes of the extreme events, not just since 1950 but from as far back as records are available.

What about the impacts of global warming on extreme events? The IPCC AR6 states that "Regional changes in the intensity and frequency of climate extremes generally scale with global warming."[85] This provides the basis for inferring that at least some extreme events will change in either frequency or intensity as warming continues.

The challenge is then to link understanding of the regional variations and change in an extreme weather event with an increase in global temperature. For regional climate change, an assessment of local trends in the frequency or intensity of extreme events of the historical data record indicates whether the regional extreme events are susceptible to global warming. However, any trend needs to be carefully interpreted in context of the multi-decadal natural climate regimes. Temperature and rainfall-related events also need to be assessed in context of land use changes (urbanization, deforestation, cultivation). Once an event type is identified that appears to be worsening with warming, physical process models and theoretical analyses can be used to assess changes in the potential magnitudes of the extreme events, relative to the historical data.

My company Climate Forecast Applications Network (CFAN) uses the following approach:

i) Interpretation of the historical variations of the particular extreme weather type in context of the overall trend, relationships to weather teleconnection regimes and climate regimes (e.g., El Nino, Atlantic Multidecadal Oscillation);
ii) Integration of scenarios of modes of natural multi-decadal variability with emissions-based scenarios of climate change;
iii) Development of extreme event scenarios from the scenarios of warming and multi-decadal variability by integrating historical climate dynamics analysis and process-based scenarios of extreme events;
iv) For each scenario of warming and natural climate variability, assess the magnitude and frequency (return times) of each extreme event type.

In all of the analyses that I have prepared of regional scenarios of extreme events for the next three decades, variations associated with natural climate variability substantially exceed any expected change from global warming. That said, it may be justifiable in some circumstances to assess that warming may have incrementally worsened a particular extreme event. In terms of longer-range scenarios of regional extreme events (say to 2100), physical process models and physical reasoning may suggest a worsening of extremes, but uncertainties about variations in multi-decadal climate regimes can be larger than incremental changes from warming. Time of Emergence (TOE) analysis provides an estimate of the point in time when the forced signal of climate change has emerged from the natural variability of the climate system with a specified level of statistical confidence.[86,87]

Rather than focusing solely on the relatively small and uncertain impacts of global warming on extreme events, a broader range of extreme weather events from the historical record (and paleo record where available) can provide a better basis for avoiding "big surprises" (see Section 9.4).

Greater attention is needed to understand the full range of climate variability that contributes to extreme weather events. For example, many of the worst US weather disasters in the historical data record occurred in the 1930s and 1950s, a period that was not significantly influenced by manmade global warming.[88] The evolution of natural multi-decadal modes of climate variability suggests that we could see another quiet period in coming decades, followed by a more active period. Until the influence of natural climate variability on extreme weather is better understood, we may be misled in our interpretations of recent trends and their attribution to manmade global warming.

In several regional climate impact assessment projects that I have been involved in, the client has hired two or three different groups to provide independent assessments. Apart from different methodologies, such assessments invariably involve expert judgment, and "which expert" matters. The bottom line is that currently there is no generally accepted best practice for making regional projections of extreme weather events on decadal time scales.

8.3.2 Scenarios for Stress Test Applications

"You know there is no constancy in human affairs, when a single swift hour can often bring a man to nothing." (Ancius Boethius[89], philosopher of the late Middle Ages)

One application of regional climate scenarios involves the use of stress tests,[90] also referred to as vulnerability analyses,[91] to evaluate specific risks and impacts of disruptive climatic conditions. In contrast to regional climate impact assessments that assess a diverse range of potential futures, stress testing involves a targeted focus on a discrete event or set of extreme conditions under which a system of interest breaks or fails.[92] Climate scenarios for stress tests are scenarios that represent disruptive weather/climate events at the level of detail necessary to identify specific adaptation actors or strategies.[93]

Creating scenarios for stress test applications requires synthesizing a variety of data sources and analytical techniques. Narrative stress test scenarios, consisting of a qualitative description of how an extreme event unfolds, are commonly used in emergency response planning exercises.[94] Quantitative stress tests are particularly beneficial for assessing climate impacts to highly managed, complex, or nonlinear systems where responses are difficult to anticipate.

Data sources for stress test applications include historical climate records, stochastically generated weather time series, paleoclimate proxies, and climate model projections. There are two paths by which climate stress tests can be developed—the storyline approach and the scenario discovery approach.[95] The storyline approach starts with a broad characterization of the types of events stakeholders are concerned about (e.g., extreme winter floods, extreme heat), followed by the identification of more specific attributes of the event or change that make it of higher risk or concern. In scenario discovery, a large ensemble of climate time series is run through an impact simulation model to identify the climatic conditions or decision alternatives under which the system fails to meet performance metrics.[96,97]

Some examples of stress testing systems with regional scenario information include: regional water resource availability, adequacy of a local sewer system, adequacy of coastal defenses to storm surge, resilience of electricity generation and transmission infrastructure.

Here is a specific example related to stress-testing electric utilities for high temperature extremes. Excessive heat challenges many aspects of the electricity generation and transmission infrastructure.[98] Heat waves affect electricity generators, potentially causing physical damage above a certain temperature threshold or forcing curtailment to avoid safety hazards. Plants may be forced to curtail operations partially or completely when the intake of cooling water exceeds operating design temperatures. Persistent extreme temperatures can lead to deterioration and abrupt failure of transformers and power lines. High temperatures also increase transmission and distribution line losses and reduce carrying capacity. Sagging power lines at high temperatures pose many risks, including fire and safety hazards.

To determine the vulnerability of different components of the electrical system to weather and climate extremes, engineering approaches predict degradation or failure of the components under adverse conditions in terms of intensity, duration, and frequency of the event. These functions are commonly referred to as fragility curves, damage functions, or dose-response functions, and give electricity system planners the ability to quantify the potential climate vulnerabilities of their energy assets. Scenario development for a regional electricity system can focus on the extreme temperature events associated with electric system stress or component failure.

Notes

1 CNA Military Advisory Board, "National Security and the Accelerating Risks of Climate Change," CNA, May 2014, https://www.cna.org/cna_files/pdf/MAB_5-8-14.pdf.

2 Christopher P. Weaver et al., "Improving the Contribution of Climate Model Information to Decision Making: The Value and Demands of Robust Decision Frameworks," WIREs Climate Change 4, no. 1 (2013): 39–60, https://doi.org/10.1002/wcc.202.

3 R. J. Lempert, "Robust Decision Making (RDM)," *Decision Making under Deep Uncertainty*, April 5, 2019, 23–51, https://doi.org/10.1007/978-3-030-05252-2_2.

4 G. A. Bradshaw and Jeffrey G. Borchers, "Uncertainty as Information: Narrowing the Science-Policy Gap," *Conservation Ecology* 4, no. 1 (March 22, 2000), https://doi.org/10.5751/es-00174-040107.

5 Erica Thompson, *Escape from Model Land: How Mathematical Models Can Lead US Astray and What We Can Do about It* (New York, NY: Basic Books, 2022).

6 Erica L. Thompson and Leonard A. Smith, "Escape from Model-Land," *Economics* 13, no. 1 (March 8, 2019), https://doi.org/10.5018/economics-ejournal.ja.2019-40.

7 "Coordinated Regional Climate Downscaling Experiment," Cordex, accessed May 9, 2022, https://cordex.org/.

8 Roman Frigg, Leonard A. Smith, and David A. Stainforth, "An Assessment of the Foundational Assumptions in High-Resolution Climate Projections: The Case of UKCP09," *Synthese* 192, no. 12 (May 7, 2015): 3979–4008, https://doi.org/10.1007/s11229-015-0739-8.

9 Hannah Nissan et al., "On the Use and Misuse of Climate Change Projections in International Development," *WIREs Climate Change* 10, no. 3 (March 14, 2019), https://doi.org/10.1002/wcc.579.

10 Theodore G. Shepherd, "Storyline Approach to the Construction of Regional Climate Change Information," *Proceedings of the Royal Society A: Mathematical, Physical and Engineering Sciences* 475, no. 2225 (May 15, 2019), https://doi.org/10.1098/rspa.2019.0013.

11 Dr. Jonathan Foley, "We Need to See the Whole Board to Stop Climate Change," Medium (GlobalEcoGuy.org, February 14, 2021), https://globalecoguy.org/we-need-to-see-the-whole-board-to-stop-climate-change-98be66412281.

12 "The Paris Agreement," UNFCCC, accessed May 9, 2022, https://unfccc.int/process-and-meetings/the-paris-agreement/the-paris-agreement.

13 "2020, One of Three Warmest Years on Record: World Meteorological Organization | | UN News," United Nations (United Nations, January 14, 2021), https://news.un.org/en/story/2021/01/1082132.

14 D. Chen et al., "Framing, Context, and Methods," in *Climate Change 2021: The Physical Science Basis. Contribution of Working Group I to the Sixth Assessment Report of the Intergovernmental Panel on Climate Change* (Geneva, CH: Intergovernmental Panel on Climate Change, 2021), 147–286.

15 Zeke Hausfather, "Analysis: When Might the World Exceed 1.5C and 2C of Global Warming?," Carbon Brief, December 4, 2020, https://www.carbonbrief.org/analysis-when-might-the-world-exceed-1-5c-and-2c-of-global-warming.

16 Nicola Maher et al., "Quantifying the Role of Internal Variability in the Temperature We Expect to Observe in the Coming Decades," *Environmental Research Letters* 15, no. 5 (September 24, 2020): 054014, https://doi.org/10.1088/1748-9326/ab7d02.

17 Eleanor Frajka-Williams, Claudie Beaulieu, and Aurelie Duchez, "Emerging Negative Atlantic Multidecadal Oscillation Index in Spite of Warm Subtropics,"

Scientific Reports 7, no. 1 (September 11, 2017), https://doi.org/10.1038/s41598-017-11046-x.

18 Ka-Kit Tung and Jiansong Zhou, "Using Data to Attribute Episodes of Warming and Cooling in Instrumental Records," *Proceedings of the National Academy of Sciences* 110, no. 6 (January 23, 2013): 2058–2063, https://doi.org/10.1073/pnas.1212471110.

19 Nicola Maher et al., "Quantifying the Role."

20 Ka-Kit Tung and Jiansong Zhou, "Using Data to Attribute Episodes."

21 Petr Chylek et al., "The Role of Atlantic Multi-Decadal Oscillation in the Global Mean Temperature Variability," *Climate Dynamics* 47, no. 9–10 (February 20, 2016): pp. 3271–3279, https://doi.org/10.1007/s00382-016-3025-7.

22 David B. Enfield and Luis Cid-Serrano, "Projecting the Risk of Future Climate Shifts," *International Journal of Climatology* 26, no. 7 (January 26, 2006): 885–895, https://doi.org/10.1002/joc.1293.

23 Marcia Glaze Wyatt and Judith A. Curry, "Role for Eurasian Arctic Shelf Sea Ice in a Secularly Varying Hemispheric Climate Signal during the 20th Century," *Climate Dynamics* 42, no. 9–10 (May 2014): 2763–2782, https://doi.org/10.1007/s00382-013-1950-2.

24 Philipp Blom, *Nature's Mutiny: How the Little Ice Age of the Long Seventeenth Century Transformed the West and Shaped the Present* (New York, NY: Liveright Publishing Corporation, a division of W.W. Norton & Company, 2020).

25 J.-Y. Lee et al., "Future Global Climate: Scenario-Based Projections and Near-Term Information," in *Climate Change 2021: The Physical Science Basis. Contribution of Working Group I to the Sixth Assessment Report of the Intergovernmental Panel on Climate Change* (Geneva, CH: Intergovernmental Panel on Climate Change, 2021), 553–672.

26 M. Sigl et al., "Timing and Climate Forcing of Volcanic Eruptions for the Past 2,500 Years," *Nature* 523, no. 7562 (July 8, 2015): 543–549, https://doi.org/10.1038/nature14565.

27 Christoph C. Raible et al., "Tambora 1815 as a Test Case for High Impact Volcanic Eruptions: Earth System Effects," *WIREs Climate Change* 7, no. 4 (June 2, 2016): 569–589, https://doi.org/10.1002/wcc.407.

28 M. Sigl et al., "Timing and Climate Forcing of Volcanic Eruptions."

29 Veronika Eyring et al., "Overview of the Coupled Model Intercomparison Project Phase 6 (CMIP6) Experimental Design and Organization," *Geoscientific Model Development* 9, no. 5 (May 26, 2016): 1937–1958, https://doi.org/10.5194/gmd-9-1937-2016.

30 Ingo Bethke et al., "Potential Volcanic Impacts on Future Climate Variability," *Nature Climate Change* 7, no. 11 (November 2017): 799–805, https://doi.org/10.1038/nclimate3394.

31 John C. Fyfe et al., "Significant Impact of Forcing Uncertainty in a Large Ensemble of Climate Model Simulations," *Proceedings of the National Academy of Sciences* 118, no. 23 (May 28, 2021), https://doi.org/10.1073/pnas.2016549118.

32 J.-Y. Lee et al., "Future Global Climate."

33 P. Foster et al., "The Earth's Energy Budget, Climate Feedbacks, and Climate Sensitivity," in *Climate Change 2021: The Physical Science Basis. Contribution of Working Group I to the Sixth Assessment Report of the Intergovernmental Panel on Climate Change* (Geneva, CH: Intergovernmental Panel on Climate Change, 2021), 923–1054.

34 Mike Lockwood, "Solar Influence on Global and Regional Climates," *Surveys in Geophysics* 33, no. 3–4 (May 16, 2012): 503–534, https://doi.org/10.1007/s10712-012-9181-3.

35 Ronan Connolly et al., "How Much Has the Sun Influenced Northern Hemisphere Temperature Trends? an Ongoing Debate," *Research in Astronomy and Astrophysics* 21, no. 6 (2021): 131, https://doi.org/10.1088/1674-4527/21/6/131.

36 Nicola Scafetta et al., "Modeling Quiet Solar Luminosity Variability from TSI Satellite Measurements and Proxy Models during 1980–2018," *Remote Sensing* 11, no. 21 (November 1, 2019): 2569, https://doi.org/10.3390/rs11212569.

37 V.M. Velasco Herrera, B. Mendoza, and G. Velasco Herrera, "Reconstruction and Prediction of the Total Solar Irradiance: From the Medieval Warm Period to the 21st Century," *New Astronomy* 34 (January 2015): 221–233, https://doi.org/10.1016/j.newast.2014.07.009.

38 Nicola Scafetta, "Discussion on Climate Oscillations: CMIP5 General Circulation Models versus a Semi-Empirical Harmonic Model Based on Astronomical Cycles," *Earth-Science Reviews* 126 (November 2013): 21–357, https://doi.org/10.1016/j.earscirev.2013.08.008.

39 Frank Stefani, "Solar and Anthropogenic Influences on Climate: Regression Analysis and Tentative Predictions," *Climate* 9, no. 11 (November 3, 2021): 163, https://doi.org/10.3390/cli9110163.

40 Gunnar Myhre et al., "Anthropogenic and Natural Radiative Forcing," *Climate Change 2013—The Physical Science Basis*, 2014, 659–740, https://doi.org/10.1017/cbo9781107415324.018.

41 S. K. Gulev et al., "Changing State of the Climate System," in *Climate Change 2021: The Physical Science Basis. Contribution of Working Group I to the Sixth Assessment Report of the Intergovernmental Panel on Climate Change* (Geneva, CH: Intergovernmental Panel on Climate Change, 2021), 287–422.

42 Katja Matthes et al., "Solar Forcing for CMIP6 (v3.2)," *Geoscientific Model Development* 10, no. 6 (June 22, 2017): 2247–2302, https://doi.org/10.5194/gmd-10-2247-2017.

43 Marcel Crok et al., "Final Evaluation Report of Climate Dialogue," PBL Netherlands Environmental Assessment Agency, February 17, 2015, https://www.pbl.nl/en/publications/final-evaluation-report-of-climate-dialogue.

44 "Climate Dialogue Offers a Platform for Discussions between Scientists," Stichting Milieu, Wetenschap en Beleid, accessed May 9, 2022, https://mwenb.nl/climate-dialogue/.

45 Ronan Connolly et al., "How Much Has the Sun."

46 I. G. Usoskin et al., "Evidence for Distinct Modes of Solar Activity," *Astronomy & Astrophysics* 562 (February 20, 2014), https://doi.org/10.1051/0004-6361/201423391.

47 S. K. Solanki et al., "Unusual Activity of the Sun during Recent Decades Compared to the Previous 11,000 Years," *Nature* 431, no. 7012 (October 28, 2004): 1084–1087, https://doi.org/10.1038/nature02995.

48 I. G. Usoskin, S. K. Solanki, and G. A. Kovaltsov, "Grand Minima and Maxima of Solar Activity: New Observational Constraints," *Astronomy & Astrophysics* 471, no. 1 (June 6, 2007): 301-309, https://doi.org/10.1051/0004-6361:20077704.

49 Ilya G. Usoskin et al., "The Maunder Minimum (1645–1715) Was Indeed a Grand Minimum: A Reassessment of Multiple Datasets," *Astronomy & Astrophysics* 581 (September 10, 2015), https://doi.org/10.1051/0004-6361/201526652.

50 Bruce Dorminey, "Why Has the Sun Gone Quiet?," Astronomy.com, August 20, 2019, https://astronomy.com/magazine/2019/08/why-has-the-sun-gone-quiet.

51 F. Inceoglu et al., "On the Current Solar Magnetic Activity in the Light of Its Behaviour during the Holocene," *Solar Physics* 291, no. 1 (November 23, 2015): 303-315, https://doi.org/10.1007/s11207-015-0805-x.

52 V.M. Velasco Herrera et al., "Reconstruction and Prediction of the Total Solar Irradiance: From the Medieval Warm Period to the 21st Century," *New Astronomy* 34 (January 2015): 221–233, https://doi.org/10.1016/j.newast.2014.07.009.

53 L. Barnard et al., "Predicting Space Climate Change," *Geophysical Research Letters* 38, no. 16 (August 19, 2011), https://doi.org/10.1029/2011gl048489.

54 Georg Feulner and Stefan Rahmstorf, "On the Effect of a New Grand Minimum of Solar Activity on the Future Climate on Earth," *Geophysical Research Letters* 37, no. 5 (March 10, 2010), https://doi.org/10.1029/2010gl042710.

55 Henrik Svensmark, *Force Majeure: The Sun's Role in Climate Change* (London, UK: The Global Warming Policy Foundation, 2019).

56 Nir J. Shaviv, "Using the Oceans as a Calorimeter to Quantify the Solar Radiative Forcing," *Journal of Geophysical Research: Space Physics* 113, no. A11 (November 4, 2008), https://doi.org/10.1029/2007ja012989.

57 Nicola Scafetta, "Discussion on Climate Oscillations."

58 Frank Stefani, "Solar and Anthropogenic Influences on Climate: Regression Analysis and Tentative Predictions," *Climate* 9, no. 11 (November 3, 2021): 163, https://doi.org/10.3390/cli9110163.

59 Katja Matthes et al., "Solar Forcing for CMIP6."

60 Georg Feulner and Stefan Rahmstorf, "On the Effect."

61 Marcel Crok, "What Will Happen during a New Maunder Minimum?," Stichting Milieu, Wetenschap en Beleid, October 15, 2014, https://mwenb.nl/what-will-happen-during-a-new-maunder-minimum/.

62 Ben Hirschler, "Davos Strives to Make Climate Talk More than Hot Air," Chicago Tribune, January 25, 2021, https://www.chicagotribune.com/lifestyles/ct-xpm-2013-01-25-sns-rt-us-davos-climatebre90o0lb-20130125-story.html.

63 "2020, One of Three Warmest Years on Record."

64 Nicola Maher et al., "Quantifying the Role."

65 Thomas R. Knutson et al., "Prospects for a Prolonged Slowdown in Global Warming in the Early 21st Century," *Nature Communications* 7, no. 1 (November 30, 2016), https://doi.org/10.1038/ncomms13676.

66 Ingo Bethke et al., "Potential Volcanic Impacts on Future Climate Variability," *Nature Climate Change* 7, no. 11 (November 2017): 799–805, https://doi.org/10.1038/nclimate3394.

67 Nicola Scafetta, "Discussion on Climate Oscillations."

68 Frank Stefani, "Solar and Anthropogenic Influences on Climate."

69 Girma J. Orssengo, "Determination of the Sun-Climate Relationship Using Empirical Mathematical Models for Climate Data Sets," *Earth and Space Science* 8, no. 9 (July 20, 2021), https://doi.org/10.1029/2019ea001015.

70 Mike Hulme, "Fetishising 'the Number': How Not to Govern Pandemics, Climate and Biodiversity," Mike Hulme, August 7, 2020, https://mikehulme.org/fetishising-the-number-how-not-to-govern-pandemics-climate-and-biodiversity/.

71 David Simpson, *Fetishism and Imagination: Dickens, Melville, Conrad* (Ann Arbor, MI: UMI, 1999).

72 Jerry Z. Muller, *The Tyranny of Metrics* (Princeton, NJ: Princeton University Press, 2019).

73 Kim Stanley Robinson, "The Coronavirus Is Rewriting Our Imaginations," The New Yorker, May 1, 2020, https://www.newyorker.com/culture/annals-of-inquiry/the-coronavirus-and-our-future.

74 O. Rössler et al., "Challenges to Link Climate Change Data Provision and User Needs: Perspective from the Cost-Action Value," *International Journal of Climatology* 39, no. 9 (April 18, 2017): 3704–3716, https://doi.org/10.1002/joc.5060.

75 Thomas Stocker, *Climate Change 2013: The Physical Science Basis: Summary for Policymakers, a Report of Working Group I of the IPCC; Technical Summary, a Report Accepted by Working Group I of the IPCC but Not Approved in Detail and Frequently Asked Questions* (Geneva, CH: Intergovernmental Panel on Climate Change, 2013).

76 "Coordinated Regional Climate Downscaling Experiment."

77 F. J. Doblas-Reyes et al., "Linking Global to Regional Climate Change," in *Climate Change 2021: The Physical Science Basis. Contribution of Working Group I to the Sixth Assessment Report of the Intergovernmental Panel on Climate Change* (Geneva, CH: Intergovernmental Panel on Climate Change, 2021), 1363–1512.

78 F. J. Doblas-Reyes et al., "Linking Global to Regional Climate Change."

79 Angela Merkel, "2014 Lowy Lecture: Angela Merkel, Chancellor of Germany," YouTube (YouTube, November 17, 2014), https://www.youtube.com/watch?v=HZYR5NVB8XA.

80 S. I. Seneviratne et al., "Weather and Climate Extreme Events in a Changing Climate," in *Climate Change 2021: The Physical Science Basis. Contribution of Working Group I to the Sixth Assessment Report of the Intergovernmental Panel on Climate Change* (Geneva, CH: Intergovernmental Panel on Climate Change, 2021), 1513–1766.

81 "It Could Happen Again. Are You Prepared?—National Weather Service," National Weather Service, October 2020, https://www.weather.gov/media/tbw/1921/1921_Hurr_99th_Anniversary.pdf.

82 David M. Ludlum, *Early American Hurricanes 1492–1870* (Boston, MA: American Meteorological Society, 1989).

83 Judith Curry, "Hurricanes and Climate Change," Climate Forecast Applications Network, June 2019, https://345f4919-32bf-4395-bc40-6489e53b0225.filesusr.com/ugd/867d28_32f52bbef6d24cbfb018540b6b8d60bd.pdf.

84 NOAA Tides & Currents, "NOAA Tides and Currents," NOAA Tides and Currents, accessed August 4, 2022, https://tidesandcurrents.noaa.gov/.

85 S. I. Seneviratne et al., "Weather and Climate Extreme Events."

86 E. Hawkins and R. Sutton, "Time of Emergence of Climate Signals," *Geophysical Research Letters* 39, no. 1 (January 7, 2012), https://doi.org/10.1029/2011gl050087.

87 Jingyuan Li et al., "Quantifying the Lead Time Required for a Linear Trend to Emerge from Natural Climate Variability," *Journal of Climate* 30, no. 24 (December 15, 2017): 10179–10191, https://doi.org/10.1175/jcli-d-16-0280.1.

88 Richard R. Heim, "An Overview of Weather and Climate Extremes – Products and Trends," *Weather and Climate Extremes* 10 (December 2015): 1–9, https://doi.org/10.1016/j.wace.2015.11.001.

89 Ancius Boethius, *The Consolation of Philosophy*, trans. Victor Watts (London, UK: Penguin, 2003).

90 Paul C. Stern et al., "Managing Risk with Climate Vulnerability Science," *Nature Climate Change* 3, no. 7 (June 25, 2013): 607–609, https://doi.org/10.1038/nclimate1929.

91 Casey Brown and Robert L. Wilby, "An Alternate Approach to Assessing Climate Risks," *Eos, Transactions American Geophysical Union* 93, no. 41 (October 9, 2012): 401–402, https://doi.org/10.1029/2012eo410001.

92 John D. Steinbruner et al., *Climate and Social Stress: Implications for Security Analysis* (Washington, DC: National Academies Press, 2013).

93 Christine M. Albano et al., "Techniques for Constructing Climate Scenarios for Stress Test Applications," *Climatic Change* 164, no. 3–4 (February 8, 2021), https://doi.org/10.1007/s10584-021-02985-6.

94 "Homeland Security Exercise and Evaluation Program (HSEEP)—FEMA," FEMA, 2020, https://www.fema.gov/sites/default/files/2020-04/Homeland-Security-Exercise-and-Evaluation-Program-Doctrine-2020-Revision-2-2-25.pdf.

95 Theodore G. Shepherd, "Storyline Approach."

96 David G. Groves and Robert J. Lempert, "A New Analytic Method for Finding Policy-Relevant Scenarios," *Global Environmental Change* 17, no. 1 (February 2007): 73–85, https://doi.org/10.1016/j.gloenvcha.2006.11.006.

97 Robert Lempert, "Scenarios That Illuminate Vulnerabilities and Robust Responses," *Climatic Change* 117, no. 4 (October 2, 2012): 627–646, https://doi.org/10.1007/s10584-012-0574-6.

98 Melissa R Allen-Dumas et al., "Extreme Weather and Climate Vulnerabilities of the Electric Grid: A Summary of Environmental Sensitivity Quantification Methods," Oak Ridge National Laboratory, August 16, 2019, https://www.energy.gov/sites/prod/files/2019/09/f67/Oak%20Ridge%20National%20Laboratory%20EIS%20Response.pdf.

Chapter Nine

WHAT'S THE WORST CASE?

"We must contemplate some extremely unpleasant possibilities, just because we want to avoid them."
—American nuclear strategist Albert Wohlstetter, writing during the Cold War[1]

Future scenarios of worst-case outcomes have an important role to play in many decision-making frameworks (Part Three). How to formulate worst-case scenarios, and assess whether they are plausible, is a substantial challenge. This chapter brings into play some complex epistemological issues, focused on formulating and assessing plausible worst-case scenarios.

Global climate models provide a coherent basis for generating scenarios of future climate change, based on emissions scenarios. The worst case is clearly associated with the 8.5 scenarios (RCP8.5, SSP5–8.5). However, Section 7.1.1 provided evidence that RCP8.5 is an implausible scenario. What exactly does implausible mean, and how do we delineate between a scenario that is improbable (<10 percent chance) versus a plausible or an implausible scenario?

Emissions-driven climate model simulations do not allow exploration of all possible future scenarios that are compatible with our background knowledge of the basic way the climate system actually behaves, as described in Chapter Eight. Some of these unexplored possibilities might turn out to be real ones.

Surprises are a class of risk that can be defined as low-likelihood but well-understood events that cannot be predicted with current understanding. Examples include: a series of major volcanic eruptions, a nuclear war, significant twenty-first-century sea level rise due to collapse of the West Antarctic Ice Sheet, and unexpected pandemics.

Black swans are a category of surprise with the following attributes.[2] A black swan is an outlier, as it lies outside the realm of regular expectations and is associated with an extreme impact. In spite of its outlier status, human nature makes us concoct explanations for its occurrence after the fact, making it explainable and predictable. Another class of surprises are known events or processes that were ignored for some reason or judged to be of negligible importance by the scientific community. These are often referred to as Pink Flamingos (also referred to as unknown knowns or known neglecteds).[3]

Four risk factors have been articulated for genuine surprise:[4]

- System complexity—risk of surprise is higher when the system under study is nonlinear and complex.
- Limited knowledge of the system's past behavior and processes that underlie that behavior.
- Past instances of genuine surprise when investigating the system.
- Novel conditions—system being subjected to boundary conditions unlike those in which it was previously studied.

Future climate change and its impacts are subject to each of these risk factors for genuine surprise. Efforts to avoid surprises begin with a fully imaginative consideration of possible future outcomes. There is value in scientific speculation on policy-relevant aspects of low probability, high-impact outcomes, even though we can neither model them realistically nor provide a precise estimate of their probability.[5]

In addition to climate model simulations, two general strategies have been employed for articulating possible black swan events related to climate change:

- Statistical extrapolation of inductive knowledge beyond the range of limited experience using fat-tailed probability distributions.[6,7]
- Physically based scientific speculation on the possibility of high impact scenarios, even if we can neither model them realistically nor provide an estimate of their likelihood ("storyline" approach).[8]

The risk management literature has discussed the need for a broad range of scenarios of future outcomes.[9,10] When background knowledge supports doing so, modifying model results to broaden the range of possibilities they represent can generate additional scenarios, including known neglecteds. Simple climate models, process models, and data-driven models can also be used as the basis for generating scenarios of future climate change outcomes. The paleoclimate record provides a rich source of information for developing future scenarios. More creative approaches, such as mental simulation and abductive reasoning, can also produce what-if scenarios.[11] Expert speculation on the influence of known neglecteds can minimize the potential for missing black swan events that are associated with known processes that were ignored for some reason.

In an influential paper, economist Martin Weitzman argued that climate policy should be directed at reducing the risks of worst-case outcomes, not at balancing the most likely values of costs and benefits.[12] It has also been argued that policy should be based on the credible worst-case outcome.[13] Worst-case scenarios of twenty-first century sea level rise are becoming anchored as outcomes that are driving local adaptation plans.[14] Projections of future extreme weather/climate events driven by the worst-case RCP8.5 scenario are highly influential in the public discourse on climate change.[15]

9.1 Scenario Probabilities and Plausibility

"Truth is so hard to tell, it sometimes needs fiction to make it plausible." (Philosopher Dagobert D. Runes)[16]

Under the auspices of the IPCC, climate scientists have been engaged in a major effort to provide probabilistic projections of climate outcomes using Bayesian and other techniques.[17] Environmental economists involved in the IPCC use probabilities to drive their impact assessment models. Energy economists use probabilities in setting emissions targets. Such probabilities are also desired by the insurance and broader financial sectors.

For a given emissions/concentration pathway, does the multi-model ensemble of simulations of the twenty-first century climate used in the IPCC assessment reports provide meaningful probabilities? The multi-model ensemble-of-opportunity developed for IPCC assessment reports (e.g., CMIP5, CMIP6) is not designed to sample uncertainty in a thorough or strategic way.[18] Model inadequacy and an insufficient number of simulations in the ensembles preclude producing meaningful probability distributions. Furthermore, the derived probability distributions are liable to be misleading because conclusions derived from them imply much greater confidence than the underlying assumptions justify. As an alternative, imprecise (interval) probabilities are used by the IPCC. For example, the *likely* range (66 percent) refers to a knowledge-based probability and indicates the analysts' judgment that more precision is not warranted.[19]

Our current inability to provide meaningful probabilities of future climate outcomes implies that we are in the regime of scenario uncertainty or deep uncertainty (Section 5.1.1). In the regime of scenario uncertainty, a collection of scenario outcomes is regarded as a range of discrete possibilities with no *a priori* allocation of likelihood—the focus is on providing bounds for the future outcomes, and perhaps providing some sense of likelihood for groups of outcomes. Deep uncertainty reflects our inability to provide plausible bounds to extreme scenario outcomes.

9.1.1 Possibility Theory

"We see the 'future' perversely tamed into numbers, with prediction and probability shielding complexity and uncertainty." (Policy scientists Cynthia Selin and Angela Guimarães Pereira)[20]

Where probabilistic prediction fails, foreknowledge is possibilistic. Possibility theory is an uncertainty theory devoted to the handling of incomplete information that can capture partial ignorance and represent partial beliefs.[21] Philosopher Gregor Betz provided a conceptual framework that distinguishes different categories of possibility and necessity to convey our uncertain knowledge about the future. Betz classified possible events to fall into two categories: (i) verified possibilities—statements that are shown to be possible; and (ii) unverified possibilities—events that are articulated, but neither shown to be possible nor impossible.[22] The epistemic status of verified possibilities is higher than that of unverified possibilities; however, the most informative scenarios for risk management may be the unverified possibilities.

A key feature of possibility theory is that it can handle incomplete information, partial ignorance, and represent partial beliefs. In evaluating scenarios, it is useful to employ the logic of partial positions for an ordered scale of events resulting in cumulative outcomes.[23] Partial positions are useful when there is deep uncertainty about the future, and decisions can or must be staged sequentially to maximize future options.[24] Adopting a partial position with a higher degree of justification provides a scenario that can be extended flexibly in many different ways when constructing a complete position and is more immune to falsification.

Sea level rise provides an easily understood example of partial positions. High end, low confidence scenarios of twenty-first century sea level rise of 2 meters or greater have been proposed, which are judged in Section 9.5 to be weakly plausible. However partial positions between 0.2 and 0.6 meters have strong justification; these lower values first must be passed (partial positions) before reaching higher values that are considered for the worst-case scenario.

9.1.2 Plausibility

"You should take the approach that you're wrong [...] Your goal is to be less wrong." (Business magnate Elon Musk)[25]

In order to be prepared for the unexpected, it is necessary to make judgments of what can plausibly happen. In contrast to probabilistic futures, plausible futures explore future scenarios that are improbable (or unlikely) but could still occur and where predictive capacity is limited due to high uncertainty, ignorance, or complexity. As an example, an event such as Superstorm Sandy that struck New York in 2012 would not be considered in future scenarios that are evaluated for their probability, since the likelihood of such a storm is ill-determined owing to the rarity of the event.

Articulation of plausible scenarios is of particular relevance in situations with large uncertainties, where the knowledge base supporting the assessment is weak and the potential for experiencing surprises relative to this knowledge is large. What does plausibility actually mean in a risk and uncertainty context? Plausibility has been defined as the quality of a scenario to hold enough evidence to be qualified as "occurrable," explicitly including scenarios that are not the most likely ones.[26] A similar definition holds the term plausible to be equivalent to "seriously possible."[27]

There is not a straightforward continuum between the plausible and implausible. Rather, there is a spectrum from the extremely plausible (norms, business as usual, foundational assumptions) to the implausible, which is nearly inconceivable or incredible.[28] Levels of plausibility are further explored in Section 9.3. However, nailing down the threshold of plausibility versus implausibility of scenarios is not necessarily the point.

Articulation of plausible scenarios provides a pathway away from prediction and probabilistic thinking that creates a more fruitful basis for making decisions for complex problems with large uncertainties.[29] The emphasis is on characterizing the strength of the knowledge supporting a scenario, as well as identification of potential surprises. Plausibility does not represent a measure of likelihood per se but constitutes part of the foundation for likelihood assignments, by providing reflections on the underlying knowledge and assumptions on which the judgments are based.

Plausibility assessments of scenarios provide a line of argument for why improbable scenarios can nevertheless be worthy of consideration and raise important questions about uncertain futures and our societal ability to manage them.[30]

9.2 Fat Tails and Tall Tales

"If we continue on a business-as-usual trajectory, then there is some non-trivial probability of a catastrophic climate outcome materializing at some future time." (Economist Martin Weitzman)[31]

While participating in a debate, I was surprised to be admonished for paying insufficient attention to uncertainty.[32] My fellow debater Lord Martin Rees[33] stated that he was very concerned about the most extreme outcomes—the fat tails of the distribution. He was apparently heavily influenced by a 2009 paper authored by economist Martin Weitzman entitled "On Modelling and Interpreting the Economics of Catastrophic Climate Change" and a subsequent book entitled *Climate Shock: The Economic Consequences of a Hotter Planet.*[34,35]

The key climate science input to the Integrated Assessment Models used by economists is the probability density function of equilibrium climate sensitivity (ECS; Section 7.2). High-end values of ECS are of considerable interest to economists. Weitzman addressed the challenge of determining high-end values of ECS by creating a precise probability distribution for ECS based upon the ECS parameters provided by the IPCC AR4—the median value and a likely range.[36,37] Weitzman argued that the tails of probability density function (pdf) of the equilibrium climate sensitivity, fattened by structural uncertainty using a Bayesian framework, can have a large effect on the cost-benefit analysis. Proceeding in the Bayesian paradigm, Weitzman produced a fat-tailed probability distribution, with a probability of 5 percent of ECS exceeding 10°C, and a 1 percent probability of exceeding 20°C.[38]

The *likely* range of ECS values cited by the IPCC AR6 is 2.5–4.0°C, and the *very likely* range is 2.0–5.0°C.[39] While these ranges are disputed (Section 7.2), there is no physical justification for values of ECS exceeding 10°C and certainly not for exceeding 20°C. The conceivable worst case for ECS is arguably ill-defined; there is no obvious way to positively infer this, and such inferences are hampered by timescale fuzziness between equilibrium climate sensitivity and the larger Earth system sensitivity associated with the long timescales of ice sheets and oceans. However, one can refute estimates of extreme values of ECS from fat-tailed distributions >10°C as arguably impossible—these values reflect the statistical manufacture of extreme values that are unjustified by either observations or theoretical understanding and extend well beyond any conceivable uncertainty or possible ignorance about the subject. Further, such extreme values of ECS imply a climate that is much less stable than observed.

The end result is that this most important part of the probability distribution that drives the economic costs of carbon is based upon a statistically manufactured fat tail whose outcome values have no scientific justification. Such statistically manufactured extreme outcomes that are not associated with any physical justification are regarded here as having no epistemic justification as a plausible scenario. Such statistical approaches are not only used by economists but have also been used in exploring extreme scenarios of sea level rise.[40]

9.3 Scenario Justification and Falsification

"Why, sometimes I've believed as many as six impossible things before breakfast." (The Queen in Lewis Carroll's *Through the Looking-Glass*)[41]

As a practical matter for considering outcomes of future climate change and its impacts, how are we to evaluate whether a scenario is plausible? In particular, how do we assess the plausibility of scenarios that could produce big surprises?

The concepts of justification and falsification can take us deep into complex debates on philosophy and epistemology. Our interest here is in the practical application of providing decision makers with a possibility distribution of justified scenarios for the twenty-first century, including the plausible worst-case scenarios. Philosopher Gregor Betz provides a useful framework for evaluating the scenarios relative to their degrees of justification and evaluating the outcomes against our background knowledge. Confirmation and refutation play complementary roles in scenario justification.[42]

Sections 9.4 and 9.5 provide examples of how worst-case scenarios can be created; but how do we approach refuting extreme scenarios or outcomes as implausible or impossible? Assessing the strength of background knowledge is an essential element in assessing the plausibility of extreme scenarios. However, the background knowledge against which extreme scenarios and their outcomes are evaluated is continually changing, which argues for frequent re-evaluation of worst-case scenarios and outcomes.

Assessing the strength of knowledge associated with a scenario can include reflections on reasonableness of the assumptions, the amount of reliable and relevant data/information, the degree of agreement among experts, phenomena understanding, the existence of accurate models, and the degree to which the knowledge base has been scrutinized with respect to surprising events.[43]

Scenarios can be evaluated based on the following criteria:

1. Evaluation of the plausibility of each link in the model or storyline used to create the scenario.
2. Evaluation of the degree to which each element contributes to the plausibility of the entire scenario.
3. Evaluation of the plausibility of the outcome and/or the inferred rate of change, in light of physical or other constraints.

A high degree of justification implies high robustness and relative immunity to falsification or rejection. Guided by the frameworks established across several different disciplines, the plausibility of scenarios of future outcomes are categorized here in terms of degrees of justification.[44,45,46,47] Below is a classification of scenarios based on the levels of justification:

1. *Strongly verified*—strongly supported by basic theoretical considerations and empirical evidence

2. *Corroborated*—empirical evidence for the outcome; it has happened before under comparable conditions

3. *Verified*—generally agreed to be consistent with relevant background theoretical and empirical knowledge

4. *Contingent*—outcome is contingent on a model simulation and the plausibility of input values such as the emissions scenarios

5. *Weakly plausible*—the scenario or scenario element is theoretically occurrable, but without empirical evidence (unverified possibility)

6. *Implausible*—logically possible based on theoretical insight but consistency with background knowledge is disputed (unverified possibility)

7. *Impossible*—internal contradictions or inconsistent with relevant background knowledge

Falsification of a scenario differs from rejecting a scenario with a low degree of justification. Falsification requires that the storyline for the scenario be in some way incoherent and contradictory, or the scenario produces unrealistic outcomes.

The plausible worst-case scenario is judged to be the most extreme scenario that is deemed occurrable and cannot be falsified as impossible based upon our background knowledge.[48] On topics where there is substantial uncertainty and/ or a rapidly advancing knowledge frontier, experts disagree on what outcomes they would categorize as a plausible worst case, even when considering the same background knowledge and the same input parameters/constraints. An example of such disagreement among experts is provided in Section 9.5 on worst-case scenarios for sea level rise.

9.4 Worst-Case Weather and Climate Events

"The future is already here—it's just not very evenly distributed." (Author William Gibson)[49]

Low likelihood, high impact weather, and climate events cause the greatest harm to vulnerable populations and assets in a particular region. Plausible worst-case scenarios can extend beyond the recent historical data record, owing to natural weather and climate variability as well as human-caused climate change.

As the climate moves away from its recent past and current states, we may experience extreme events that are unexpected in magnitude, frequency, timing, or location. Alternatively, a warmer climate may reduce some types of extreme events (Section 7.4).

This section illustrates the storyline approach for generating worst-case scenarios for different types of compound weather and climate events. The event-based storyline approach breaks down each of the drivers of the event in terms of the atmospheric weather, seasonal climate drivers, multi-decadal climate regimes, and global warming.[50]

The event types considered here are selected from cases that I have considered in the context of adaptation concerns raised by the clients of my company. These worst-case scenarios illustrate a range of event types and scenario construction approaches:

- Worst-case Florida landfalling hurricane
- Worst-case flooding in California from successive atmospheric river events
- Extended drought of the South Asian monsoon.

9.4.1 Florida Landfalling Hurricanes

The sound of the wind, that's what you never forget. The initial whisper. The growing mewing that turns into a howl. Then the cry of glass shattering. The snap of trees breaking. The grumbling of a roof peeling apart. (Writer Ana Veciana-Suarez Remembering Hurricane Andrew as 2004s Hurricane Ivan threatens)[51]

The state of Florida in the United States has encountered 41 percent of all historical landfalling hurricanes striking the United States. A total of 38 Florida landfalling hurricanes since 1850 have been major hurricanes (Category 3 or greater), with wind speeds exceeding 110 mph (177 km/hr).[52] Because of its location directly between the Atlantic Ocean and Gulf of Mexico, Florida is susceptible to hurricanes that come from many directions.

Damage from landfalling hurricanes is caused by winds, storm surge, and rainfall. As a hurricane travels inland, its wind speeds are reduced owing to land friction and loss of the oceanic moisture supply. However, tornadoes can be generated by the hurricane well inland of the landfall location, and the number of hurricane-induced tornadoes is influenced by the hurricane maximum wind speed and horizontal size.[53] Storm surge is determined by the hurricane intensity (maximum sustained wind speed), horizontal size, and forward speed of motion. Storm surge is also heavily influenced by landfall location (slope of the continental shelf), tidal conditions, and trajectory of the hurricane relative to the shape of the coast. Rainfall accumulation does not scale with hurricane intensity, but is influenced by the horizontal size of the hurricane and its forward speed.

The historical record provides a basis for constructing worst-case scenarios for Florida landfalling hurricanes. The complexity of determining maximum landfall wind speeds is described in an article by scientists from the National Hurricane Center on revising the data for Hurricane Andrew (1992).[54] The strongest landfall winds in the United States occurred with the Labor Day Hurricane of 1935, which struck the Florida Keys and the Gulf Coast of Florida.[55] The estimated maximum landfall wind speed based on recorded atmospheric pressure is believed to be approximately 165 mph. More recently, Hurricane Andrew (1992) and Hurricane Michael (2018) registered sustained winds at landfall of 150 mph and 160 mph, respectively.[56] The 1935 Labor Day hurricane produced a storm surge on the Florida Keys that was estimated to exceed 18 feet. The Great Miami hurricane of 1926 produced a storm surge estimated at 14–15 feet.[57] The largest rainfall was 45 inches produced by Hurricane Easy in 1950.[58]

Two additional factors that influence hurricane impacts include the forward track speed and overall wind field radius (the horizontal size of the hurricane). Since 1995, the Florida Commission on Hurricane Loss Projection Methodology has compiled potential hurricane impacts based on modeling.[59] Recent model reports reflect potential hurricane forward speeds ranging from 1.0 to 19.8 m/s.[60] The horizontal extent of gale-force winds (>40 mph) appears greatest for Category 3 and 4 hurricanes (rather than the strongest hurricanes).[61]

A synthetic hurricane can be constructed using a radial wind model that inputs maximum sustained winds or minimum surface pressure, horizontal extent of the radius of gale force winds, and storm forward speed.[62] Storm surge models further require information on local coastal bathymetry and tides. The input variables from historical storms, combined with a radial wind speed model and storm surge model, provide the basis for developing storylines for worst-case hurricanes.

How will global warming influence these worst-case scenarios, relative to the historical values? The IPCC AR6 (Chapter Eleven) summarizes the challenges in interpreting trends of extreme weather events from historical data that are associated with inhomogeneous data sets and large regional signals of natural climate variability.[63] Simulation of hurricanes using climate models is very challenging, owing to the high model resolution that is required to actually produce a strong hurricane. Hence, inferences about hurricane changes in a warmer climate are made from historical data sets, high-resolution process models, and theoretical considerations.

The key conclusions from the IPCC AR6, of relevance to the hurricane parameters of interest here, are listed below:[64]

- Peak wind speeds of the most intense tropical cyclones (TCs) are projected to increase at the global scale with increasing global warming (*high confidence*). The increase in global TC maximum surface wind speeds is about 5 percent for a 2°C global warming across a number of high-resolution multi-decadal studies
- It is *very likely* that heavy precipitation events will intensify and become more frequent in most regions with additional global warming. At the global scale, extreme daily precipitation events are projected to intensify by about 7 percent for each 1°C of global warming (*high confidence*).
- It is *more likely than not* that the slowdown of hurricane translation speed over the US has contributions from anthropogenic forcing.
- The projected change in both magnitude and sign of hurricane size is uncertain.
- Projected increases in sea level, average hurricane intensity, and hurricane rainfall rates each generally act to further elevate future storm surge and fresh-water flooding.

What is the impact of an additional 2°C warming (relative to 2020) on the worst-case historical hurricanes, as per the conclusions of the IPCC AR6? A 5 percent increase in maximum sustained winds would increase the worst-case winds of 165 mph to 173 mph (note that 5 percent is well within the measurement error of determining the actual landfall winds). A 14 percent increase in the worst-case rainfall scenario of 45 inches would become 51 inches. An increase in sea level of 3 feet would increase a storm surge of 18 feet to 21 feet.

9.4.2 ARkStorm

> "Such a desolate scene I hope never to see again. Most of the city is still under water, and has been for three months." (Ecologist William Brewer, reporting from Sacramento California in 1862)[65]

Sporadic atmospheric rivers that occur in relatively narrow regions in the atmosphere are responsible for most of the horizontal transport of water vapor outside of the tropics. Strong atmospheric river events can produce hurricane-force winds and copious amounts of rainfall.

The ARkStorm is a hypothetical but scientifically realistic megastorm scenario developed by the United States Geological Survey, based on historical occurrences in the United States state of California.[66] The event is similar to the exceptionally intense California storms that occurred between December 1861 and January 1862, which dumped nearly 10 feet of rain in parts of California and submerged the entire Central Valley under as much as 15 feet of water.[67,68] Six megastorms more severe than 1861–1862 have occurred in California during the last 1800 years, occurring at intervals of 200 years or so.[69]

While the ARkStorm is patterned after the 1861–62 historical events, it also uses data from large storms in 1969 and 1986. Based on these events, the ARkStorm scenario was created using sophisticated weather models and expert analysis.[70] The scenario developers characterized the resulting floods, landslides, coastal erosion, and inundation that translate into infrastructural, environmental, agricultural, social, and economic impacts.

The ARkStorm scenario occurring in the twenty-first century would flood thousands of square miles of urban and agricultural land, resulting in thousands of landslides, and disrupt lifelines throughout the state of California for days or weeks. The cost of such an event has been estimated to be on the order of US $725 billion, and such an event could require the evacuation of 1,500,000 people.[71]

So, how might the ARkStorm scenario be altered by global warming? While the total volume of California precipitation is not likely to change significantly under continued global warming, less snowpack is expected, and snowmelt is expected to occur earlier in the year. Rainfall from individual atmospheric river events is expected to increase, with a much lower proportion falling as snow.[72] The IPCC AR6 concluded that there is high confidence that the magnitude and duration of atmospheric rivers will increase in the future, leading to increased precipitation.[73] With regards to the frequency of atmospheric river events striking the US west coast, there is low confidence in projecting any trend.

How much more rain can we expect in a warmer climate, relative to the ARkStorm scenario based on the 1861–62 event? The estimation of rainfall from atmospheric rivers has large uncertainties, especially as they hit topographically complex coastal regions. A study using a regional climate model under the extreme emissions scenario RCP8.5 found increases in California rainfall from atmospheric rivers ranging from 15 to 50 percent in different topographic environments.[74] More realistic emissions scenarios would be associated with a smaller increase in rainfall.

The bottom line is that the occurrence of such an extreme storm in California during the twenty-first century is *as likely as not*. Plausible worst-case scenarios could be 50 percent worse than the 1862 event, associated either with warming or simply a worse case analogous to those observed earlier in the paleoclimate record.

9.4.3 South Asian Monsoon Failure

"It was June in Maharashtra, and the monsoon would not come. The whole district lay panting in the heat, the burning sky clapped tight overhead like the lid of a tandoor oven. Lean goats stumbled down the narrow alleyways, udders hanging slack and dry beneath them; beggars cried for water in every village." (Writer Arinn Dembo)[75]

Much of the world's population lives in monsoon Asia, depending on summer monsoon rainfall to provide over 80 percent of the region's precipitation. The Asian monsoon impacts water resources for drinking, sanitation, industry, agriculture, hydropower production, ecosystem health, and overall socioeconomic well-being for India, Pakistan, Bangladesh, and surrounding countries. At least 60 percent of the agriculture in this region is rain fed (not irrigated).[76] Hence, the Asian monsoon is one of the most anticipated, tracked, and studied weather systems in the world. There is a pressing need to understand how the monsoon will change in the future, to understand the interplay between anthropogenic and natural drivers, and to assess the worst-case scenario for a multi-year monsoon drought or failure.

Monsoon rainfall varies from year to year as a result of variations of sea surface temperatures in the Indian and Pacific Oceans, volcanic eruptions, and land snow cover and soil moisture over the Asian continent. The relevant drivers for decadal to century changes in the Indian summer monsoon are summarized by the IPCC AR6:[77]

- Increased greenhouse gas concentrations (chiefly CO_2) are a strong contributor to changes in the monsoon, support an increase in rainfall.
- Industrial emissions of aerosol particles are understood to weaken the monsoon.
- Massive expansion of agriculture at the expense of forest and shrublands, with widespread irrigation, contribute to drying.
- Decadal modes of climate variability such as the Pacific Decadal Oscillation and Atlantic Multidecadal Oscillation cause decadal modulation of the monsoon.

The interplay of these external and internal drivers is key to understanding past and future monsoon change. Since the 1950s, the observed monsoon rainfall has gradually declined over India—this decline is not in line with climate model simulations of increasing monsoon rainfall in a warming world. The observed decline has been interpreted to result from the rapid warming in the Indian Ocean (decreasing the land-ocean temperature gradient), changes in land use and land cover, and increased emissions of aerosol particles. The IPCC AR6 has high confidence that anthropogenic aerosol emissions have dominated the observed declining trends of Indian summer monsoon rainfall.[78] Apparent increasing Indian monsoon rainfall since 2002 is due either to a change in dominance of a particular forcing (for example from aerosol to greenhouse gases) or to a change in phase of the Pacific Decadal Oscillation.[79]

The IPCC AR6 has high confidence that South Asia monsoon precipitation will increase during the twenty-first century in response to continued global warming across the higher emissions scenarios, mostly in the mid and long terms. The AR6 assesses that monsoon precipitation will *likely* increase by 1.3–2.4 percent per °C of warming.

With regards to worst-case outcomes, severe and persistent droughts in monsoon Asia are evident in tree-ring paleoclimate reconstructions, particularly from the Monsoon Asia Drought Atlas (MADA).[80] The MADA utilizes a network of more than 300 annually resolved tree-ring width time series covering the last 700 years to reconstruct the seasonalized Palmer Drought Severity Index (PDSI) for the summer (June-July-August) monsoon season.[81] Monsoon failures and megadroughts have repeatedly affected the agrarian peoples of Asia over the past millennium. Epochal events have been associated with extended monsoon droughts, such as the fall of the Ming Dynasty in China in 1644, substantial upheaval in Southeast Asia that coincided with a multi-decadal drought 1756–68, and widespread regional famines during the East India drought of 1790–96 and the late Victorian Drought of 1876–78.[82]

The so-called Strange Parallels drought (1756–68) is a good candidate to provide the basis for constructing a worst-case scenario.[83] MADA reveals that Southeast Asia and much of India were affected by this multidecadal drought.[84] This spatially broad and persistent megadrought is one of the most important periods of monsoon failure found in the MADA.

To interpret the causes of the Strange Parallels monsoon drought, a pre-industrial control run of 1300 years length from a global climate model was used. This analysis revealed the influence of two different types of El Niño (canonical El Niño and El Niño Modoki) on drought conditions over monsoon Asia. Canonical El Niño events with warming in the eastern equatorial Pacific, have a substantial influence on interannual variability in monsoon rains.[85] In contrast, multi-year drought periods, resembling those sustained during the Strange Parallels drought, feature anomalous Pacific warming around the dateline, typical of El Niño Modoki events. El Niño Modoki (Japanese for "similar but different") is associated with warming in the central equatorial Pacific, which exhibits a larger decadal signal.[86]

With regards to the worst-case scenario in the twenty-first century of a multi-decadal monsoon, do we have any reasons to reject a future scenario analogous to the Strange Parallels monsoon drought, or anticipate that such a scenario would be worse in the twenty-first century? Here are the factors to consider, with "positive" denoting an ameliorating effect:

- Continued warming is expected to overall increase monsoon rainfall. (positive, depending on the magnitude of the warming)
- A decrease in aerosol particles from burning fossil fuels would increase rainfall. The direction of this trend depends on government policies in the region. (positive)
- The IPCC regards any changes in El Niño characteristics associated with emissions-forced warming to be highly uncertain. Further, changes in the rate of warming and temperature distributions in the Indian Ocean also remains uncertain. (uncertain)
- Variations in multi-decadal modes of ocean circulations. (e.g., Pacific Decadal Oscillation) (alternating positive and negative)
- Land use changes towards more cultivated lands. (negative)

While overall monsoon rainfall is expected to increase in a warmer climate provided that pollution aerosol does not increase, there is no basis for rejecting the worst-case scenario based on the Strange Parallels drought. This drought was apparently caused by unusual circulation patterns in the Pacific. Based on our current lack of understanding of how global warming might alter such circulation regimes in the future, we have no basis for rejecting this scenario. However, a large amount of warming makes such a scenario of multidecadal monsoon drought less likely.

A warming climate is unambiguously associated with increased global rainfall, notably in the regions impacted by the Asian monsoon. About half of the global population lives in the region impacted by the Asian monsoon—nearly all of this region is water stressed, particularly in the face of rapidly growing populations. I have often wondered if global CO_2 emissions policy was decided by one person/one vote, whether a majority of the global population would vote for more CO_2 emissions (provided that aerosol emissions are reduced) so that there would be greater water availability, a factor that dominates much of the overall well-being in the region.

9.5 Sea Level Rise

"What would an ocean be without a monster lurking in the dark? It would be like sleep without dreams." (German film director Werner Herzog)[87]

Sea level rise is an issue of significant concern, given the large number of people who live in coastal regions and the value of coastal infrastructure and property. The concern over sea level rise is not about the 20 centimeters (8 inches) or so that global mean sea level has risen since 1900. Rather, the concern is about projections of twenty-first century sea level rise based on climate model-based projections of human-caused global warming.

Upper limit scenarios for twenty-first century sea level rise are an essential tool for scientists, engineers, and policy analysts tasked with designing responses and adaptation strategies (e.g., planning for coastal safety in cities and long-term investment in critical infrastructure). Given the current state of knowledge, the specific challenges for identifying worst-case sea level rise scenarios are the deep uncertainty surrounding the stability of marine ice sheets in Antarctica and clarification of geological constraints on extreme values of sea level rise.

The focus of this section is on global mean sea level rise—regional sea level rise is further influenced by regional/local vertical land motion and ocean circulation patterns. The plausible worst-case scenario for global mean sea level rise for the twenty-first century is the most complicated case considered in this chapter. There is no shortage of worst-case sea level rise scenarios in the published literature; the challenge is to assess plausibility of these scenarios and clarify our ignorance.

9.5.1 Storylines of West Antarctic Ice Sheet collapse

"Huge blocks of ice, weighing many tons, were lifted into the air and tossed aside as other masses rose beneath them. We were helpless intruders in a strange world, our lives dependent upon the play of grim elementary forces that made a mock of our puny efforts." (Explorer Sir Ernest Shackleton)[88]

Over the past century, melting of mountain glaciers has been the main contributor to increasing ocean water mass. However, most land ice is stored in the Antarctic and Greenland ice sheets. Nearly all worst-case scenarios of twenty-first century sea level rise are associated with accelerated loss from the West Antarctic Ice Sheet (WAIS). Estimates of the contribution to global sea level rise from a complete collapse of the WAIS are 3.2–4.2 meters.[89,90]

The Western Antarctic Ice Sheet is classified as a marine ice sheet, meaning that its bed lies well below sea level. The weight of the ice has caused the underlying rock to sink by between 0.5 and 1 kilometer (1,500–3,000 feet) below sea level. Under the force of its own weight, the ice sheet deforms and flows over the bedrock. When an ice stream reaches the coast, it either breaks away or continues to flow outward onto the water, resulting in a floating ice shelf extending from the continent.

A major uncertainty in future West Antarctic mass losses is the possibility of rapid ice loss through instability of the marine ice sheet via the mechanisms of Marine Ice Sheet Instability and Marine Ice Cliff Instability (MICI).[91]

Because the West Antarctic Ice Sheet rests on bedrock below sea level, the ice sheet is vulnerable to melting from the ocean. If these marine ice shelves—the floating extensions of glacial ice flowing into the ocean—lose mass, their buttressing capacity is reduced, accelerating the ice flow seaward. This self-sustaining process is known as Marine Ice Sheet Instability (MISI).[92]

The disappearance of ice shelves allows formation of ice cliffs, which may be inherently unstable if they are tall enough to generate stresses that exceed the strength of the ice. Ice cliff failure can lead to ice sheet retreat via a process called MICI, that has been hypothesized to cause partial collapse of the West Antarctic Ice Sheet within a few centuries.[93] The IPCC AR6 assesses that there is low agreement on the exact MICI mechanism and limited evidence of its occurrence in the present or the past.[94] Thus, the potential of MICI to impact the future sea level remains speculative.[95]

Another storyline relates to geologically-induced melting at the base of the ice sheet—a factor not mentioned in the IPCC assessment reports.[96] The West Antarctic Ice Sheet lies atop a major volcanic rift system with 138 documented volcanoes that are widely distributed throughout West Antarctic.[97,98] These volcanoes produce a steady flux of heat, but episodic active volcanoes can potentially produce significant local melting,[99] particularly under the vulnerable Thwaites Glacier.[100] The discovery of high geothermal heat flux and volcanoes beneath the West Antarctic Ice Sheet means that there is an additional source of heat that melts the ice and lubricate its passage toward the sea. The impact of under ice volcanoes in terms of producing significant and varying under ice melting, is as of yet an undetermined effect on the glacier mass balance.[101]

9.5.2 Candidate Worst-Case Scenarios

"The point of predicting the future is that we should not be too surprised when it arrives." (Environmental scientist Michael Oppenheimer et al.)[102]

Worst-case scenarios for twenty-first century sea level rise have been developed in different ways: convening an expert committee to develop extreme scenarios,[103] conducting a large expert assessment survey,[104] combining process models of ice sheet contribution with climate model projections,[105] or semi-empirical approaches based on past relationships of sea level rise with temperature.[106]

Between the period 2008 and 2013, estimates of the worst-case or high-end sea level rise scenarios by 2100 ranged from 1.1 to 2 meters, showing an increase over the period.[107] In 2017, the US NOAA published a report entitled *Global and Regional Sea Level Rise Scenarios for the United States.* This report presented a worst-case scenario of 2.5 meters (8.2 feet).[108] This represents an increase from a previous 2012 NOAA report that denoted the worst-case global sea level rise scenario to be 2.0 meters.[109] The rationale for increasing sea level rise to 2.5 meters was the inclusion of MICI.[110]

A 2018 expert elicitation provided high-end estimates of sea level rise for 2100 relative to 2005, under a high-emissions warming scenario close to RCP8.5, of 3.29 meters at the 99th percentile and 2.38 meters at the 95th percentile. For a moderate emissions scenario (slightly below RCP4.5), the high-end projections for 2100 are 1.63 meters at the 99th percentile and 1.26 meters at the 95th percentile. This study concluded that experts' judgments of uncertainties in projections of the ice sheet contribution to sea level rise increased following the publication of the IPCC AR5 (2013).[111]

The IPCC AR6 presented three categories of ice sheet projections: (1) projections from ice sheet models that represent processes in which there is at least medium confidence, (2) projections from an Antarctic ice-sheet model that incorporates the MICI, and (3) projections based on structured expert judgment (SEJ). The IPCC AR6 ascribes *low confidence* to projections incorporating MICI because there is *low confidence* in the current ability to quantify MICI. *Low confidence* is also ascribed to projections based on SEJ, because individual experts participating in the SEJ study may have incorporated processes in whose quantification there is low confidence, and the experts' reasoning has not been examined in detail.[112]

Considering only sea level projections for 2100 representing processes in whose quantification there is at least *medium confidence* (1), the 95th percentile projections are 1.0 meters under RCP4.5, and 1.6 meters under RCP8.5. Considering also projections incorporating MICI or SEJ (low confidence), the 95th percentile projections for 2100 are 1.6 meters under RCP4.5 and 2.4 meters under RCP8.5.[113]

9.5.3 Scenario Falsification and the Plausible Worst Case

"In my expert opinion, based on the historic record, the rapid pulses, and current rates of sea level rise acceleration, I project a 4.6–9.1 meter (15–30 foot) rise in sea level by 2100 if current trends continue." (Harold Wanless, Professor, and Chair of the Department of Geological Sciences, University of Miami, in written testimony provided in Juliana versus the United States climate change litigation)[114]

Values of projected sea level rise for 2100 that exceed 1 meter require either the high emissions scenario (RCP8.5/SSP5–8.5) or the inclusion of the MICI process. Poorly understood processes of ice sheet instabilities and geological processes, characterized by deep uncertainty, have the potential to strongly increase Antarctic mass loss on century timescales—hence these scenarios shouldn't be completely ignored even if they are judged to be at best weakly plausible based on our current background knowledge. However, scenarios of twenty-first century sea level rise as high as 9 meters have been put forward, by invoking MICI and paleoclimate analogs.

For physical reasons, there should exist an upper limit to sea-level rise by 2100.[115] However, the exact upper limit is debatable. The upper limits are related to the nature of the current reservoirs of land ice, the achievable rates of sea level rise, and limits associated with ice sheet processes. As a check on scenarios developed from process models and/or more speculative methods, integral constraints on basic physical processes provide a rationale for potentially falsifying extreme scenarios.

The scenario of a 2.5 meter sea level rise in the twenty-first century requires a rate of sea level rise of 25–44 millimeters per year over the second half of the twenty-first century.[116] For reference, the average rate of global sea level rise for the last several decades is 3.3 millimeters per year (note: 3 millimeters is about the thickness of two stacked pennies).[117]

Are these high rates of sea level rise in the twenty-first century plausible? Projected rates of sea level rise required to achieve an integral sea level rise above 1.8 meters by 2100 are larger than the rates at the onset of the last deglaciation.[118] Additional insights are provided from a previous interglacial period (about 120,000 years ago, the Eemian). The late Eemian sea level is estimated to have exceeded present values by 6.6 meters and is unlikely to have exceeded 9.4 meters.[119] It was concluded that present ice sheets could sustain a rate of global sea level rise of about 5.6–9.2 millimeters per year for several centuries, with these rates potentially spiking to higher values for shorter periods.[120] Starting from present-day conditions, such high rates of sea level rise would require unprecedented ice-loss mechanisms without interglacial precedents, such as catastrophic collapse of the West Antarctic Ice Sheet or activation of major East Antarctic Ice Sheet retreat.[121]

An alternative strategy for falsifying ice loss scenarios relates to identifying physical constraints on specific ice loss mechanisms. Extreme scenarios have been falsified based on kinematic constraints on marine ice sheet contributions to twenty-first century sea level rise. It was found that a total sea level rise of about 2 meters by 2100 could occur under physically possible glaciological conditions, but only if all variables are quickly accelerated to extremely high limits.[122] It is concluded that increases in excess of 2 meters are physically untenable.[123]

An additional brake on ice loss from the West Antarctic Ice Sheet is the recent finding that the ground under the rapidly melting Amundsen Sea Embayment of West Antarctica is rising at the astonishingly rapid rate of 41 millimeters per year as an adjustment to reduced ice mass loading, which acts to stabilize the ice sheet. Models that include this feedback find that much of the West Antarctic Ice Sheet is preserved for moderate climate warming.[124]

In summary, scenarios of twenty-first century sea level rise exceeding about 1.8 meters require conditions without natural interglacial precedents. These extreme scenarios require a cascade of extremely unlikely events and parameters. The joint likelihood of these extremely unlikely events arguably crosses the threshold to implausibility.

Notes

1 Albert Wohlstetter, "No Highway to High Purpose," *Rand Corporation*, 1960, https://doi.org/10.7249/p2084.

2 Nassim Nicholas Taleb, *The Black Swan: The Impact of the Highly Improbable* (New York, NY: Random House, 2012).

3 Frank Hoffman, "Black Swans and Pink Flamingos: Five Principles for Force Design," War on the Rocks, August 19, 2015, https://warontherocks.com/2015/08/black-swans-and-pink-flamingos-five-principles-for-force-design/.

4 Wendy S. Parker and James S. Risbey, "False Precision, Surprise and Improved Uncertainty Assessment," *Philosophical Transactions of the Royal Society A: Mathematical, Physical and Engineering Sciences* 373, no. 2055 (November 28, 2015): 20140453, https://doi.org/10.1098/rsta.2014.0453.

5 Leonard A. Smith and Nicholas Stern, "Uncertainty in Science and Its Role in Climate Policy," *Philosophical Transactions of the Royal Society A: Mathematical, Physical and Engineering Sciences* 369, no. 1956 (December 13, 2011): 4818–4841, https://doi.org/10.1098/rsta.2011.0149.

6 Martin L Weitzman, "On Modeling and Interpreting the Economics of Catastrophic Climate Change," *Review of Economics and Statistics* 91, no. 1 (February 2009): 1–19, https://doi.org/10.1162/rest.91.1.1.

7 T. Wahl et al., "Understanding Extreme Sea Levels for Broad-Scale Coastal Impact and Adaptation Analysis," *Nature Communications* 8, no. 1 (July 7, 2017), https://doi.org/10.1038/ncomms16075.

8 James Hansen et al., "Ice Melt, Sea Level Rise and Superstorms: Evidence from Paleoclimate Data, Climate Modeling, and Modern Observations That 2 °C Global Warming Could Be Dangerous," *Atmospheric Chemistry and Physics* 16, no. 6 (March 22, 2016): 3761–3812, https://doi.org/10.5194/acp-16-3761-2016.

9 David G. Groves and Robert J. Lempert, "A New Analytic Method for Finding Policy-Relevant Scenarios," *Global Environmental Change* 17, no. 1 (February 2007): 73–85, https://doi.org/10.1016/j.gloenvcha.2006.11.006.

10 Evelina Trutnevyte et al., "Reinvigorating the Scenario Technique to Expand Uncertainty Consideration," *Climatic Change* 135, no. 3–4 (January 22, 2016): 373–379, https://doi.org/10.1007/s10584-015-1585-x.

11 Susan J. Debad, *Learning from the Science of Cognition and Perception for Decision Making: Proceedings of a Workshop* (Washington, DC: The National Academies Press, 2018).

12 Martin L Weitzman, "On Modeling and Interpreting."

13 Frank Ackerman, *Worst-Case Economics: Extreme Events in Climate and Finance* (London, UK: Anthem Press, 2017).

14 Caroline A. Katsman et al., "Exploring High-End Scenarios for Local Sea Level Rise to Develop Flood Protection Strategies for a Low-Lying Delta—the Netherlands as an Example," *Climatic Change* 109, no. 3–4 (February 24, 2011): 617–645, https://doi.org/10.1007/s10584-011-0037-5.

15 David Wallace-Wells, *The Uninhabitable Earth: Life after Warming* (New York, NY: Tim Duggan Books, 2019).

16 Dagobert D. Runes, *A Dictionary of Thought* (New York, NY: Philosophical Library, 1959).

17 Wendy S. Parker, "Predicting Weather and Climate: Uncertainty, Ensembles and Probability," Studies in History and Philosophy of Science Part B: Studies in History and Philosophy of Modern Physics 41, no. 3 (September 2010): 263–272, https://doi.org/10.1016/j.shpsb.2010.07.006.

18 D.A Stainforth et al., "Confidence, Uncertainty and Decision-Support Relevance."

19 Terje Aven and Ortwin Renn, "An Evaluation of the Treatment of Risk and Uncertainties in the IPCC Reports on Climate Change," *Risk Analysis* 35, no. 4 (April 2015): 701–712, https://doi.org/10.1111/risa.12298.

20 Cynthia Selin and Ângela Guimaraes Pereira, "Pursuing Plausibility," *International Journal of Foresight and Innovation Policy* 9, no. 2/3/4 (2013): 93–109, https://doi.org/10.1504/ijfip.2013.058616.

21 Didier Dubois and Henry Prade, "Possibility Theory and Its Applications: Where Do We Stand?," *Springer Handbook of Computational Intelligence*, 2015, 31–60, https://doi.org/10.1007/978-3-662-43505-2_3.

22 Gregor Betz, "What's the Worst Case."

23 Gregor Betz, "On Degrees of Justification," *Erkenntnis* 77, no. 2 (August 27, 2011): 237–272, https://doi.org/10.1007/s10670-011-9314-y.

24 Jonathan Rosenhead, "Robustness Analysis: Keeping Your Options Open," in *Rational Analysis for a Problematic World Revisited: Problems Structuring Methods for Complexity, Uncertainty, and Conflict* (Chichester, NY: Wiley, 2001), 181–207.

25 *Elon Musk—Starting a Business, YouTube*, 2014, https://youtu.be/0Bo-RA0sGLU.

26 Arnim Wiek et al., "Plausibility Indications in Future Scenarios," *International Journal of Foresight and Innovation Policy* 9, no. 2/3/4 (2013): 133, https://doi.org/10.1504/ijfip.2013.058611.

27 Alfred Nordmann, "(IM)plausibility²," *International Journal of Foresight and Innovation Policy* 9, no. 2/3/4 (2013): 125, https://doi.org/10.1504/ijfip.2013.058612.

28 Cynthia Selin and Ângela Guimaraes Pereira, "Pursuing Plausibility."

29 Ibid.

30 Arnim Wiek et al., "Plausibility Indications."

31 Martin Weitzman, "The Odds of Disaster: An Economist's Warning on Global Warming," PBS (Public Broadcasting Service, May 23, 2013), https://www.pbs.org/newshour/economy/the-odds-of-disaster-an-econom-1.

32 "The next Environmental Crisis," IAI TV—Changing how the world thinks, November 15, 2021, https://iai.tv/live/the-next-environmental-crisis.

33 "Martin Rees," Welcome to the Royal Society, accessed May 17, 2022, https://royalsociety.org/people/martin-rees-12156/.

34 Martin L Weitzman, "On Modeling and Interpreting."

35 Gernot Wagner and Martin L. Weitzman, *Climate Shock: The Economic Consequences of a Hotter Planet* (Princeton, NJ: Princeton University Press, 2015).
36 Susan Solomon et al., eds., *Climate Change 2007: The Physical Science Basis: Contribution of Working Group I to the Fourth Assessment Report of the Intergovernmental Panel on Climate Change* (Cambridge, UK: Cambridge Univ. Press, 2007).
37 Martin L Weitzman, "On Modeling and Interpreting."
38 Ibid.
39 D. Chen et al., "Framing, Context, and Methods Supplementary Material," in *Climate Change 2021: The Physical Science Basis. Contribution of Working Group I to the Sixth Assessment Report of the Intergovernmental Panel on Climate Change* (Geneva, CH: Intergovernmental Panel on Climate Change, 2021), 147–286.
40 Jonathan L. Bamber et al., "Ice Sheet Contributions to Future Sea-Level Rise from Structured Expert Judgment," *Proceedings of the National Academy of Sciences* 116, no. 23 (May 20, 2019): 11195–11200, https://doi.org/10.1073/pnas.1817205116.
41 Lewis Carroll, *Through the Looking-Glass* (London, UK: Macmillian, 1872).
42 Gregor Betz, "On Degrees of Justification," *Erkenntnis* 77, no. 2 (August 27, 2011): 237–272, https://doi.org/10.1007/s10670-011-9314-y.
43 Terje Aven and Shital Thekdi, *Risk Science an Introduction* (London, UK: Routledge, Taylor & Francis Group, 2022).
44 Gregor Betz, "What's the Worst Case? the Methodology of Possibilistic Prediction," *Analyse &Amp; Kritik* 32, no. 1 (January 2010): 87–106, https://doi.org/10.1515/auk-2010-0105.
45 Nir Friedman and Joseph Halpern, "Plausibility Measures: a User's Guide," in *Uncertainty in Artificial Intelligence: Proceedings of the Eleventh Conference (1995) ; August 18–20, 1995 ; Eleventh Conference on Uncertainty in Artificial Intelligence, McGill University, Montreal, Quebec, Canada,* ed. Phillipe Besnard and Steve Hanks (San Francisco, CA: Morgan Kaufmann, 1995), 175–184.
46 Franz Huber, "The Plausibility-Informativeness Theory," in *New Waves in Epistemology* (Aldershot, UK: Palgrave Macmillan, 2008), 164–191.
47 Arnim Wiek et al., "Plausibility Indications."
48 Gregor Betz, "What's the Worst Case."
49 Scott Rosenberg, "Virtual Reality Check Digital Daydreams, Cyberspace Nightmares," *San Francisco Examiner,* April 19, 1972, C1.
50 Emanuele Bevacqua et al., "Guidelines for Studying Diverse Types of Compound Weather and Climate Events," *Earth's Future* 9, no. 11 (October 25, 2021), https://doi.org/10.1029/2021ef002340.
51 Christopher W. Landsea et al., "A Reanalysis of Hurricane Andrew's Intensity," *Bulletin of the American Meteorological Society* 85, no. 11 (November 2004): 1699–1712, https://doi.org/10.1175/bams-85-11-1699.
52 "US Hurricane Landfalls," Atlantic Oceanographic and Meteorological Laboratories, accessed May 18, 2022, https://www.aoml.noaa.gov/hrd/hurdat/All_U.S._Hurricanes.html.
53 James I. Belanger et al., "Variability in Tornado Frequency Associated with U.S. Landfalling Tropical Cyclones," *Geophysical Research Letters* 36, no. 17 (September 3, 2009), https://doi.org/10.1029/2009gl040013.
54 Christopher W. Landsea et al., "A Reanalysis of Hurricane Andrew's Intensity."
55 Christopher W. Landsea and James L. Franklin, "Atlantic Hurricane Database Uncertainty and Presentation of a New Database Format," *Monthly Weather Review* 141, no. 10 (October 2013): 3576–3592, https://doi.org/10.1175/mwr-d-12-00254.1.

56 Ibid.

57 Jay Barnes, *Florida's Hurricane History* (Chapel Hill, NC: The University of North Carolina Press, 2007).

58 David Roth, " Tropical Cyclone Rainfall," Weather Prediction Center (WPC) home page, August 11, 2021, https://www.wpc.ncep.noaa.gov/tropical/rain/tcrainfall.html.

59 "About the FCHLPM," Florida Commission on Hurricane Loss Projection Methodology, accessed May 18, 2022, https://www.sbafla.com/methodology/AbouttheFCHLPM.aspx.

60 "Florida Commission on Hurricane Loss Projection Methodology—Sbafla. com," Hurricane Model Submissions—Applied Research Associates, Inc., accessed May 18, 2022, https://www.sbafla.com/method/LinkClick.aspx?fileticket=mJTehUBMlD4%3d&tabid=1448&portalid=8&mid=3958.

61 "Modeler Submissions," Florida Commission on Hurricane Loss Projection Methodology, accessed May 18, 2022, https://www.sbafla.com/methodology/ModelerSubmissions.aspx.

62 Greg Holland et al., "A revised model for radial profiles of Hurricane Winds," Monthly Weather Review 138, no. 12 (December 2010): 4393–4401, https://doi.org/10.1175/2010MWR3317.1.

63 S. I. Seneviratne et al., "Weather and Climate Extreme Events in a Changing Climate," in *Climate Change 2021: The Physical Science Basis. Contribution of Working Group I to the Sixth Assessment Report of the Intergovernmental Panel on Climate Change* (Geneva, CH: Intergovernmental Panel on Climate Change, 2021), 1513–1766.

64 S. I. Seneviratne et al., "Weather and Climate Extreme Events."

65 William Brewer, *Up and Down California in 1860–1864 The Journal of William H. Brewer* (Berkley, CA: University of California Press, 1930).

66 Keith Porter et al., *Overview of the Arkstorm Scenario* (Reston, VA: Dept. of the Interior. U.S. Geological Survey, 2011).

67 Jan Null and Joelle Hulbert, "California Washed Away: The Great Flood of 1862," *Weatherwise* 60, no. 1 (2007): 26–30, https://doi.org/10.3200/wewi.60.1.26-30.

68 Michael D. Dettinger and B. Lynn Ingram, "The Coming Megafloods," *Scientific American* 308, no. 1 (2013): 64–71, https://doi.org/10.1038/scientificamerican0113-64.

69 Keith Porter et al., *Overview of the Arkstorm Scenario.*

70 Ibid.

71 Ibid.

72 Alexander Gershunov et al., "Precipitation Regime Change in Western North America: The Role of Atmospheric Rivers," *Scientific Reports* 9, no. 1 (July 9, 2019), https://doi.org/10.1038/s41598-019-46169-w.

73 H. Douville et al., "Water Cycle Changes," in *Climate Change 2021: The Physical Science Basis. Contribution of Working Group I to the Sixth Assessment Report of the Intergovernmental Panel on Climate Change* (Geneva, CH: Intergovernmental Panel on Climate Change, 2021), 1055–1210.

74 Xingying Huang, Daniel L. Swain, and Alex D. Hall, "Future Precipitation Increase from Very High Resolution Ensemble Downscaling of Extreme Atmospheric River Storms in California," *Science Advances* 6, no. 29 (July 2020), https://doi.org/10.1126/sciadv.aba1323.

75 Arinn Dembo, *Monsoon and Other Stories* (Vancouver, BC: Kthonia Press, 2011).

76 Margo Weiss, "In India, Reducing the Dependency on Monsoon Precipitation," State of the Planet, May 28, 2014, https://news.climate.

columbia.edu/2014/05/28/in-india-reducing-the-dependency-on-monsoon-precipitation/.

77 F. J. Doblas-Reyes et al., "Linking Global to Regional Climate Change," in *Climate Change 2021: The Physical Science Basis. Contribution of Working Group I to the Sixth Assessment Report of the Intergovernmental Panel on Climate Change* (Geneva, CH: Intergovernmental Panel on Climate Change, 2021), 1363–1512.

78 V. Eyring et al., "Human Influence on the Climate System," in *Climate Change 2021: The Physical Science Basis. Contribution of Working Group I to the Sixth Assessment Report of the Intergovernmental Panel on Climate Change* (Geneva, CH: Intergovernmental Panel on Climate Change, 2021), 423–552.

79 Qinjian Jin and Chien Wang, "A Revival of Indian Summer Monsoon Rainfall since 2002," *Nature Climate Change* 7, no. 8 (July 24, 2017): 587–594, https://doi.org/10.1038/nclimate3348.

80 Monsoon Asia Drought Atlas, accessed May 17, 2022, http://drought.memphis.edu/MADA/.

81 Edward R. Cook et al., "Asian Monsoon Failure and Megadrought during the Last Millennium," *Science* 328, no. 5977 (April 23, 2010): 486–489, https://doi.org/10.1126/science.1185188.

82 Ibid.

83 Ibid.

84 Monsoon Asia Drought Atlas.

85 Manuel Hernandez et al., "Multi-Scale Drought and Ocean–Atmosphere Variability in Monsoon Asia," *Environmental Research Letters* 10, no. 7 (July 2015): 074010, https://doi.org/10.1088/1748-9326/10/7/074010.

86 Hengyi Weng et al., "Impacts of Recent El Niño Modoki on Dry/Wet Conditions in the Pacific Rim during Boreal Summer," *Climate Dynamics* 29, no. 2–3 (March 6, 2007): 113–129, https://doi.org/10.1007/s00382-007-0234-0.

87 "Our Deep Need for Monsters That Lurk in the Dark," BBC News (BBC, June 23, 2015), https://www.bbc.com/news/magazine-33226376.

88 Sir Ernest Shackelton, *South: The Story of Shackleton's Last Expedition 1914–1917* (William Heinemann, 1919).

89 Jonathan L. Bamber et al., "Reassessment of the Potential Sea-Level Rise from a Collapse of the West Antarctic Ice Sheet," *Science* 324, no. 5929 (May 15, 2009): 901–903, https://doi.org/10.1126/science.1169335.

90 Linda Pan et al., "Rapid Postglacial Rebound Amplifies Global Sea Level Rise Following West Antarctic Ice Sheet Collapse," *Science Advances* 7, no. 18 (April 30, 2021), https://doi.org/10.1126/sciadv.abf7787.

91 Robert M. DeConto and David Pollard, "Contribution of Antarctica to Past and Future Sea-Level Rise," *Nature* 531, no. 7596 (March 30, 2016): 591–597, https://doi.org/10.1038/nature17145.

92 Ibid.

93 Ibid.

94 B. Fox-Kemper et al., "Ocean, Cryosphere and Sea Level Change," in *Climate Change 2021: The Physical Science Basis. Contribution of Working Group I to the Sixth Assessment Report of the Intergovernmental Panel on Climate Change* (Geneva, CH: Intergovernmental Panel on Climate Change, 2021), 1211–1362.

95 Tamsin L. Edwards et al., "Revisiting Antarctic Ice Loss Due to Marine Ice-Cliff Instability," *Nature* 566, no. 7742 (February 6, 2019): 58–64, https://doi.org/10.1038/s41586-019-0901-4.

96 Mareen Lösing, Jörg Ebbing, and Wolfgang Szwillus, "Geothermal Heat Flux in Antarctica: Assessing Models and Observations by Bayesian Inversion," *Frontiers in Earth Science* 8 (April 21, 2020), https://doi.org/10.3389/feart.2020.00105.

97 Yasmina M. Martos et al., "Heat Flux Distribution of Antarctica Unveiled," *Geophysical Research Letters* 44, no. 22 (November 6, 2017), https://doi.org/10.1002/2017gl075609.

98 Maximillian van Wyk de Vries, Robert G. Bingham, and Andrew S. Hein, "A New Volcanic Province: An Inventory of Subglacial Volcanoes in West Antarctica," *Geological Society, London, Special Publications* 461, no. 1 (May 29, 2017): 231–248, https://doi.org/10.1144/sp461.7.

99 Brice Loose et al., "Evidence of an Active Volcanic Heat Source beneath the Pine Island Glacier," *Nature Communications* 9, no. 1 (June 22, 2018), https://doi.org/10.1038/s41467-018-04421-3.

100 Ricarda Dziadek et al., "High Geothermal Heat Flow beneath Thwaites Glacier in West Antarctica Inferred from Aeromagnetic Data," *Communications Earth &Amp; Environment* 2, no. 1 (August 18, 2021), https://doi.org/10.1038/s43247-021-00242-3.

101 Alex Burton-Johnson et al., "Review Article: Geothermal Heat Flow in Antarctica: Current and Future Directions," *The Cryosphere* 14, no. 11 (November 10, 2020): 3843–3873, https://doi.org/10.5194/tc-14-3843-2020.

102 Michael Oppenheimer et al., "Expert Judgement and Uncertainty Quantification for Climate Change," Nature Climate Change 6, no. 5 (April 27, 2016): 445–451, https://doi.org/10.1038/nclimate2959.

103 Caroline A. Katsman et al., "Exploring High-End Scenarios."

104 Benjamin P. Horton et al., "Expert Assessment of Sea-Level Rise by AD 2100 and AD 2300," *Quaternary Science Reviews* 84 (January 2014): 1–6, https://doi.org/10.1016/j.quascirev.2013.11.002.

105 J. L. Bamber and W. P. Aspinall, "An Expert Judgement Assessment of Future Sea Level Rise from the Ice Sheets," *Nature Climate Change* 3, no. 4 (January 6, 2013): 424–427, https://doi.org/10.1038/nclimate1778.

106 Stefan Rahmstorf, "A Semi-Empirical Approach to Projecting Future Sea-Level Rise," *Science* 315, no. 5810 (January 19, 2007): 368–370, https://doi.org/10.1126/science.1135456.

107 S Jevrejeva et al., "Upper Limit for Sea Level Projections by 2100," Environmental Research Letters 9, no. 10 (October 10, 2014): 104008, https://doi.org/10.1088/1748-9326/9/10/104008.

108 William V. Sweet et al., "Global and Regional Sea Level Rise Scenarios for the United States," NOAA Technical Report NOS CO-OPS 083, (Silver Springs, MD: National Oceanic and Atmospheric Administration, 2017).

109 Adam Parris et al., "Global Sea Level Rise Scenarios for the United States National Climate Assessment," NOAA Technical Report OAR CPO-1, (Silver Springs, MD: National Oceanic and Atmospheric Administration, 2012).

110 William V. Sweet et al., "Global and Regional Sea Level Rise Scenarios."

111 Jonathan L. Bamber et al., "Ice Sheet Contributions to Future Sea-Level Rise from Structured Expert Judgment," *Proceedings of the National Academy of Sciences* 116, no. 23 (May 20, 2019): 11195–11200, https://doi.org/10.1073/pnas.1817205116.

112 B. Fox-Kemper et al., "Ocean, Cryosphere and Sea Level Change."

113 Ibid.
114 Juliana v United States (Expert Report of Dr. Harold R. Wanless June 28, 2018).
115 S Jevrejeva et al., "Upper Limit for Sea Level Projections."
116 William V. Sweet et al., "Global and Regional Sea Level Rise Scenarios."
117 "Rising Waters: How NASA Is Monitoring Sea Level Rise," NASA (NASA), accessed May 19, 2022, https://www.nasa.gov/specials/sea-level-rise-2020/.
118 Eelco J. Rohling et al., "A Geological Perspective on Potential Future Sea-Level Rise," *Scientific Reports* 3, no. 1 (December 12, 2013), https://doi.org/10.1038/srep03461.
119 Robert E. Kopp et al., "Probabilistic Assessment of Sea Level during the Last Interglacial Stage," Nature 462, no. 7275 (December 17, 2009): 863–867, https://doi.org/10.1038/nature08686.
120 Robert E. Kopp et al., "Probabilistic Assessment of Sea Level."
121 Eelco J. Rohling et al., "A Geological Perspective."
122 W. T. Pfeffer et al., "Kinematic Constraints on Glacier Contributions to 21st-Century Sea-Level Rise," Science 321, no. 5894 (September 5, 2008): 1340–1343, https://doi.org/10.1126/science.1159099.
123 W. T. Pfeffer et al., "Kinematic Constraints."
124 Valentina R. Barletta et al., "Observed Rapid Bedrock Uplift in Amundsen Sea Embayment Promotes Ice-Sheet Stability," *Science* 360, no. 6395 (June 22, 2018): 1335–1339, https://doi.org/10.1126/science.aao1447.

Part Three

CLIMATE RISK AND RESPONSE

"The future will be neither a nirvana nor a hell on earth, but an evolution of the past, a combination of our best endeavours hindered by obstacles and aided by serendipity."

—Physicist and engineer Michael J. Kelly[1]

Climate change is a risk because it may affect prosperity and security, and because its consequences are uncertain. The way we understand and describe a risk strongly influences the way in which it is analyzed, with implications for risk management and decision-making. By characterizing climate change as a well-understood problem with a strong consensus, traditional risk management approaches assume that climate change can and ought to be rationally managed, or at the very least contained, and preferably eliminated. However, the diversity of climate-related impact drivers and their complex linkages, various inherent and irreducible uncertainties, ambiguities about the consequences of climate change, and the unequal distribution of exposure and effects across geography and time, confound any simple or uncontested application of traditional risk management approaches.

Characterization of climate change as a simple, tame hazard risk of dose-response (such as regulation of food additives or use of antibiotics in feedstocks) to be controlled via the Precautionary Principle has torqued both the science and the policy process in misleading directions. As a result, the policy process that has evolved over the past several decades is not only inadequate to deal with the risks associated with climate change, but has fueled societal controversies around climate risk.

Human-caused climate change has become a topic of contested politics, with great economic stakes associated with both the problem and its proposed solutions. Guided by the analyses in Parts One and Two on the nature of the climate change problem and scenarios of future climate outcomes, Part Three presents a framework for analyzing climate risks in all of their complexity and ambiguity, toward formulating pragmatic and adaptable policies. Integrative thinking in the context of the tension associated with different perspectives, best practices from risk science and decision-making under deep uncertainty, and focusing on resilience and antifragility can lead to broader risk management frameworks that are politically viable and support human well-being, both now and in the future.

Note

1 Michael J Kelly, "How the World Really Works: A Scientist's Guide to Our Past, Present and Future," Net Zero Watch, January 29, 2022, https://www.netzerowatch.com/how-the-world-really-works-a-scientists-guide-to-our-past-present-and-future/.

Chapter Ten

RISK AND ITS ASSESSMENT

"[I]n itself, nothing is a risk, there are no risks in reality. Inversely, anything can be a risk; it all depends on how one frames the danger, considers the event."
—Philosopher François Ewald[1]

The concept of risk is an outgrowth of concerns about coping with dangers. For most of human history, the assessment and management of risk occurred informally by trial and error. More than 2400 years ago, Athenians articulated the capacity of assessing risk before making decisions.[2] As the influence and scale of technology expanded, it became evident that society needed to assess risks proactively. Science has increasingly allowed us to recognize and measure more subtle hazards, which has been coupled with a general decreasing tolerance for risk in modern industrialized society.[3]

Risk assessment and management as a scientific field is only about 40–50 years old. There are many different perspectives on risk, and some of these perspectives represent substantially different frameworks. However, there is broad agreement on the basic ideas and principles of risk understanding, assessment, communication, and management.[4]

Risk may have positive or negative outcomes or may simply be associated with uncertainty. Fire and accidents only have negative outcomes, and they are often referred to as hazard risks. Taking a risk can also result in a positive outcome. Risk can also be related to uncertainty of outcome. However, most applications of risk analysis focus on adverse outcomes.

Climate change presents a challenge to risk assessment that is uniquely complex, uncertain, and ambiguous. This chapter provides perspectives on risk and its perception, how risk is characterized, and why climate change is such a challenging problem for risk analysis.

10.1 Risk and Perception

"You are so convinced that you believe only what you believe that you believe, that you remain utterly blind to what you *really* believe without believing you believe it." (Writer Orson Scott Card)[5]

Climate risk is generally regarded as a hazard risk, although there are potential positive outcomes as well. The related terms "threat" and "hazard" refer to something that could cause harm, with the possibility of trouble, danger, or ruin.[6] The hazard may be something that is impending and imminent, or something that is likely, or merely possible. A forecasted extreme weather event is an imminent, impending hazard. For climate change supported by scenarios from global climate models or expert assessments, *likely* is the descriptor in IPCC parlance. Possibility is used for military threats and threats to cybersecurity, based on creatively imagined scenarios.

Risk has often been characterized as some type of statistical variance—the product of the likelihood of occurrence and the impact if it occurs. However, such a characterization is appropriate only for simple or tame problems. Broader definitions of risk integrate specified consequences of an event or actions, a measure of uncertainty associated with the consequences, and the strength of the background knowledge that supports the assessment.[7]

The IPCC defines climate change risk through the dynamic association of the three core elements of risk: hazard, exposure, and vulnerability. Hazard captures the frequency and intensity of extreme weather and climate events or slowly emerging change such as sea level rise. Exposure refers to the elements of populace, infrastructures, and social welfare at risk. Vulnerability refers to the exposures' potential to suffer as well as its ability to manage, withstand and rebound.[8]

Climate risk focuses on negative concerns surrounding the consequences of manmade global warming:[9]

1. Risk of death, injury, ill health, or disrupted livelihoods in low lying coastal zones and small island developing states due to storm surges, coastal flooding, and sea level rise;
2. Risk of severe ill health and disrupted livelihoods for large urban populations due to inland flooding;
3. System risks due to extreme weather events leading to breakdown of infrastructure networks and critical services such as electricity, water supply, and health and emergency services;
4. Risk of mortality and morbidity during periods of extreme heat, particularly for vulnerable urban populations and those working outdoors in urban or rural areas;
5. Risk of food insecurity and the breakdown of food systems linked to warming, drought, flooding, and precipitation variability and extremes, particularly for poorer populations in urban and rural settings;

6. Risk of loss of rural livelihoods and income due to insufficient access to drinking and irrigation water and reduced agricultural productivity, particularly for farmers and pastoralists with minimal capital in semi-arid regions;
7. Risk of loss of marine and coastal ecosystems, biodiversity and ecosystem goods, functions and services they provide for coastal livelihoods, especially for fishing communities in the tropics and the Arctic; and
8. Risk of loss of terrestrial and inland water ecosystems, biodiversity, and the ecosystem goods, functions, and services they provide for livelihoods.

As evidenced from this list, risks from human-caused climate change are convoluted with natural weather and climate variability, and are dominated by societal vulnerabilities of developing states and poorer populations.

10.1.1 Risk Perceptions

"Something frightening poses a perceived risk. Something dangerous poses a real risk." (Physician Hans Rosling et al.)[10]

Apart from the objective facts about a risk, the social sciences find that our interpretation of those facts is ultimately subjective. Risk science makes a clear distinction between professional judgments about risk versus the public perception of risk. Risk perception is a person's subjective judgment or appraisal of risk, which can involve social, cultural, and psychological factors.

No matter how strongly we feel about our perceptions of risk, we often get risk wrong. People worry about some things more than the evidence warrants (e.g., nuclear radiation, genetically modified food), and less about other threats than the evidence warrants (e.g., obesity, using mobile phones while driving). This gap in risk perception produces social policies that protect us more from what we are afraid of than from what actually threatens us the most. Understanding the psychology of risk perception is important for rationally managing the risks that arise when our subjective risk perception system gets things dangerously wrong.[11]

The Psychometric Paradigm research of psychologist Paul Slovic and collaborators describes a suite of psychological characteristics that make risks feel more or less frightening, relative to the actual facts[12,13,14]

- Natural versus manmade risks
- Risks that are detectable versus undetectable (without special instrumentation)
- Controllable versus uncontrollable risks

- Voluntary versus imposed risks
- Risks with benefits versus uncompensated risks
- Known risks versus vague risks
- Risks central to people's everyday lives versus uncommon risks
- Future versus immediate risks
- Equitable versus asymmetric distribution of risks.

In each of these pairs, the first risk type is generally preferred to the second risk type. For example, risks that are common, self-controlled, and voluntary, such as driving, generate the least public apprehension. Risks that are rare and imposed that lack potential upside, like terrorism, invoke the most dread.[15]

Communicators hoping to spur action on climate change emphasize the manmade aspects of climate change, the unfair burden of risks on undeveloped countries and poor people, and the more immediate risks of severe weather events. The recent occurrence of infrequent events (e.g., a hurricane or flood) produces disparate perceptions of the risk of low probability events, that translate into perceptions of overall climate change risk.

The cultural theory of risk contends that individual views on risk are filtered through cultural world views about how society should operate.[16] People tend to relate their position on the environmental risk spectrum with how they perceive nature. Risk seekers view nature as robust, while risk avoiders see nature as fragile. In between can be found risk regulators that see a robust nature with bounds. Those who are indifferent to risk view nature as unpredictable and view risk truly as just chance.[17]

Even if the initial harm is small, social risk may be greatly amplified by the collective response or irrational behaviors of individuals. The response to climate risk, driven by alarmism and "extinction" rhetoric, has arguably crossed the threshold to actually increasing the social risk associated with climate change.

10.1.2 Risk Characterization

"We must see these things objectively, as we do a tree; and understand that they exist whether we like them or not. We must not try and turn them into something different by the mere exercise of our own minds, as if we were witches." (Writer and theologian G. K. Chesterton)[18]

The concept of hazard risk captures both the threat contribution and vulnerability to the threat. Risk can be split into two main components: (i) occurrence of the triggering events and associated uncertainties; and (ii) consequences, given the events, and the associated uncertainties.

The concept of resilience is closely related to vulnerability. Resilience is defined as the ability of the system to maintain functionality and recover, given that one or more triggering events occur. The ability—or lack of ability—of impacted systems to maintain performance and recover from the event will influence the consequences of the event.

An important element of characterizing risk is evaluating the strength of knowledge (Section 9.3). Concerns about strength of the climate change knowledge base are raised by people questioning aspects of the IPCC's assessment that are used to infer climate risk. The IPCC approach is based on judgment of the available evidence and agreement among experts. More sophisticated knowledge characterizations for risk management include:[19]

(i)   The degree to which the assumptions made are reasonable/realistic based on our background knowledge—growing concern about the focus on implausible emissions scenarios RCP8.5/SSP5–8.5 (Section 7.1)

(ii) The degree to which data/information exists and are reliable and relevant—the historical and paleo database is inadequate for a complete characterization of natural climate variability on multi-decadal to millennial time scales

(iii) The degree to which there is disagreement among experts (including those from different environments or academic fields)—attempts to suppress disagreement and alternative perspectives among experts (Section 2.4)

(iv) The degree to which the phenomena involved are understood and accurate models exist—concerns about the fidelity and utility of climate models (Chapter Six)

(v)   The degree to which the knowledge has been thoroughly examined with respect to unknown knowns—neglect of the unknown knowns associated with natural climate variability (Section 8.2).

Accepting the IPCC's assessments as the best available knowledge base is not inconsistent with acknowledging significant weaknesses in the knowledge base in the context of climate risk analysis.

There are various types of risk, independent of the nature of the triggering event or the consequence. In 1921, economist Frank Knight articulated a definition of risk that distinguished risk from uncertainty.[20] Knightian risk implies that robust probability information is available about future outcomes, allowing risks to be treated in terms of probability and effects. Knightian risk underlies the technocratic, decisionistic, and economic models of risk assessment and management.[21] Similarly, mathematician Benoit Mandelbrodt differentiates between "mild" and "wild" risk.[22] Systems that are predictable, have large quantities of sample cases to explore and that conform to normal style distributions would be considered mild. In contrast, random behavior, unusual distributions, and largely unpredictable systems fall into the wild realm.[23] The danger lies in making the assumption a risk is mild when it is indeed wild.

Knightian-based, or mild, risk assessment and management is suited to simple, or tame, problems (Section 3.4). For simple risks, the cause for the risk is well known, the negative consequences are obvious and definite, and randomness is low. The risk consequences are reversible and are well understood in terms of the science and regulation. Simple risks are recurrent and not affected by ongoing or expected major changes; hence, assessing the risks in statistical terms is meaningful. Examples include automobile accidents and regularly recurring weather hazards—hazards that insurance companies have traditionally covered.

Simple risks are characterized by low complexity, low uncertainty, and low ambiguity. However, the most worrisome risks are not associated with simple problems. Wild risks are associated with high uncertainty and even ignorance. Complexity gives rise to systemic risk. Related to the term "mess" that was introduced in Section 3.4, "ambiguity" means that there are different legitimate viewpoints from which to evaluate whether there are, or could be, adverse effects and whether these risks are tolerable. Ambiguity results from divergent and contested perspectives on the justification, severity, or wider meanings associated with a perceived threat.[24]

Some example types of risk that are not simple:[25]

- Combination of low probability with high extent of damage. This includes technological risks associated with nuclear power plants or dams.
- Disaster potential is high and relatively well known, but little is known about causal factors. Earthquakes and volcanoes are examples.
- Human interventions in the environment that cause wide-ranging, persistent, and irreversible changes, which are discovered after the interventions. Examples include large infrastructure projects such as dams and river engineering.
- There is a considerable delay between the triggering event and the occurrence of damage. Fossil fuel emissions are an example.

Risk conceptions can also be usefully categorized based on future predictability— the extent to which historical records are useful in assessing the risk of a future outcome. When there is a clear causal relationship, the past can be a very good predictor of the future. However, when uncertainty prevails, the past is not a good indicator of the future, and unprecedented outcomes can occur.[26]

The concept of emerging risk is gaining increasing attention. Emerging risk is related to an activity when the background knowledge is weak but contains indications/justified beliefs that a new type of event could occur in the future and potentially have severe consequences to something humans value. The weak background knowledge results in difficulty specifying consequences as well as in specifying scenarios.[27] Intergenerational impacts are a concern with emerging risks. Global warming is clearly an emerging risk. An important consideration when thinking about emerging risks is the speed at which they can become significant, which is referred to as the risk velocity.

Climate change risk includes elements of both incremental risk and emergency risk. Incremental risk displays creeping characteristics and the "fat tail" effect. Since changes take place rather slowly, the adverse consequences take a long time to emerge. Some climate impacts are expected to accumulate and worsen over time and across space. The slow creep of sea level rise is an example of incremental risk. A more complex example is the incremental risk of long-term water shortage in Africa, which reduces food production over time and aggravates malnutrition and other health problems, leading to migration and regional violent conflicts. The potential impacts and long-term consequences can easily be underestimated in the early stage. Once incremental risks cross a critical point, they are difficult to manage.

Emergency risks are associated with extreme weather events; technically these are weather risks and not climate risks, even if global warming could be shown to incrementally worsen the weather hazard. Weather risk can become climate risk if global warming causes the event to exceed a vulnerability threshold that otherwise wouldn't have been exceeded by the weather event. Attempts are also made to assess incremental costs/damages associated with extreme weather events. Such assessments are very challenging to make against the background of natural weather and climate variability.

10.1.3 Direct versus Systemic Risk

"A dike ten thousand feet long begins its crumbling with holes made by ants." (Chinese philosopher (Zhou Dynasty) Han Fei Tzu)[28]

Direct risks from a hazard emerge from a direct impact arising from the driver or event. Examples of direct risk from weather and climate include: heat stress to people, crop production failures, damage from unusual river flooding, sea level rise for coastal cities. Direct risks can be non-linear—while average conditions may change gradually, the risks can increase more rapidly if critical vulnerability thresholds are exceeded.

The risks of climate change to human interests depend not only on the direct impacts from changes in the physical climate, but also on the response of complex human systems such as the global economy, food markets, and the system of international security. In complex systems, small changes can sometimes lead to large divergences in future state. Systemic risks are characterized by a high degree of complexity, uncertainty, and ambiguity.[29] Many problems categorized as a "wicked mess" (Section 3.4) can also be categorized as systemic risks.

The greatest risks associated with climate change are systemic. Systemic risks are much less straightforward to assess than direct risks. Systemic risks may be initiated by a local stressor such as an extreme weather event, which can have impacts extending far beyond the impacted location. An example from national and international security, systemic risk is triggered by regional drought, resulting in migration and displacement, food price spikes, civil unrest, state failure, terrorism, humanitarian crises, and transboundary water disputes.[30]

Hazard events that previously might have caused localized impacts can now have cascading and even global impacts. Consider the severe, widespread flooding in Thailand during 2011. The flooding was triggered by the landfall of a tropical storm, spreading through much of Thailand along the Mekong and Chao Phraya River basins. The systemic nature of this event arose from flooding of the industrial estates and production plants around Bangkok.[31] Apart from cascading effects on Thailand's economy, this local flood ended up having systemic impacts across countries, regions, and economic sectors, affecting supply chains as far away as Japan and the United States. Key manufacturing sectors such as the automobile, electronics, and electrical appliances industries experienced abrupt declines in production and exports. The disruption caused a global shortage of hard disk drives, which lasted throughout 2012.[32] Components manufactured in Thailand were essential for products finalized in other countries, causing disruption or collapse of entire production chains.

10.2 Risk Assessment

"Risk is a function of how poorly a strategy will perform if the 'wrong' scenario occurs." (Management and competitiveness expert Michael E. Porter)[33]

Risk assessment is the systematic process of: identifying risk sources, threats, hazards, and opportunities; understanding how these can materialize/occur; identifying specific trigger events and event sequences, and what their consequences could be; representing and expressing uncertainties; and determining the significance of the risk.[34] The role of risk assessment is to support understanding, communication, and decision-making processes.

Traditional risk assessment has been based on historic data—assessing probabilities of severity, frequency, and impact based on experience from past events. In times of global change, this approach is no longer adequate to capture future risks. Scenarios of future climate and extreme events are generated by models and by storyline approaches (Chapters Seven to Nine).

Apart from the likelihood and impacts of the risk, values and ancillary issues that contribute to risk perception should be integrated into the risk assessment. Further, risk evaluation requires assessment of risk(s)–benefit(s) and risk–risk trade-offs. Risks need to be assessed in context of all of the relevant dimensions that matter to the affected populations.[35] A broad assessment of the risk provides a knowledge base so that decision makers have better guidance on how to select measures for managing the risk.

Narrow risk assessments of well-understood phenomena with ample data might be uncontested. However, for contentious issues, the outcome of a risk assessment may be strongly influenced by the many inherent value judgments (often unknowingly) made by the analyst. Scientific studies have trouble resolving differences when the value assumption process is murky. Wise policymakers are likely to dismiss overly certain assessments, properly recognizing their biases or naivete. Trust is built via an open dialog around value assumptions, reducing the concerns of hidden agendas.[36]

10.2.1 Acceptable versus Intolerable Risk

"Time heals nothing. It only brings other issues and tissues, and takes what is incurable or unacceptable out of the center of our attention." (Ana Claudia Antunes, author of The Tao of Physical and Spiritual)[37]

An essential aim of risk assessment is to qualify the risks in terms of social acceptability and tolerability. The "traffic light model" identifies three categories for risks: normal/acceptable risks, intermediate/tolerable risks, and intolerable risks. Normal risks are characterized by low complexity, little statistical uncertainty, and low catastrophic potential (Knightian or mild risk), and are well understood by science and easily managed by regulatory mechanisms.[38]

Activities are tolerable if they are considered as worth pursuing for the associated benefits. For tolerable risks, efforts for risk reduction or coping are welcomed provided that the benefits of the activities are not lost. Burning fossil fuels has historically been considered a tolerable risk.[39] Genuinely intolerable risks include existential threats—such as portrayed by the earth-impacting comet in the movie Don't Look Up—or "ruin" problems.[40,41]

How to draw the lines between "acceptable," "tolerable" and "intolerable" is one of the most controversial tasks in the risk governance process for complex risks. Ambiguity results from divergent and contested perspectives on the justification, severity, or wider meanings associated with a perceived threat.[42] Climate change risks have been characterized as acceptable, tolerable, and intolerable by different individuals and constituencies—clearly an ambiguous situation.

Using the example of extreme weather risks, fairly routine destruction by hurricanes in certain coastal regions is generally categorized as tolerable, with some locations working to decrease vulnerability over time by improving building codes and engineering the coast to minimize the impact of storm surge. By contrast, a rare but extreme event such as Superstorm Sandy striking New York City was judged to be associated with intolerable risk and a large adaptation deficit, given the large population and property value that was impacted and high vulnerability of the infrastructure.

For complex problems, often neither the risks nor the benefits can be clearly identified or agreed upon. Multiple dimensions and values need to be considered. A heavily contested assessment of the (in)tolerability or acceptability of risks is highly relevant input to the decision-making process, in the context of the ambiguity of the risk.

10.2.2 Assessment of Systemic Risks

"[I]t is not a calculated risk if you haven't calculated it."[43] (Naved Abdali, author of *Investor*)

In most problems of systemic risk, there are numerous ambiguous and unknown risks. Due to the complexity of systemic risk and the limited understanding of the risk propagation paths, there are substantial uncertainties in terms of the cascading effect of the systemic risk on the influence domain.

Key characteristics of systemic risk are broadly categorized under five themes: the scale of the system, the relationship of the elements within a system, the level of system understanding, the transboundary effects, and the outcomes of systemic risk.[44]

An important tool in characterizing systemic risk is forensic analysis of previous disasters. Forensic analysis combines retrospective longitudinal analysis, disaster scenarios, comparative case analysis, and meta-analysis research, along with enhanced involvement of development stakeholders.[45] Forensic analysis can help us better understand how and why disasters occur and their complex linkages with natural, infrastructure, and societal systems.[46]

Recognizing the depth of uncertainty about the future state of complex systems, assessment of systemic risk relies on the tools of scenarios and wargaming to help us think about what might happen. A key challenge for systemic risk assessment is to identify key linkages and the potential for critical transitions, in the context of how systemic risks cascade through interdependent networks in both physical and social systems.[47] Complex challenges like pandemics or climate change should be assessed in the global context of megatrends such as urbanization, growing populations, demographic changes, environmental degradation, and technological change.

In risks that go beyond ordinary dimensions, such as COVID-19 and climate change, the reliability of any risk assessment is low, uncertainty is high, catastrophic potential can possibly reach alarming dimensions and systematic knowledge about the distribution of consequences is missing. Such risks may generate global, irreversible damages, which may accumulate during a long time or mobilize or frighten the population; or alternatively, the risks may turn out to relatively benign. An unequivocal conclusion about the degree of validity associated with the scientific risk evaluation isn't feasible. For such problems, risk assessments are most effectively used as a tool to represent and describe knowledge and lack of knowledge, and to formulate narratives of possible danger.

10.3 Climate Change Risk

"I have long argued climate messaging needs to be about here and now, kitchen table issues […] not 8 decades from now or way in the Arctic." (Meteorologist Marshall Shepherd)[48]

Many public organizations, ranging from the United Nations to local municipalities, have conducted risk assessments of climate change impacts. There is also a growing trend for private companies with responsibility to shareholders to conduct climate risk assessments. Virtually all of these climate risk assessments are based on assessment reports from the IPCC.

Cumulative risks at national and global levels are linked to the principle of avoiding dangerous climate change as defined by the UNFCCC, which is tied to assumed risk thresholds of 1.5°C and 2°C of global warming above pre-industrial levels. The most recent IPCC reports have formulated a risk-based approach for understanding the risk factors that lead to regional/local climate change impacts, and the role of adaptation initiatives in managing these risk factors. The importance of regional climate change risk assessments is recognized politically through the UNFCCC in the context of Warsaw International Mechanism for Loss and Damage and the Sendai Framework for Disaster Risk Reduction under the auspices of the UN Office for Disaster Risk Reduction.[49,50]

Thus far, government-sponsored climate change assessments have primarily been rooted in communicating a scientific consensus around anticipated changes in the physical climate system, along with some impacts that are generally implied by those climate changes. The assessments have not characterized or analyzed specific societal risks due to climate change, including socio-environmental causes of vulnerability related to climate hazards,[51] although the WGII Report in the AR6 took steps in this direction.

Climate risk assessments for industry sectors and infrastructure sectors reveal challenges from interconnectivities, interdependencies across value chains, and the impacts of policy and regulations.[52] International dimensions include mechanisms such as prices, material flows, movement of people, and political stability, with specific emphasis on food security and geopolitical risks.[53]

The complexity, uncertainty, and ambiguity associated with climate risk point to a number of challenges and frontier issues for assessing the risks associated with climate change in a policy-based context. These include the issue of dealing with systemic risks, critical knowledge gaps regarding Earth system processes, and the complexity of societal responses—which could collectively lead to extremely challenging outcomes. Much of the uncertainty in climate change assessments is irreducible; this uncertainty should be treated as a source of actionable knowledge through its influence on the viability of different decision strategies. While the UNFCCC has framed climate risk in terms of dangers, risk assessment should evaluate both positive and negative aspects of change to support a balanced appraisal of alternative decision options. Other issues relate to risk assessment processes themselves, such as managing uncertainty and clarifying the hidden values assumptions of the risk analysts.[54]

This section presents the case that it is time for a shift in the objectives and implementation of climate change assessments—from making what amounts to a general case for action regarding eliminating fossil fuel emissions, to characterizing specific regional risks and vulnerabilities in a manner that helps people develop, select, carry out, and monitor actions that ultimately have greater benefits than costs.[55]

10.3.1 How We have Mischaracterized Climate Risk

"The global climate change debate has gone badly wrong. Many mainstream environmentalists are arguing for the wrong actions and for the wrong reasons, and so long as they continue to do so they put all our futures in jeopardy." (Philosopher Thomas Wells)[56]

The communication and handling of the climate change risk is strongly dependent on the way risk is conceptualized and described.

In the early 1980s, the United Nations Environmental Program (UNEP) became bullish on the idea that fossil fuels would produce dangerous climate change. The prospect of eliminating fossil fuels was congruent with UNEP's broader interests in environmental quality and world governance. At Villach, Austria, in 1985 at the beginning of the climate treaty movement, the policy movement to eliminate fossil fuels became detached from its moorings in the science—the rhetoric of precaution argued that we should act anyway to eliminate fossil fuels, just in case.[57,58] This perspective became codified by the 1992 UN Framework Convention on Climate Change (UNFCCC) Treaty in 1992, the Kyoto Protocol in 1997, and the 2015 Paris Climate Agreement.

The fundamental presupposition of the UNFCCC climate treaties and agreements is that human-caused climate change associated with fossil fuel emissions is dangerous—natural climate variability and most other human impacts on climate were essentially ignored. There was an implicit assumption that warming was dangerous, ignoring any potential benefits. In attempting to build political will for the international agreements, the adverse impacts of fossil-fuel driven climate change were exaggerated—more frequent and intense severe weather/climate events, sea level rise, and many adverse ecosystem, health, economic and geopolitical impacts, with all of their complex causes, were conflated with fossil-fuel driven warming (Section 1.1).

The torquing of climate science and the manufacture of a consensus around human-caused climate change (Chapter Two) not only oversimplified the scientific and social challenges, but led to the adoption of a "predict then act" strategy to manage and control the climate. This strategy is based on the belief that climate change is a simple or tame problem, with science and all its uncertainties trumping all practical questions and conflicting values and purposes.

One of the biggest problems associated with climate change assessments is that there is no simple way to articulate the danger associated with a warmer climate (Section 1.3). To circumvent this challenge, the problem of climate change has been reduced to one of the CO_2 emissions, with the elimination of human-caused emissions as the primary strategy for dealing with the problem. This strategy ignores the fact that there is no precise link between CO_2 emissions and global mean surface temperature, owing to uncertainties in climate sensitivity to CO_2 (Section 7.2) and the poorly constrained natural variations of the climate system (Section 8.2). An alternative approach focuses on a specified "dangerous" surface temperature threshold.[59] This approach also requires accurate accounting and projections of natural climate variability from solar variability, volcanic eruptions, and multi-decadal to millennial scale ocean variations (Section 8.2). An additional complication is that the temperature targets of 1.5 and 2.0°C do not have a sound scientific basis as thresholds for danger (Section 1.3).

Further, the temperature-based approach of specifying a dangerous threshold ignores nonlinearities in the climate system. Climatic "tipping points" (Section 1.3), which have been described as "real critical thresholds," do not all occur at the same temperature threshold. If unknown climate tipping points exist that are associated with natural climate processes, there can be no certainty that the risk of climate catastrophe would be eliminated even if emissions fell to zero.[60]

The end result of the way in which the IPCC/UNFCCC have characterized the climate change problem is that we are focused on meeting emissions targets that are arguably unachievable in the near term, rather than on obtaining a deeper understanding of the climate system and the broader causes of vulnerabilities of human and natural systems. Further, the slow incremental risks of warming have been mischaracterized as urgent, leading to rapid implementation of policies that are not only costly and suboptimal, but also arguably reduce societal resilience to weather and climate variability, whatever their causes.

The politics of international climate governance has produced systematic biases in the kinds of expertise and evidence that are deemed appropriate for consideration.[61] The UNFCCC and IPCC have characterized climate change as an environmental and economic problem, and geoscientists and economists have dominated the assessment and policy making process.

However, the issues with increasing CO_2 and warming are primarily social, not environmental. The Earth has undergone geological periods of higher temperatures and atmospheric CO_2 concentrations, during which life thrived. Characterization of climate change as an environmental problem has downplayed the cultural and political dimensions of the issue.[62] In fact, many social scientists have argued that climate change is a values problem, and that the disciplinary constrictions imposed by the IPCC and UNFCCC have neglected many important insights arising from a wide range of expert and unaccredited sources.[63]

Clarifying the multitude of values in play would help illuminate the ambiguity surrounding interpretations of climate risk. However, characterizing climate change as a values problem can also lead into blind alleys. Many environmentalists have moralized the climate change problem as a simple, righteous values choice: Are you for the planet or against it?[64] This moralizing neglects to understand that people engage in activities that are of value to them that happen to emit carbon as a byproduct.[65] Further, this narrow moralizing systematically excludes important ethical values, such as improving the lives of the billion people presently living in unacceptable poverty or protecting other aspects of the environment.[66]

The end result is that after 30 years of the UNFCCC/IPCC, we remain fixated on the minutiae of greenhouse gas emissions levels and the abstract and impossible problem of constraining the atmospheric CO_2 concentration—while ignoring natural climate variability and drastically simplifying the human impacts and responses. As long as the current situation prevails, the IPCC's assessments of anthropogenic climate change and the UNFCCC recommendations for action will remain seriously inadequate.

This fixation on global CO_2 levels alone as the problem and solution has led risk scientist Terje Aven to conclude:

> "The current thinking and approaches guiding this conceptualization and description have been shown to lack scientific rigour, the consequence being that climate change risk and uncertainties are poorly presented. The climate change field needs to strengthen its risk science basis, to improve the current situation."[67]

The UNFCCC is promoting a solution to an exceedingly complex, uncertain, and ambiguous problem, without reference to wider ethical issues and political and practical feasibility. We have neglected to truly understand the problem and systematically and broadly evaluate the feasible policy space.

10.3.2 Reframing the Assessment of Climate Risk

"[A] problem well put is half solved." (Philosopher John Dewey)[68]

Climate risks are exceedingly complex and scientific forecasts are subject to large uncertainties. As a result, there is deep uncertainty within climate change risk assessments. In the face of this deep uncertainty, a conventional predict-then-act approach is paralyzed by limits to prediction and lack of clarity regarding the potential consequences. Estimating the severity of the consequences based only on the potential economic losses ignores a series of impacts that are not easily quantified. In summary, we are far from a thorough assessment of full climate risk that would allow for efficient and targeted policies and support rational risk taking.[69]

With this context, the risks from climate change need to be reframed more broadly, in the context of advances in risk science.[70]

There is an emerging literature on characterizing systemic risks from climate change. Systemic risks induced by climate change have three basic characteristics: (1) the wide scope of impacts, affecting a wide range of natural, economic, and social systems; (2) the long duration of impacts; and (3) cumulativeness of these impacts that grow over time.[71]

Assessment of systemic risk must account for the complexity of the climate system and the integrated effects on politics, the economy, society, culture, and the environment. Both emergency risks associated with extreme weather events and the slow creep of incremental risks have characteristics of systemic risk. Interaction of climate with a variety of processes of global environmental change (both natural and human systems) represents an increased need to identify and characterize multiple stressors and their interactions across different scales.[72]

How does a categorization of systemic risk actually help with the practical assessment of climate risk? Is such a diagnosis similar to a categorization of a problem as a wicked mess (Section 3.4), which may lead people to throw up their hands and shrug their shoulders? It is sometimes argued that a full assessment of the risks of climate change would be counterproductive, because the risks may be so large and the solutions so difficult that people will be overwhelmed with a feeling of helplessness, and will look the other way. However, rather than stymieing a solution, such categorizations help identify situations where simplistic solutions won't work and suggest paths that may lead to partial solutions and reductions in vulnerability.

There are a number of strategies that can be helpful for assessing systemic risks. These include portraying the full range of extreme future scenarios (global and regional; natural and human caused), including extreme weather and climate events and worst-case scenarios (Chapters Eight and Nine). These scenarios can be used to drive narratives of danger and to stress test social and infrastructure systems. It is important to differentiate between acute impacts associated with extreme weather events versus incremental risks that develop over time, based on different latent periods and manifestations. Forensic analysis of previous disasters and their complex effects can provide important insights in developing scenarios of systemic risks.

Judgments of intolerable risks from climate change relate to mistakenly conflating the slow creep of global warming (an emerging risk) with consequences from extreme weather and climate events (emergency risks) and concerns about inequitable risk exposure to poorer populations. Removing the risks associated with extreme weather and climate events from the consequences of global warming diminishes the perceived urgency for reducing fossil fuel emissions. The poorest populations would benefit far more from access to grid electricity and help in reducing vulnerability from extreme weather events, than from reductions to the amount of CO_2 in the atmosphere (see Chapter Thirteen).[73]

The uncertainties and complexity of the forces driving the social, biological, and physical dimensions of climate change ensure that we will most likely be surprised by the eventual outcomes. However, there are several features of the climate problem that suggest that climate change related risks will increase in the future. These factors are:[74]

- There is a substantial time lag between the emission of greenhouse gases and the related impacts, and the link to many impacts is not yet fully understood.
- Science is unable to predict long-term impacts accurately enough to provide a strong signpost for action.
- There is large heterogeneity in historical, current, and expected future emissions among countries and sectors, and there is no single culprit.
- Variance and heterogeneity of climate related hazards and benefits are large, and winners and losers may change over time.
- There is large heterogeneity in vulnerability among countries, sectors, and ecosystems.

The question of intergenerational equity is of special importance for emerging risks, owing to the lag between the emissions of greenhouse gases and the expected occurrence of the damage. There is no simple way to decide what duty of care we owe to future generations, but the IPCC's socioeconomic pathways for the twenty-first century all have the world being better off by 2100, even under the most extreme emissions scenarios (Section 7.1).

A key to securing meaningful action to reduce climate change risk is to clearly separate the incremental, emerging risks associated with long-term changes in average conditions from the emergency risks associated with extreme weather and climate events. Risks from extreme weather and climate events need to be addressed in both the present and future climate. Any assessment of climate risk needs to approach these two kinds of risk differently, depending on the nature of the decisions it aims to inform.

Once the incremental risks are separated from the emergency risks, the perception of urgency in reducing emissions is diminished. It becomes far more important to develop strategies for managing atmospheric greenhouse gases in a way that supports a transition that does not have adverse impact on security, the economy, industry, and agriculture. The challenges for risk assessment are then to identify transition risks associated with proposed solutions to mitigate the rising atmospheric CO_2 content.

At this point, assessing the transition risks of eliminating fossil fuels is arguably more important than attempting to refine our assessments of the deeply uncertain incremental risks associated with fossil fuel emissions (see Chapter Fourteen). Consequences of the transition are associated with a fairly solid knowledge base, leading many people to be more concerned about transition risks than they are about the more uncertain risks from climate change itself that are associated with a far weaker knowledge base.[75,76,77]

Assessments of emergency risk associated with extreme weather and climate events generally focus on event type, sectoral impacts or regional assessments. Too many of these assessments are seemingly intended to build political will for eliminating fossil fuel emissions rather than conducting a serious assessment of emergency risk including systemic aspects.

A laudable example of weather emergency risk assessment is the ARkStorm scenario described in Section 9.4.2, related to extreme winter flooding in central California.[78] The state of California's ShakeOut Earthquake Scenario to support preparedness for a major earthquake demonstrated how postulating a hypothetical but plausible catastrophe provides a basis for better examining the interdependencies in social structure and infrastructure and expose choke-points and vulnerabilities.[79] The ARkStorm risk assessment project engaged emergency planners, businesses, universities, government agencies, and others. Following a thorough development of the storm scenario and its direct impacts on inundation and landslides, they assessed secondary impacts on highways, electricity systems, wastewater treatment, water supply, dams, levees, telecommunications, agriculture, buildings, evacuation and sheltering strategies, business interruption, truck transport, and environmental and health issues associated with hazardous materials. The ARkStorm scenario and risk assessment informed community decision-making and is helping communities increase their resilience to severe California winter storms.[80]

In assessing emergency risk from extreme weather/climate events, there are some key principles to follow:[81]

1. Emergency risk from extreme weather/climate events is dominated by plausible worst-case events. Looking at extreme event data back to 1950, and calculating return periods, isn't useful here. Global climate models are of no direct use for extreme event scenarios. Of the three extreme events selected in Section 9.4, the Florida hurricane scenario was based on data from 1935, the ARkStorm scenario was based on data from 1861–62, and the south Asian monsoon failure was based on data from 1756–68—the latter two cases were informed not only by historical records but also by paleoclimate reconstructions.

2. Consider all events, even if their probability of occurrence in a particular location is low. The low probability/high impact events, provided that the events are judged to be plausible or have occurred in the past, are associated with the greatest risks. Nevertheless, the cumulative impact of weaker events can also be significant.

3. Be holistic. Account for all relevant factors, as far as possible—including human behavior and the complex interactions between different parts of a system. Evaluate both direct and systemic risks across the relevant time and space windows appropriate to the decision process.
4. Assess the current risk from weather/climate extremes, based on historical extreme events and today's societal vulnerabilities. Additional scenarios can be created with incremental effects from global warming.

The seemingly overwhelming challenge of climate risk assessment and the associated political conflicts become more tractable by separating the incremental, emerging risk from the emergency risk associated with weather/climate extremes. By regionalizing risk governance, the management approach does not depend on an impossible global agreement for a dubious "perfect" solution that requires moral and political coercion. The regional approach offers a feasible pathway that respects the existing interests and values of the human beings concerned. This approach works within our existing political institutions and offers transparent arguments for action within our present valuational framework, rather than requiring us all to assume a new and narrow set of values.[82]

10.3.3 Climate Change versus COVID-19 Risk

"There is no power for change greater than a community discovering what it cares about." (Writer Margaret Wheatley)[83]

The outbreak of the COVID-19 pandemic at the end of 2019 has generated or contributed to a series of interconnected financial, societal, and political crises on a global scale. This has triggered comparative studies of systemic risks of pandemics and climate change, which share many commonalities despite obvious differences.[84]

Both pandemic and climate risks are systemic, in that their impacts propagate rapidly across an interconnected world. Both COVID-19 and climate change risks are nonstationary, with past probabilities and distributions of occurrences rapidly shifting and proving to be inadequate for future projections. Both risks are nonlinear, in that their socioeconomic impact grows disproportionally and even catastrophically once certain thresholds are breached (such as hospital capacity to treat pandemic patients). Both COVID-19 and climate change are risk multipliers, that highlight and exacerbate hitherto untested vulnerabilities inherent in the financial, energy, food, and healthcare systems and the overall economy. Both risks affect disproportionally the world's most vulnerable populations.[85]

However, the timescales of both the occurrence and the resolution of pandemics and climate hazards are different. Pandemic risk is measured in weeks, months, and years; climate risk is measured in decades and centuries. A pandemic presents imminent, discrete, and directly discernable dangers. By contrast, the risks from climate change, are gradual, cumulative, and distributed, manifesting themselves over time and in spatially heterogeneous way.[86] The primary difference between the pandemic and climate hazards is the salience of the threat and the perceived urgency of dealing with it.[87]

Notes

1 Graham Burchell and François Ewald, "Insurance and Risk," in *The Foucault Effect: Studies in Governmentality; with Two Lectures and an Interview with Michel Foucault* (Chicago, IL: Univ. of Chicago Press, 1991).
2 Joshua P. Nudell, "An Ancient Greek Approach to Risk and the Lessons It Can Offer the Modern World," The Conversation, February 22, 2021, https://theconversation.com/an-ancient-greek-approach-to-risk-and-the-lessons-it-can-offer-the-modern-world-154139.
3 Daniel J. Rozell, "Values in Risk Assessment," in *Dangerous Science: Science Policy and Risk Analysis for Scientists and Engineers* (London, UK: Ubiquity Press, 2020), 29–56.
4 "Resources," Society for Risk Analysis, accessed May 24, 2022, https://www.sra.org/resources/.
5 Orson Scott Card, *Shadow of the Hegemon* (New York, NY: Tor, 2009).
6 Ortwin Renn, *Risk Governance Coping with Uncertainty in a Complex World* (London, UK: Routledge, 2008).
7 Terje Aven, "Risk Assessment and Risk Management: Review of Recent Advances on Their Foundation," *European Journal of Operational Research* 253, no. 1 (August 16, 2016): 1–13, https://doi.org/10.1016/j.ejor.2015.12.023.
8 Christopher B. Field, *Managing the Risks of Extreme Events and Disasters to Advance Climate Change Adaptation: Special Report of the Intergovernmental Panel on Climate Change* (Cambridge, UK: Cambridge University Press, 2012).
9 Brian C. O'Neill et al., "IPCC Reasons for Concern Regarding Climate Change Risks," *Nature Climate Change* 7, no. 1 (January 4, 2017): 28–37, https://doi.org/10.1038/nclimate3179.
10 Hans Rosling et al., *Factfulness: Ten Reasons We're Wrong about the World—and Why Things Are Better than You Think* (New York, NY: Flatiron Books, 2018).
11 David Ropeik, *How Risky Is It Really?: Why Our Fears Don't Always Match the Facts* (New York, NY: McGraw-Hill, 2010).
12 Paul Slovic, "Perception of Risk," *Science* 236, no. 4799 (April 17, 1987): 280–285, https://doi.org/10.1126/science.3563507.
13 Carl Cranor, "A Plea for a Rich Conception of Risks," in *The Ethics of Technological Risk*, ed. L Asveld and S Roeser (London, UK: Routledge, 2008).

14 Nicolas Espinoza, "Incommensurability: The Failure to Compare Risks," in *The Ethics of Technological Risk*, ed. L Asveld and S Roeser (London, UK: Routledge, 2008).

15 Daniel J. Rozell, *Dangerous Science* (London, UK: Ubiquity Press, 2020).

16 Mary Douglas and Aaron Wildavsky, *Risk and Culture an Essay on the Selection of Technological and Environmental Dangers* (Berkeley, CA: Univ. of California Press, 1983).

17 Elizabeth Fisher and Judith S. Jones, *Implementing the Precautionary Principle: Perspectives and Prospects* (Cheltenham, UK: Edward Elgar Publishing Limited, 2006).

18 G K Chesterton, "The Fulfillment of Wishes," Chesterton Digital Library, November 22, 1913, https://library.chesterton.org/the-fulfillment-of-wishes-40578/.

19 Terje Aven, "Improving Risk Characterisations in Practical Situations by Highlighting Knowledge Aspects, with Applications to Risk Matrices," *Reliability Engineering & System Safety* 167 (November 2017): 42–48, https://doi.org/10.1016/j.ress.2017.05.006.

20 Frank H. Knight, *Risk, Uncertainty and Profit* (Boston, MA: Houghton, 1921).

21 Ortwin Renn et al., "Coping with Complexity, Uncertainty and Ambiguity in Risk Governance: A Synthesis," *AMBIO* 40, no. 2 (February 3, 2011): 231–246, https://doi.org/10.1007/s13280-010-0134-0.

22 Benoit B. Mandelbrot and Richard L. Hudson, *The (Mis)Behaviour of Markets: A Fractal View of Risk, Ruin, and Reward* (London, UK: Profile Books, 2010).

23 Ibid.

24 Andy Stirling, "Risk, Uncertainty and Precaution: Some Instrumental Implications from the Social Sciences," *Negotiating Environmental Change*, 2003, https://doi.org/10.4337/9781843765653.00008.

25 Ortwin Renn and Andreas Klinke, "Systemic Risks: A New Challenge for Risk Management," *EMBO Reports* 5, no. S1 (November 2004), https://doi.org/10.1038/sj.embor.7400227.

26 Emmy Wassénius and Beatrice I. Crona, "Adapting Risk Assessments for a Complex Future," *One Earth* 5, no. 1 (January 21, 2022): 35–43, https://doi.org/10.1016/j.oneear.2021.12.004.

27 R. Flage and T. Aven, "Emerging Risk – Conceptual Definition and a Relation to Black Swan Type of Events," *Reliability Engineering & System Safety* 144 (December 2015): 61–67, https://doi.org/10.1016/j.ress.2015.07.008.

28 Han Fei Tzu and W. K. Liao, *The Complete Works of Han Fei Tzu: A Classic of Chinese Political Science* (London, UK: Arthur Probsthain, 1959).

29 Ortwin Renn and Andreas Klinke, "Systemic Risks."

30 David King et al., "Climate Change: A Risk Assessment" (Cambridge, UK, 2015).

31 Aekapol Chongvilaivan, "Thailand's 2011 Flooding: Its Impact on Direct Exports, and Disruption of Global Supply Chains," *ARTNeT Working Paper Series* 113 (2012).

32 Ibid.

33 Michael E. Porter, *Competitive Advantage: Creating and Sustaining Superior Performance* (New York, NY: Free Press, 1985).

34 Terje Aven, "Risk Assessment and Risk Management."

35 Ortwin Renn et al., "Coping with Complexity."

36 Daniel J. Rozell, *Dangerous Science*.

37 Ana Claudia Antunes, *The Tao of Physical and Spiritual* (Lulu, 2008).

38 Ortwin Renn and Andreas Klinke, "Systemic Risks."

39 Alex Epstein, *The Moral Case for Fossil Fuels* (New York, NY: Portfolio/Penguin, 2015).

40 "Don't Look Up," Don't Look Up (Netfilx, December 24, 2021), https://www.netflix.com/title/81252357.

41 Nassim Nicholas Taleb et al., "The Precautionary Principle (with Application to the Genetic Modification of Organisms)," October 2014, https://doi.org/https://doi.org/10.48550/arxiv.1410.5787.

42 Andy Stirling, "Keep It Complex."

43 Naved Abdali, *Investing—Hopes, Hypes, & Heartbreaks* (Mississauga, ON: Rosehurst Publishing, 2021).

44 J Sillmann et al., "ISC-UNDRR-RISK KAN Briefing Note on Systemic Risk" (Paris, FR: International Science Council, 2022).

45 Anthony Oliver-Smith et al., "Forensic Investigations of Disasters (FORIN): a Conceptual Framework and Guide to Research," *Integrated Research on Disaster Risk*, 2016.

46 Arabella Fraser et al., "Developing Frameworks to Understand Disaster Causation: From Forensic Disaster Investigation to Risk Root Cause Analysis," *Journal of Extreme Events* 03, no. 02 (2016): 1650008, https://doi.org/10.1142/s2345737616500081.

47 W. Neil Adger et al., "Advances in Risk Assessment for Climate Change Adaptation Policy," *Philosophical Transactions of the Royal Society A: Mathematical, Physical and Engineering Sciences* 376, no. 2121 (2018): 20180106, https://doi.org/10.1098/rsta.2018.0106.

48 Marshall Shepherd (@DrShepherd2013), "I have long argued climate messaging needs to be about here and now, kitchen table issues…not 8 decades from now or way in the Arctic," Twitter, February 12, 2022, https://mobile.twitter.com/DrShepherd2013/status/1492506606583656450.

49 "Warsaw International Mechanism for Loss and Damage Associated with Climate Change Impacts," United Nations Framework Convention on Climate Change, accessed May 26, 2022, https://unfccc.int/topics/adaptation-and-resilience/workstreams/loss-and-damage/warsaw-international-mechanism.

50 "Sendai Framework for Disaster Risk Reduction 2015–2030," World Health Organization, 2015, https://www.who.int/publications/m/item/sendai-framework-for-disaster-risk-reduction-2015-2030.

51 C P Weaver et al., "Reframing Climate Change Assessments around Risk: Recommendations for the US National Climate Assessment," *Environmental Research Letters* 12, no. 8 (2017): 080201, https://doi.org/10.1088/1748-9326/aa7494.

52 Swenja Surminski et al., "Assessing Climate Risks across Different Business Sectors and Industries: An Investigation of Methodological Challenges at National Scale for the UK," *Philosophical Transactions of the Royal Society A: Mathematical, Physical and Engineering Sciences* 376, no. 2121 (April 30, 2018): 20170307, https://doi.org/10.1098/rsta.2017.0307.

53 Andy J. Challinor et al., "Transmission of Climate Risks across Sectors and Borders," *Philosophical Transactions of the Royal Society A: Mathematical, Physical and Engineering Sciences* 376, no. 2121 (April 30, 2018): 20170301, https://doi.org/10.1098/rsta.2017.0301.

54 Ibid.
55 C P Weaver et al., "Reframing Climate Change Assessments."
56 Thomas Rodham Wells, "Debating Climate Change: The Need for Economic Reasoning," The Philosopher's Beard, September 10, 2012, http://www. philosophersbeard.org/2012/09/debating-climate-change-need-for.html.
57 "The Origins of the IPCC: How the World Woke up to Climate Change," International Science Council, March 10, 2018, https://council.science/current/blog/the-origins-of-the-ipcc-how-the-world-woke-up-to-climate-change/.
58 Bernie Lewin, *Searching for the Catastrophe Signal: The Origins of the Intergovermental Panel on Climate Change* (Melbourne, AU: The Global Warming Policy Foundation, 2017).
59 Adam Lucas, "Risking the Earth Part 1: Reassessing Dangerous Anthropogenic Interference and Climate Risk in IPCC Processes," *Climate Risk Management* 31 (2021): 100257, https://doi.org/10.1016/j.crm.2020.100257.
60 Ibid.
61 Adam Lucas, "Risking the Earth Part 2: Power Politics and Structural Reform of the IPCC and UNFCCC," *Climate Risk Management* 31 (2021): 100260, https://doi.org/10.1016/j.crm.2020.100260.
62 Mike Hulme and Martin Mahony, "Climate Change: What Do We Know about the IPCC?," *Progress in Physical Geography: Earth and Environment* 34, no. 5 (June 18, 2010): 705–718, https://doi.org/10.1177/0309133310373719.
63 Adam Lucas, "Risking the Earth Part 2."
64 Thomas Rodham Wells, "Debating Climate Change."
65 Alex Epstein, *The Moral Case for Fossil Fuels.*
66 Rafaela Hillerbrand and Michael Ghil, "Anthropogenic Climate Change: Scientific Uncertainties and Moral Dilemmas," *Physica D: Nonlinear Phenomena* 237, no. 14–17 (August 15, 2008): 2132–2138, https://doi.org/10.1016/j.physd.2008.02.015.
67 Terje Aven, "Climate Change Risk – What Is It and How Should It Be Expressed?," *Journal of Risk Research* 23, no. 11 (November 9, 2019): 1387–1404, https://doi.org/10.1080/13669877.2019.1687578.
68 John Dewey, *Logic: The Theory of Inquiry* (New York, NY: Holt, Reinhart and Winston, 1938).
69 Terje Aven, "Climate Change Risk."
70 Terje Aven, "Risk Assessment and Risk Management."
71 Hui-Min LI et al., "Understanding Systemic Risk Induced by Climate Change," *Advances in Climate Change Research*, June 2021, https://doi.org/10.1016/j.accre.2021.05.006.
72 Ibid.
73 Bjørn Lomborg, *False Alarm: How Climate Change Panic Costs Us Trillions, Hurts the Poor, and Fails to Fix the Planet* (New York, NY: Basic Books, an imprint of Perseus Books, LLC., a subsidiary of Hachette Book Group, Inc., 2020).
74 Michael Obersteiner et al., "Managing Climate Risk" (Laxenburg, AT: International Institute for Applied Systems Analysis, 2001).
75 Mark Carney, "Breaking the Tragedy of the Horizon – Climate Change and Financial Stability," Bank for International Settlements, September 29, 2015, https://www.bis.org/review/r151009a.pdf.
76 Christa Clapp et al., "Shades of Climate Risk. Categorizing Climate Risk for Investors," CICERO Research Archive (CICERO Center for International

Climate and Environmental Research—Oslo, 2017), https://pub.cicero.oslo.no/cicero-xmlui/handle/11250/2430660.

77 Matthew Scott et al., "The Bank's Response to Climate Change," Bank of England, June 16, 2017, https://www.bankofengland.co.uk/quarterly-bulletin/2017/q2/the-banks-response-to-climate-change.

78 Keith Porter et al., *Overview of the Arkstorm Scenario* (Reston, VA: Dept. of the Interior. U.S. Geological Survey, 2011).

79 "The Great California Shakeout," Great ShakeOut Earthquake Drills, 2008, https://www.shakeout.org/california/scenario/.

80 Keith Porter et al., *Overview of the Arkstorm Scenario.*

81 David King et al., "Climate Change: A Risk Assessment."

82 Thomas Rodham Wells, "Debating Climate Change."

83 Margaret J. Wheatley, *Turning to One Another: Simple Conversations to Restore Hope to the Future* (San Francisco, CA: Berrett-Koehler, 2002).

84 Dickon Pinner et al., "Addressing Climate Change in a Post-Pandemic World," McKinsey & Company (McKinsey & Company, April 7, 2020), https://www.mckinsey.com/business-functions/sustainability/our-insights/addressing-climate-change-in-a-post-pandemic-world.

85 Ibid.

86 Ibid.

87 Geoffrey Saville, "The Wicked Problems of Pandemics and Climate Change," Willis Towers Watson, June 3, 2020, https://www.wtwco.com/en-US/Insights/2020/05/the-wicked-problems-of-pandemics-and-climate-change.

Chapter Eleven

RISK MANAGEMENT

"[O]ne could prepare, one could strive, one could make choices, but ultimately life was an elaborate game of providence and probability."

—Author Daniel Silva[1]

Managing risks involves balancing amongst concerns, safety, profits, and reputation in the context of the assessed risks. In general, risk management evaluates a collection of alternatives, considers pros and cons, and then reaches a decision that best achieves the decision-makers' priorities.[2]

Risk management is closely related to policy analysis and creation. A policy acts as a belief or plan to guide decision-making toward favored outcomes for international organizations, governments, private sector organizations, or individuals. The development and operation of policies are often influenced by risk management practices and decision theory.[3]

Risk assessments often include a combination of strategic risks (long term), tactical risks (medium term), and operational risks (short term). Management priorities and strategies for risks on each of these time scales are typically different, with conflicts arising when near- and long-term strategies have the potential to worsen either the long-term or near-term impacts.

Risks associated with climate change are characterized by large uncertainties and emergence, with complex interplays between emergency and incremental, long-term risks. For such situations, there is a growing focus on dynamic risk assessment and management rather than on conventional risk management and control.[4] Particularly for emergent risks, the risks as well as the risk management responses need to be monitored and adjusted.

11.1 Risk Management Principles

"The unfortunate reality is that efforts to regulate one risk can create other, often more dangerous risks." (Legal scholar Jonathan Adler)[5]

Risk management starts with reviewing the results generated by risk estimation, characterization, and assessment. Acceptable risk requires no further management. With intolerable risk, benefits aside, management involves phasing out or eliminating the risk if at all possible. If that is impossible, then mitigation and increasing resilience come into focus. With tolerable risk, while there are perceived benefits, there remains a call for risk reduction. Public risk management for such risks should focus on designing and implementing actions that result in either acceptable risk or create sustainable long-term tolerance via reduction, mitigation, or increasing resilience. For contested risks, risk management focuses on methods to create consensus, designing actions that increase tolerability among the parties most concerned. If needed, alternative courses of action can be developed for those who view the risk as acceptable or tolerable considering benefits compared to other risks.[6]

A successful risk management initiative should be "**P**roportionate, **A**ligned, **C**omprehensive, **E**mbedded and **D**ynamic (PACED)."[7] *Proportionate* balances the level of effort put into risk management with the level of risk. Risk management activities should be *aligned* with other policies and priorities. Risk management should be *comprehensive*, so that initiatives consider all the aspects of the system at risk. Risk management activities should be *embedded* with other relevant policies. Finally, risk management activities should be *dynamic* and responsive to the changing political, economic, and technological environment.[8]

When the uncertainties are large, adaptive risk management is a strategy whereby different decision alternatives are considered, one is selected and observations are registered, learning is achieved, and adjustments and adaptations are made over time. Effectively, adaptive risk management is about learning from well-designed and analyzed trial and error.[9] The challenge is to provide a risk understanding and characterization that is able to incorporate changes in the knowledge base over time.[10]

In risk management, there is invariably tension caused by different perspectives, traditional risk analysis, resilience, and antifragility. Integrative approaches to risk management have a perspective of governance and combine scientific evidence with economic considerations as well as social concerns and societal values.[11]

11.1.1 Risk Responses

"We are living in a world that is beyond controllability." (Sociologist Ulrich Beck)[12]

There are four major categories of risk response, that are suited to different types of risk situations.[13]

- Avoidance (eliminate)
- Reduction (optimize—mitigate)
- Sharing (transfer—outsource or insure)
- Retention (accept and budget)

Alternative framings of risk response include ACAT—Avoid, Control, Accept, or Transfer.[14] Analogously, opportunities have specific response strategies: exploit, share, enhance, ignore.

Risk avoidance includes not performing an activity that could present risk.[15] COVID-19 lockdowns and elimination of fossil fuel emissions are examples of risk avoidance. Avoidance is an answer only for simple risks, since avoiding risks also means losing out on the potential gain that accepting (retaining) the risk may have allowed. For example, there are serious downsides and costs to COVID-19 lockdowns and rapid elimination of fossil fuels.

Risk reduction or optimization involves lowering the severity of a loss or the chance of its occurrence. Acknowledging that risks can be positive or negative, optimizing risks means finding a balance between negative outcomes and the benefit of the operation or activity; and between risk reduction and the level of effort and cost applied.[16] The ALARP principle is to reduce the exposure to hazards (risks) to **A**s **L**ow **A**s **R**easonably **P**ractical. Continuous improvement is a key element to ALARP.[17] Beyond lowering the risk to what is reasonably possible, exposure is reduced continuously in an iterative, reasonable process.

Risk sharing involves sharing the burden of loss or the benefit of gain and the measures to reduce a risk.[18] Insurance and risk retention pools are examples of risk sharing.

Risk retention involves accepting the loss when a hazard occurs. All risks that are not avoided, reduced, or transferred are retained, by default. Risk retention is a viable strategy for small risks where the cost of insuring against the risk would be greater over time than the total losses sustained. Risk retention may also be acceptable if the chance of a very large loss is small or if the cost to insure is so great that it would impact other objectives of value.[19]

11.1.2 Risk Management Strategies

"Many of today's ecological policy issues are contentious, socially divisive, and full of conundrums." (Political scientist Robert Lackey)[20]

The risk analysis process provides essential input for evaluating risk management strategies. Uncertainty, complexity, ambiguity, and strength of the knowledge base, along with a contingent evaluation of the level of risk (uncertain, complex, and/or ambiguous), provide some guidance for risk management. Risk management strategies are about how to design a process to develop sensible options for the particular risk.

Three major strategies are used for managing or governing risk:[21]

(I) Risk-informed strategies—cost/benefit and expected utility for well-characterized risks

(II) Cautionary/precautionary strategies—precautionary principle; robustness and resilience for risks with uncertainties, complexity, and potential surprises

(III) Discursive strategies—ambiguity of different views related to the relevant values.

Simple risks with relatively low uncertainty and low stakes are a comfortable domain for applied science and engineering, where the application of cost/benefit and expected utility analysis is straightforward. If uncertainty is high and/or the stakes are high, the decision-making environment is much more volatile, and different risk management frameworks can provide conflicting recommendations. However, many risks in categories (II) or (III) are managed as if they are in category (I). The consequences of this mismanagement range from social amplification or irresponsible attenuation of the risk, sustained controversy, deadlocks, legitimacy problems, conflicts, and expensive rebound measures.[22] We see all of these consequences in the context of mismanagement of climate risk by assuming that it is a simple category (I) risk.

When complexity is dominant and uncertainty and ambiguity are low, or for risk problems that are highly ambiguous, discourse-based management is useful. Discursive strategies are based on a participative process that involves stakeholders as well as the affected public. The aim of such a process is to produce a collective understanding on how to interpret the situation and how to design procedures of justifying collectively binding decisions on acceptability and tolerability. The risk managers' task is to create a situation in which those who believe that the risk is worth taking and those who believe that the pending consequences do not justify the potential benefits are willing to construct and create strategies that are acceptable to the various stakes and interests.[23] The disagreements surrounding the tolerability of climate risks versus the proposed solutions imply the need for discursive strategies, although polarization on both sides of the climate risk debate makes finding common ground very challenging.

The main focus of managing hazard risks is on loss prevention, damage limitation, and cost containment. Reduction in overall hazard risk severity can be achieved by reducing both the impact and consequences. For example, the seat belt in a car can reduce the impact of an accident, but has no effect on the likelihood of having an accident.[24]

An important consideration when evaluating the effectiveness of risk management is the level of confidence that should be placed in a particular control. Particularly for long-term emerging risks such as climate change, there is low confidence that controls will be fully implemented or that they will be as effective as expected or required. If effectiveness of a control is uncertain, a greater variability of outcomes may be expected.[25]

It has been assumed that rapid elimination of CO_2 emissions is the necessary control to mitigate climate risk. However, there is no strategy in place to implement such controls globally or to monitor compliance. Apart from taking the edge off of possible tail risks associated with tipping points that may be related to a global temperature increase, there is low confidence that such measures will actually improve the weather/climate or human well-being in the twenty-first century.

In evaluating risk management strategies, assessment is needed of the risk magnitude versus risk capacity, and the feasibility and costs of controlling the risk. A practical difficulty is being forced to retain a risk that is recognized as being beyond the risk appetite, or even the risk capacity, if the feasibility of controlling this risk is low and costs are high. It is exactly this difficulty that is at the heart of the debate over emissions reductions in response to climate change.

11.2 Principles of Precaution

"Precautions are always blamed. When successful they are said to be unnecessary." (Nineteenth-century scholar and theologian Benjamin Jowett)[26]

Caution is a normal human emotional reaction, prioritizing carefulness and avoidance of danger or mistakes. Examples include being careful when walking on a wet floor or crossing a busy highway, installing a smoke detector, purchasing fire insurance. It is considered wise to be cautious, even when it does not seem necessary, to avoid problems and future losses.

Precaution is a subset of caution, associated with preparing for something that may have severe consequences, but for which there is substantial scientific uncertainty about the severity of the causal mechanisms.[27] Adopting an appropriate degree of precaution with respect to feared health and environmental hazards is fundamental to risk management.

For well understood risks, the advantages of cautionary measures are obvious. The real problem is in deciding how precautionary to be for poorly understood risks in the face of scientific uncertainties, opportunity losses, and generation of additional risks from the precautionary measures.

With regard to climate change, owing to large scientific uncertainties, we are in the realm of precaution rather than caution. An essential objective of risk management is to find the right balance between protection and precaution versus development and growth. For systemic risks, unintended consequences of applying the precautionary principle can be profound.

11.2.1 Precautionary Principle

"If every conceivable precaution is taken at first, one is often too discouraged to proceed at all." (Nobel Laureate chemist Archer J. P. Martin)[28]

The precautionary principle has emerged from the acknowledgment that scientific uncertainty can persist when action may be needed. The precautionary principle pertains to uncertain risks for which quantitative risk assessments are meaningless. Although there are suspicions of danger, the probability of occurrence or the effect in terms of damage cannot be estimated, and even the potential danger and the relevant causalities may be unclear.

It is important to distinguish between the cautionary principle and the precautionary principle. The Society for Risk Analysis provides these definitions:[29,30]

- Cautionary principle: "If the consequences of an activity could be serious and subject to uncertainties, then cautionary measures should be taken, or the activity should not be carried out."
- Precautionary principle: "If the consequences of an activity could be serious and subject to *scientific* uncertainties, then precautionary measures should be taken, or the activity should not be carried out."

Both the cautionary and precautionary principles appeal to the common-sense idea that "it is better to be safe than sorry." The key difference between the cautionary and the precautionary principle is that the latter refers to "scientific uncertainties," whereas the former just refers to "uncertainties."[31] As such, the precautionary principle is a special case of the cautionary principle that is associated with a particular class of uncertainties.[32]

Consider the following example that illustrates the distinction between caution and precaution: the German decision to phase out their nuclear power plants by the end of 2022, in response to the 2011 Fukushima nuclear disaster caused by a tsunami.[33] Judgments were made that the risks from nuclear accidents and nuclear waste were unacceptable, independent of the probability of large accidents occurring and also independent of nuclear power's economic benefit to society. This decision gave very strong weight to the cautionary principle. In this example, there aren't scientific uncertainties, and hence the precautionary principle does not apply.

When there are scientific uncertainties associated with environmental issues, the precautionary principle has become a binding element of international laws and agreements, and has been embraced by many countries. The precautionary principle has appeared in international conventions on ozone, global climate, and biodiversity. The precautionary principle was adopted in 2002 by the European Commission, with context provided by the European Environment Agency's 2001 Report *Late Lessons from Early Warnings.*[34]

The precautionary principle is based on the assumption that a false prediction that a human activity *will not* result in significant environmental harm will typically be more harmful to society than a false prediction that it *will* result in significant environmental harm.

The precautionary principle is not a decision rule, but rather a meta-level principle that provides guidance on how we should think in relation to risk and protect something of value. Despite its broad international acceptance, the precautionary principle and its role in risk management is controversial in the risk science community.

The precautionary principle is about avoiding possible harm, rather than respecting a wider set of values—an easier life, greater health, innovation, economic prosperity, or other values.[35] At the heart of the climate policy debate is whether the precautionary "cure" in terms of rapid elimination of CO_2 emissions is worse than the climate "disease," in terms of the costs of eliminating of CO_2 emissions, concerns about energy security, and broader opportunity costs.

Government interventions based on the precautionary principle typically impose costs, burdens, and their own harm. Further, well-intended interventions into complex systems invariably have unintended consequences. A single-minded focus on the precautionary principle associated with a single risk distracts decision makers from actually examining whether the precautionary intervention is likely to make us better off overall. "System neglect" involves a failure to attend to the systemic effects of regulation. The precautionary principle blinds us to many aspects of risk-related situations and focuses only on a narrow segment of what is at stake.

Legal scholar and behavioral economist Cass Sunstein argues that the precautionary principle gives a misleading appearance of being workable only because of cognitive mechanisms that lead people to have a narrow rather than wide view of the risk landscape. This cognitive bias makes it possible to ignore or neglect some of the risks that are actually at stake. Risks of one kind or another are on all sides of regulatory choices, and it is typically impossible to avoid running afoul of contradictory choices in the context of the precautionary principle.[36]

Germany's cautionary ban on nuclear power provides an example of unintended consequences. This ban contributed to the energy impoverishment of their citizens from increased prices that led to an increase in mortality during very cold temperatures.[37] The ban increased fossil fuel emissions when Germany became more reliant on natural gas. In 2022, when Germany's supply of natural gas was threatened by Russia's war on Ukraine, Germany increased its dependence on coal. The increased reliance on fossil fuels resulting from Germany's ban on nuclear power has not only destabilized their energy supply and increased its cost, but has also caused a spike in greenhouse gas emissions and air pollution, and has contributed to geopolitical instability in the context of the Russian war on Ukraine.[38]

The precautionary principle effectively shifts the burden of proof away from the traditions of common law, by requiring innovators and producers to prove their innocence when anyone raises the specter of threats of harm. The precautionary principle makes it very easy to rationalize restrictive measurements in the absence of fully established cause-effect relationships. To avoid absurdity, the idea of "potential risk" is understood to require a certain threshold of scientific plausibility.[39] All legal formulations of the precautionary principle include a knowledge condition—some level of proof needed to trigger application. While this knowledge condition is often vague or ambiguously formulated, scientists and experts are expected to provide proof of plausibility.[40]

Under conditions of deep uncertainty, the precautionary principle is not the only way to frame the decision-making problem—in fact, the precautionary principle actually perverts the risk management process if risk cannot be eliminated.[41] Risk management has many tools that can be applied under such conditions, as described elsewhere in this chapter and in Chapter Twelve.

It has been argued that the precautionary principle should come at the end of the risk management process if our capacity to prevent harm has failed or the value of the benefits could not be justified. In this case, the precautionary principle should be applied with care taken for the consequences of any lost benefits. In effect, the precautionary principle should be regarded as the safety net when human ingenuity fails. However, the European Environment Agency's version of precaution works in reverse, demanding the risk management process begin with the application of the precautionary principle.[42]

The growing influence of the precautionary principle in policy making for health and environmental issues, particularly in Europe, reflects a shift from a risk-taking age to a risk-prevention era, with substantial implications for human development.[43]

11.2.2 *Proportionary and Proactionary Principles*

"[T]he question of whether risk should be avoided or embraced may come to be a defining feature of future ideological struggles." (Philosopher Steve Fuller)[44]

The proportionary and proactionary principles put some brakes on the precautionary principle.

Proportionality is a general principle of European Union law requiring that action taken does not go beyond what is necessary to achieve the objective(s) aimed for. In particular, action must not pose an unnecessary burden on those affected by it or interfere too extremely with fundamental rights. Proportionality provides a framework to guide action when there are competing demands on public policy decisions.[45]

Proportionality restricts authorities in the exercise of their powers by requiring them to strike a balance between the means used and the intended aim. A pre-condition is that the measure is adequate to achieve the envisaged objective. The principle of necessity is a sub-dimension of the broader principle of proportionality, which provides a mechanism for controlling measures that restrict rights and liberties.

Additional limits to the precautionary principle are provided by the innovation principle and the proactionary principle—these are informal principles that are not canonized in law. Most of us value both of these:

1. Protecting our freedom to innovate technologically
2. Protecting ourselves and our environment from excessive collateral damage[46]

The precautionary principle focuses on #2, whereas the innovation and proactionary principles seek a balance between #1 and #2.

By concentrating only on a single set of risks, such as uncertainties associated with the introduction of a new technology, the precautionary principle neglects the negative implications caused by the lack of technological development. Further, efforts to regulate one risk can create other, often more dangerous, risks. If regulations divert resources away from potentially life-saving or safety-enhancing activities or activities that enhance human well-being, the regulations can make people worse off.[47]

Consider the regulation of biotechnology. Calls for regulation of biotechnology are precautionary, since the concerns are driven by scientific uncertainty. Regulation of biotechnology may seem compelling, until one considers the trade-offs. Excessive regulation may sacrifice the benefits of innovation in the interest of safety. For example, excessive precautionary regulation could limit the introduction of high-yield crops, nutritionally enhanced foodstuffs, or new vaccines, leaving the world less safe and prosperous than it would be otherwise. This example describes how regulation of a new technology aligned with the precautionary principle could make us more sorry than safe, by eliminating opportunity benefits.[48]

The innovation principle is normally stated as:

> "Whenever legislation is under consideration its impact on innovation should be assessed and addressed."[49]

The proactionary principle is designed to bridge the gap between no caution and the precautionary principle. The precautionary principle enforces a static world view that attempts to eliminate risk, whereas the proactionary principle promotes a dynamic worldview that promotes human development and the risk-taking that produces the leaps in knowledge that have improved our world. The proactionary principle allows for handling the mixed effects of any innovation through compensation and remediation instead of prohibition.[50] Rather than attempting to avoid risk, the risk is embraced and managed. The proactionary principle valorizes calculated risk-taking as essential to human progress.[51]

11.3 Applications of the Precautionary Principle

> "The torment of precautions often exceeds the dangers to be avoided. It is sometimes better to abandon one's self to destiny." (French military and political leader Napolean Bonaparte)[52]

Consideration of COVID-19 and climate change provides contrasting and insightful examples of the application of the precautionary principle to deeply uncertain, complex, and systemic risks.

11.3.1 COVID-19

> "Instead of adapting the protection to the actual level of risk, [the precautionary principle] tends to adapt the perception of risk to the growing need for protection—making protection itself one of the major risks." (Philosopher Roberto Esposito)[53]

The systemic risks surrounding COVID-19 provides many illuminating perspectives on risk management, caution, and precaution.

In the United States, the premise of individual responsibility dominates policies for many health issues. However, the potential harm to public health from COVID-19 was sufficiently grave that states were obligated to try to guard against them. The immediacy of the threat of grave harm obligated states to act immediately, rather than waiting for the threat to develop further and the availability of additional information.

Since the danger from COVID-19 was real and present, some have argued that the threat could be managed by cautionary and traditional risk management measures, and that there was no need to resort to the precautionary principle and lockdowns.[54] While the COVID-19 hazard was proven in the sense that the disease was contagious and that people have died and would continue to do so, the projected course of the pandemic was fraught with many uncertainties that made the precautionary principle relevant in the early stages of the disease. Rapid advances in knowledge about transmission, the impacts of the disease, and how to treat it have made applying the precautionary principle less relevant over the course of time.[55]

At the level of national and state governments, many different approaches to dealing with COVID-19 were taken. No government could say that it hadn't been warned about a coming pandemic. Some governments including Taiwan and South Korea, who speedily acknowledged the threat and had learned from their experience with SARS in the early 2000s, undertook risk management measures. These measures included rapid and extensive testing, contact tracing, masking, and social distancing. Although individuals were quarantined, there were no lockdowns.

In the United States and UK, there was an astonishing lack of preparedness, with inadequate testing and even essential medical supplies.[56] Then in mid-March with the realization of a serious threat, these countries applied the precautionary principle, mandating extensive measures to avoid overwhelming hospitals. These measures included isolation, quarantine, social distancing requirements, curfews, lockdowns, shutdowns, and the closure of internal and external borders.

During COVID-19's second wave, most European countries opted for less stringent measures, since the fear of an economic disaster imposed many exceptions to the "stay at home" rule. Quarantines and lockdowns were particularly extensive and of long duration in China, Australia, and New Zealand. Prior to COVID-19, the precautionary principle had never before been applied so comprehensively, impacting everyone's private life in almost every way.[57]

The relevant scientific basis for managing the risk from COVID-19 was deeply uncertain in the beginning, but nevertheless, prominent epidemiologists, scientists, and public health experts all spoke with one voice. These experts argued that we must act as if two or three out of a hundred infected people will die; the disease is spread primarily by droplets and on surfaces; there is no immunity after infection; and everyone, no matter what age, is equally at risk of hospitalization and death after infection. The tragedy is that as the worst suppositions about the virus turned out to be wrong, lockdown policies continued to be enforced.[58]

Influential scientists, journalists, and public health officials compounded the problem by militarizing the precautionary principle.[59] Scientists who called for more investigation about epidemiological facts about the virus and economists who raised the possibility of economic collateral harm were viciously attacked. Prominent experts who had reservations about the rush to lockdown or dared to question the assumptions underlying lockdown policies were de-platformed from Facebook and Twitter.

The threat of serious damage coupled with scientific uncertainty early in the pandemic provided a rationale for invoking the precautionary principle. However, scientific uncertainty was quickly reduced in the early months, and it became increasingly acknowledged that extended lockdowns were causing more harm than they were preventing.[60] In many countries and states, the lockdowns developed a life of their own, well after the time when hospital capacity became manageable and uncertainty had been reduced.

Unintended negative consequences of lockdowns that flowed from the precautionary principle left a trail of missed cancer surgeries, lost livelihoods, mental struggles, elevated numbers of suicides in young people, learning losses for school children, and restrictions that dehumanized the elderly in their final months. Businesses failed, there was catastrophic economic damage, market signals were distorted. In 2022 there are still global shortages and major supply chain issues as well as inflation.

International IDEA, a democracy advocacy organization, recently concluded that many countries had become more authoritarian in an effort to contain COVID-19.[61] Measures to enforce lockdowns were particularly severe in some countries, notably China, with extensive surveillance and severe punishments for non-compliance.[62] Isolation is more easily justified for individuals that are actually ill; however mandatory quarantines that restrict physical liberty of healthy people are more controversial. The imposed restrictions were considered legitimate as they sought to safeguard the public welfare.[63] However, there have been massive demonstrations against COVID-19 mandates, particularly in Canada and the United States.[64]

An analysis of over a hundred studies on the costs and benefits of lockdowns concluded that they have had, at best, a marginal effect on the number of COVID-19 deaths. The ineffectiveness of lockdowns could not have been predicted at the time the lockdowns were implemented, simply because lockdowns on such a large scale had not been previously attempted.[65]

If policymakers had considered all aspects associated with lockdowns, they might have concluded that the precautionary principle is not particularly useful to help decide on the wisdom of lockdowns, particularly for extended periods. Both sides of lockdown policies had the potential for catastrophic harm, with no viable way to compare the risks and consequences under the precautionary principle. Under such conditions, policymakers should have looked to other risk management methods that have helped the world cope with previous epidemics.[66]

While public health is a major value to protect and a pandemic gives us only bad choices, we should make every effort to determine and deploy the least bad one—what David Katz, a doctor specializing in preventive medicine, calls "total harm minimization."[67] Proportionality and necessity can help minimize adverse impacts of precautionary measures, including minimizing the costs to individual rights and liberties.

A key element of risk management and precaution is to learn from our mistakes. We need to evaluate whether our response matched or exceeded the COVID-19 threat, at every stage of the pandemic. We need to evaluate the costs and benefits of the response, at every stage of the pandemic. We need to plan for the next pandemic and develop risk management strategies that are more effective than a panicked application of the precautionary principle.

11.3.2 Climate Change

"[P]rudence consists in knowing how to distinguish the character of troubles, and for choice to take the lesser evil." (Renaissance Italian diplomat and philosopher Niccolò Machiavelli)[68]

Like COVID-19, human-caused climate change is a deeply uncertain global risk. The threat from COVID-19 was immediate, with a major loss of life being the unambiguous near-term outcome. By contrast, the dangers are less clear cut for human-caused climate change. There are long time lags between forcing and response so that climate threats are in the future, with large geographic heterogeneity in the benefits versus dangers of warming. Further, any threats from climate change are convoluted with natural climate variability and energy systems that have enabled rapid human development and well-being over the past century.

The precautionary principle is invoked by the United Nations Framework Convention on Climate Change (UNFCCC) in two places. Article 3(3) states that:[69]

"Parties should take precautionary measures to anticipate, prevent or minimize the causes of climate change and mitigate its adverse effects."

"Where there are threats of serious or irreversible damage, lack of full scientific certainty should not be used as a reason for postponing such measures, taking into account that policies and measures to deal with climate change should be cost-effective so as to ensure global benefits at the lowest possible cost."

As set forth in Article 2, the Convention's ultimate purpose is to stabilize greenhouse gas concentrations at a level that would "prevent dangerous anthropogenic interference with the climate system." The appropriate level should be "achieved within a time-frame sufficient to allow ecosystems to adapt naturally to climate change, to ensure that food production is not threatened and to enable economic development to proceed in a sustainable manner."[70]

The phrase "policies and measures to deal with climate change should be cost-effective so as to ensure global benefits at the lowest possible cost" and "to enable economic development to proceed in a sustainable manner" places economics in a central role in application of the precautionary principle to human-caused climate change.

While uncertainty does not preclude action, the degree of uncertainty might be relevant in determining what measures should be considered cost-effective. A weather forecast with a high chance of rain and possible thunderstorms does not justify taking refuge in the basement but it does justify carrying an umbrella.

Cost-benefit analysis has been used by the IPCC and national governments to justify the costs of emissions reductions. Three elements are critical to proper cost-benefit analysis of any climate policies: (i) how will economic impacts be forecasted, (ii) how to effectively balance tradeoffs between far future harm and present expenses associated with mitigation (the social discount rate), and (iii) how to account for potential catastrophic outcomes.[71] Even if science could provide a perfect forecast of global temperatures, each element of the economic impact analysis for climate change is the subject of radical uncertainties and serious debate.[72]

Despite pervasive uncertainties, in 2009 the US Interagency Working Group on Social Cost of Carbon was convened to provide a uniform estimate of the social cost of carbon using Integrated Assessment Models (IAMs). This exercise engendered substantial criticism from legal scholars and economists. Criticisms included weakly defended assumptions about future economic growth and technological change, arbitrarily damage functions that were essentially guesses, and a very large range of plausible parameter values that can be used to obtain almost any result.[73,74]

It is difficult to avoid the conclusion that the IAMs are simply too weakly justified and too varied to provide confidence in their estimates. However, the economic models do confirm the crucial role of three factors in determining the social cost of carbon: the magnitude of climate sensitivity (Section 7.2), the size of the social discount rate, and the handling of possible catastrophic outcomes. Economist Robert Pindyck concludes that possible catastrophic outcomes are the dominant factor in determining the social cost of carbon.[75]

Social discount rates seek to put a present value on future costs and benefits. With respect to climate change, the discount rate determines how much today's society should invest in trying to limit the impacts of climate change in the future.[76]

A high discount rate implies that people put less weight on the future and therefore that less investment is needed now to guard against future costs.[77] The use of a low discount rate supports the view that we should act now to protect future generations from climate change impacts, given more importance to future generations' well-being in cost-benefit analyses. The discount rate is not just an issue for economists—it is an ethical issue for which there is no objective way to determine a value. There is no right answer, only a judgment call. However, a key argument for a low discount rate relates to the possibility of catastrophic climate change, which in any event would override the effect of discounting, however low the probability of such an event.

This leads us to the second factor that dominates the social cost of carbon— the likelihood of catastrophic outcomes. The issues of worst-case weather and climate events and sea level rise were addressed in Chapter Nine. The prospect of tipping points in the Earth's climate—whether caused by natural and/or human-caused climate change—were discussed in Section 1.3.2.

In response to the challenges presented to the precautionary principle by low likelihood catastrophic outcomes, Cass Sunstein has articulated several different versions of a catastrophic risk precautionary principle.[78] The least stringent version requires only that regulators take into account even highly unlikely catastrophes. Another version suggests that society should be willing to pay insurance against possible catastrophe. The most extreme version selects the worst-case scenario and attempts to eliminate it.[79] The catastrophic risk precautionary principle doesn't really help with assessing whether mitigatory measures are cost-effective, since values beyond economic ones come into play for genuinely catastrophic outcomes.

In summary, beyond identifying climate change as a serious enough problem to justify some response, the precautionary principle and cost/benefit analysis provides essentially no guidance on the appropriate magnitude of the response or the best mitigatory actions.

While the concept of precaution is important in managing risks from climate change, the precautionary principle in itself is of little help. Further, applications of the precautionary principle to manmade climate change, in context of setting strict targets and timelines for emissions reductions, are themselves causing harm. The rapid transition away from fossil fuels has substantial costs, is reducing energy availability and security in some countries, and is changing geopolitical dynamics in negative ways as related to supplies of natural resources (see Chapter Fourteen).

In January 2022, the World Economic Forum's Global Risks Report identified the three most severe risks on a global scale over the next 10 years as biodiversity loss, extreme weather and, in top place, "climate action failure."[80] How has "climate action failure"—failure of a specific risk management strategy targeting the elimination of fossil fuels—become an independent risk, apparently exceeding the risk from climate change itself? By attempting to deal with the risks of human-caused climate change in the context of the precautionary principle, the unknown and uncertain dangers from climate change (both natural and human-caused) have been displaced by the precautionary urgency of implementing a specific policy solution—rapid reductions of emissions from fossil fuels.

11.4 Resilience and Robustness

"[N]o precautions, and no precautionary principle, can avoid problems that we do not yet foresee. Hence, we need a stance of problem-fixing, not just problem-avoidance." (Physicist David Deutsch)[81]

Risk assessment has limitations in being able to capture all aspects of risk and uncertainties, particularly under conditions of high levels of randomness and complexity (including potential surprises), and therefore needs to be supplemented with measures of robustness and resilience. Resilience and robustness fall under the category of cautionary and precautionary measures of risk management.

Resilience is a generic term that has been most used in the safety domain, whereas robustness is most commonly referred to in business and operational research contexts.[82] There is a growing interest in applying both concepts to complex human and environmental systems, including sustainability and climate change. Robustness and resilience-based measures are seen as fundamental building blocks for risk management and risk governance.

With regard to hazard risks, resilience, and robustness strategies are targeted at reducing impacts of a hazard event, in stark contrast to the precautionary principle that focuses on eliminating or at least reducing the magnitude of the hazard itself.

11.4.1 Resilience

"Man plans and God laughs." (Old Yiddish adage)[83]

Ecological resilience can be defined as "the capacity of a system to absorb disturbance and reorganize while undergoing change so as to still retain essentially the same function, structure, identity, and feedbacks."[84]

Analogously in the context of engineered systems and organizations, resilience has been defined as the ability to maintain functionality and recover, given that one or more adverse events or disruptions occur. The resilience concept is closely related to vulnerability. The ability—and lack of ability—of affected systems to maintain performance and recover determine the frequency and magnitude of adverse impacts.

There has been a growing interest in the topic of resilience in the context of government and local or municipal authorities. This interest initially arose in relation to civil emergencies and natural hazards such as earthquakes and extreme weather events. Application of resilience concepts to complex human and natural systems, including sustainability and climate change issues, is at the forefront of risk science research.

Resilience within a system or organization is linked to capability of sustaining or restoring core functionality in response to a stressor.[85] "Sustain" and "restore" present two different connotations to resilience. The first is an operational connotation that relates to the operational ability to respond and restore. The second connotation is to sustain functionality, which relates to the infrastructure of the system itself.

Resilience can be further divided between that which is general or specific. A system's general resilience allows for it to effectively respond to any type of shock or stress including those which are unexpected or unknown. Whereas specific resilience reflects the system's ability to deal with a distinct stress or shock while maintaining a specific function. Building general resilience, with its focus on preparing the system for unknown disturbances, is particularly beneficial in situations of large randomness or high complexity within the system.[86]

Resilience engineering scholars refer to the inherent complexities of modern socio-technical systems that make these systems inherently risky. The goal of resilience fundamentally becomes the capability to adapt to emerging risks in order to ensure the success of an inherently risky system.[87]

A resilient system or organization is characterized by:[88]

- awareness of changes in the external, internal, and risk management environments, so that constant attention to resilience is ensured;

- 'prevent, protect, and prepare' in relation to all types of resources; and
- 'respond, recover, and review' in relation to disruptive events, including the ability to respond rapidly, review lessons learnt, and adapt.

Concepts of resilience underlie adaptation to climate change (Chapter Thirteen).

11.4.2 Robustness

"We live in a world with fundamental disagreements related to policy—not just choices, but about ends, means, and problem framings." (Policy and risk analyst Robert Lempert)[89]

Robustness is the quality of being strong and unlikely to fail—the antonym of vulnerability. In context of a system, robustness refers to the ability to tolerate perturbations that might affect the system's function. Robustness can be defined as "the ability of a system to resist change without adapting its initial stable configuration."[90]

It is reasonable to consider a system, policy, action, or decision highly robust if most or all of the following attributes are present, and inadequately robust if weak across all the attributes:[91]

- *Resilience* of a system is the attribute of rapid recovery of critical functions. A system or policy is robust against uncertainty if it can rapidly recover from adverse surprise and achieve critical outcomes.
- *Redundancy* is the attribute of providing multiple alternative solutions. Robustness to surprise can be achieved by having alternative responses available.
- *Flexibility* is the ability for rapid modification of tools and methods. Adaptiveness is the ability to adjust goals and methods in the mid- to long-term. A system is robust if it can be adjusted as information and understanding change.
- *Comprehensiveness* of a system or policy is its interdisciplinary system-wide coherence to address the multi-faceted nature of the problem. A system is robust if it integrates relevant considerations from technology, organizational structure and capabilities, cultural attitudes and beliefs, historical context, economic mechanisms, and forces, or other factors.

Stabilization of greenhouse gas concentrations at a target level is a non-robust strategy in an environment that is extremely uncertain and nonlinear.[92] The stabilization concept implicitly assumes that there is high confidence in setting the target level, and that there is no or little risk associated with a target level. Further, it assumes that the risks of having a safety margin in the target level is lower than the risks and costs of complying with the target.

The concept of robustness with regard to climate change argues for building a broad technological portfolio of mitigation and adaptation measures that might be useful/effective, given the uncertainties of the physical and socioeconomic systems.[93]

11.5 Managing Systemic Risk

"Some life dilemmas cannot be solved by study or rational thought. We just live with them, struggle with them, and become one with them [...] To live, we must die every instant. We must perish again and again in the storms that make life possible." (Vietnamese spiritual leader Thich Nhat Hanh)[94]

Insights into managing systemic risk are provided by a series of reports issued by the UN Office for Disaster Risk Reduction.[95]

Managing systemic risks and wicked problems is complicated not only owing to high levels of complexity, but also by the lack of a determinable final point of stability. Given the extremely intricate dependencies among the many moving components within a complex dynamic system, resolving one piece of a problem may actually create more problems elsewhere in the system.[96]

There are certain actions policymakers and analysts can take to better understand and devise solutions to managing systemic risk.[97] Systemic risk cannot be eliminated, but it can be reduced and addressed more effectively. The challenge is to use the connections and dynamism in complex systems to reduce risk and increase resilience.

There are three emerging strategies for understanding and managing systemic risk:[98]

(a) Apply systems-based approaches to address the dynamic drivers of risk
(b) Integrate diverse knowledge sources and mobilize the collective intelligence
(c) Acknowledge deep uncertainty (Chapter Twelve) and employ adaptive planning, evolutionary development, and continual improvement to encourage flexible responses.

Addressing systemic risk requires developing enhanced approaches to enabling systems thinking and systems approaches. The challenge is to identify and connect the relevant information, which requires facilitating collective intelligence.[99] Systems thinking involves the following:[100]

- Making meaningful connections within and between systems
- Seeking to understand the big picture
- Changing perspectives to increase understanding
- Evaluating how our mental models shape our views and actions
- Observing how elements within systems change over time
- Challenging our beliefs and assumptions
- Recognizing that a system's structure generates its behavior
- Identifying the circular nature and feedbacks in complex cause and effect relationships
- Understanding the impact of time delays when exploring cause and effect relationships
- Considering short-term, long-term, and unintended consequences of actions
- Considering the issue fully and resisting the impulse to come to a quick conclusion
- Using understanding of system structure to identify possible leverage actions
- Checking results and changing actions if needed

Notes

1 Daniel Silva, *The Cellist: A Novel* (New York, NY: Harper, 2021).
2 Terje Aven, "Risk Assessment and Risk Management: Review of Recent Advances on Their Foundation," *European Journal of Operational Research* 253, no. 1 (August 16, 2016): 1–13, https://doi.org/10.1016/j.ejor.2015.12.023.
3 Ibid.
4 Ibid.
5 Jonathan H Adler, "More Sorry than Safe: Assessing the Precautionary Principle and the Proposed International Biosafety Protocol," *Texas International Law Journal* 35 (2000): 173–205.
6 Ortwin Renn et al., "Coping with Complexity, Uncertainty and Ambiguity in Risk Governance: A Synthesis," *AMBIO* 40, no. 2 (February 3, 2011): 231–246, https://doi.org/10.1007/s13280-010-0134-0.
7 Paul Hopkin, *Fundamentals of Risk Management: Understanding, Evaluating and Implementing Effective Risk Management* (New York, NY: Kogan Page, 2017).
8 Ibid.
9 Louis Anthony Cox, "Confronting Deep Uncertainties in Risk Analysis," *Risk Analysis* 32, no. 10 (April 10, 2012): 1607–1629, https://doi.org/10.1111/j.1539-6924.2012.01792.x.
10 Terje Aven, "Risk Assessment and Risk Management."
11 Ibid.
12 Stuart Jeffries, "Interview: Stuart Jeffries Meets Sociologist Ulrich Beck," The Guardian (Guardian News and Media, February 11, 2006), https://www.theguardian.com/books/2006/feb/11/society.politics.

13 Mark S Dorman, *Introduction to Risk Management and Insurance* (Upper Saddle River, NJ: Pearson-Prentice Hall, 2008).

14 Diomidis H. Stamatis, *Risk Management Using Failure Mode and Effect Analysis (FMEA)* (Milwaukee, WI: ASQ Quality Press, 2019).

15 Ibid.

16 Ibid.

17 *Reducing Risks, Protecting People: HSE's Decision Making Process* (London, UK: Great Britain, Health and Safety Executive, 2001).

18 *ISO/IEC GUIDE 73:2002 Risk Management—Vocabulary* (Geneva, CH: International Organization for Standardization, 2002).

19 Diomidis H. Stamatis, *Risk Management Using Failure Mode and Effect Analysis.*

20 Robert T Lackey, "Ecological Policy," 2019, https://osu-wams-blogs-uploads. s3.amazonaws.com/blogs.dir/2961/files/2019/07/Ecological-Policy-Book-2019.pdf.

21 Terje Aven, "Risk Assessment and Risk Management."

22 Ortwin Renn et al., "Coping with Complexity, Uncertainty and Ambiguity."

23 Andreas Klinke and Ortwin Renn, "A New Approach to Risk Evaluation and Management: Risk–Based, Precaution-Based, and Discourse-Based Strategies," *Risk Analysis* 22, no. 6 (December 11, 2002): 1071–1094, https://doi. org/10.1111/1539-6924.00274.

24 Ibid.

25 Ibid.

26 Robert S. Rait, ed., *Memorials of Albert Venn Dicey; Being Chiefly Letters and Diaries* (London, UK: Macmillan, 1925).

27 Terje Aven, "The Cautionary Principle in Risk Management: Foundation and Practical Use," *Reliability Engineering & System Safety* 191 (November 1, 2019): 106585, https://doi.org/10.1016/j.ress.2019.106585.

28 Archer J. P. Martin, "The Development of Partition Chromatography," The Nobel Prize, December 12, 1952, https://www.nobelprize.org/uploads/2018/06/martin-lecture.pdf.

29 Terje Aven, "The Cautionary Principle in Risk Management."

30 Terje Aven et al., "Society for Risk Analysis Glossary," Society for Risk Analysis, 2018, https://www.sra.org/wp-content/uploads/2020/04/SRA-Glossary-FINAL.pdf.

31 Terje Aven, "The Cautionary Principle in Risk Management."

32 Ibid.

33 Lars Kramm, "The German Nuclear Phase-Out After Fukushima: A Peculiar Path or an Example for Others?," *Renewable Energy Law and Policy Review* 3, no. 4 (2012): 251–262.

34 Poul Harremoës et al., *Late Lessons from Early Warnings: The Precautionary Principle 1896–2000* (Copenhagen, DK: European Environment Agency, 2001).

35 Cass R. Sunstein, "The Catastrophic Harm Precautionary Principle," *SSRN Electronic Journal*, 2007, https://doi.org/10.2139/ssrn.2532598.

36 Ibid.

37 Matthew Neidell et al., "Be Cautious with the Precautionary Principle: Evidence from Fukushima Daiichi Nuclear Accident," *National Bureau of Economic Research Working Papers*, October 2019, https://doi.org/10.3386/w26395.

38 Ibid.

39 Cass R. Sunstein, "Precautions & Nature," *Daedalus* 137, no. 2 (2008): 49–58.

40 Max More, "Perils, Part 5: Failures of the Precautionary Principle," Max More's Strategic Philosophy, August 23, 2010, http://strategicphilosophy.blogspot.com/2010/08/perils-part-5-failures-of-precautionary.html.

41 RiskMonger, "The Top Ten Keystone Corona Moments of 2020: Part 5/10 – the Perversion of Precaution," The Risk-Monger, December 30, 2020, https://risk-monger.com/2020/12/30/the-top-ten-keystonecorona-moments-of-2020-part-5-10-the-perversion-of-precaution/.

42 RiskMonger, "The Post-Covid-19 Blueprint (Conclusion): A Risk Management Process Fit for the 21st Century," The Risk-Monger, July 17, 2020, https://risk-monger.com/2020/07/17/a-risk-management-process-fit-for-the-21st-century/.

43 Ulrich Beck, *Risk Society: Towards a New Modernity* (London, UK: Sage Publ., 1992).

44 Steve Fuller, "The Proactionary Principle," The Breakthrough Institute, August 8, 2013, https://thebreakthrough.org/articles/the-proactionary-principle.

45 Ibid.

46 Max More, "The Perils of Precaution," Max More's Strategic Philosophy, August 22, 2010, http://strategicphilosophy.blogspot.com/2010/08/perils-of-precaution.html.

47 Jonathan H Adler, "More Sorry than Safe."

48 Ibid.

49 Paul Leonard, "The Innovation Principle," Encompass, 2016, https://encompass-europe.com/comment/the-innovation-principle.

50 Max More, "The Proactionary Principle," Wayback Machine, July 29, 2005, https://web.archive.org/web/20120202021736/https://www.maxmore.com/proactionary.htm.

51 Steve Fuller, "The Proactionary Principle."

52 Napoleon Bonaparte, *Napoleon in His Own Words from the French of Jules Bertaut* (Chicago, IL: A.C. McClurg & Co, 1916).

53 Roberto Esposito, *Immunitas: The Protection and Negations of Life*, trans. Zakiya Hanafi (Cambridge, UK: Polity Press, 2011).

54 Klaus Meßerschmidt, "Covid-19 Legislation in the Light of the Precautionary Principle," *The Theory and Practice of Legislation* 8, no. 3 (June 19, 2020): 267–292, https://doi.org/10.1080/20508840.2020.1783627.

55 Jonathan Birch, "Applying the Precautionary Principle to Pandemics," PhilArchive, August 19, 2021, https://philarchive.org/rec/BIRATP.

56 David E. Sanger et al., "Before Virus Outbreak, a Cascade of Warnings Went Unheeded," The New York Times (The New York Times, March 19, 2020), https://www.nytimes.com/2020/03/19/us/politics/trump-coronavirus-outbreak.html.

57 Harry Eyres, "How Coronavirus Has Led to the Return of the Precautionary Principle," New Statesman, April 7, 2020, https://www.newstatesman.com/politics/2020/04/how-coronavirus-has-led-return-precautionary-principle.

58 Jay Bhattacharya, "Editor's Note—on the Catastrophic Misapplication of the Precautionary Principle," Collateral Global, June 14, 2021, https://collateralglobal.org/article/misapplication-of-the-precautionary-principle/.

59 Martin Kulldorff and Jay Bhattacharya, "One of the Lockdowns' Greatest Casualties Could Be Science," The Federalist, March 18, 2021, https://thefederalist.com/2021/03/18/one-of-the-lockdowns-greatest-casualties-could-be-science/.

60 Vera Lúcia Raposo, "Quarantines: Between Precaution and Necessity. A Look at Covid-19," *Public Health Ethics*, January 25, 2021, https://doi.org/10.1093/phe/phaa037.

61 Gabrielle Bauer, "Danger: Caution Ahead," The New Atlantis, February 4, 2022, https://www.thenewatlantis.com/publications/danger-caution-ahead.

62 Pratik Jakhar, "Coronavirus: China's Tech Fights Back," BBC News (BBC, March 3, 2020), https://www.bbc.com/news/technology-51717164.

63 Vera Lúcia Raposo, "Quarantines: Between Precaution and Necessity."

64 Paula Newton, "How the Canadian Covid-19 Protests Could Play Out," CNN, February 11, 2022, https://www.cnn.com/2022/02/11/americas/canada-covid-protests-outcome-explainer/index.html.

65 Douglas W. Allen, "Covid-19 Lockdown Cost/Benefits: A Critical Assessment of the Literature," *International Journal of the Economics of Business* 29, no. 1 (September 29, 2021): 1–32, https://doi.org/10.1080/13571516.2021.1976051.

66 Jay Bhattacharya, "Editor's Note."

67 David Katz, "A National Response to the COVID-19 Pandemic for #TotalHarmMinimization," True Health Initiative, April 8, 2020, https://www.truehealthinitiative.org/wp-content/uploads/2020/04/TotalHarmMinimization-2020-04-08.pdf.

68 Niccolò Machiavelli, *The Prince*, trans. William K. Marriott (New York, NY: E.P. Dutton & Co., 1958).

69 United Nations, "United Nations Framework Convention on Climate Change," (New York: United Nations, General Assembly, 1992), https://unfccc.int/resource/docs/convkp/conveng.pdf.

70 Ibid.

71 Daniel A. Farber, "Coping with Uncertainty: Cost-Benefit Analysis, the Precautionary Principle, and Climate Change," *Washington Law Review* 90, no. 4 (2015), https://doi.org/dx.doi.org/10.2139/ssrn.2637105.

72 Ibid.

73 Eric Posner and Johnathan Masur, "'Climate Regulation and the Limits of Cost-Benefit Analysis," *University of Chicago Public Law & Legal Theory Working Paper*, no. 315 (2010).

74 Robert S Pindyck, "Climate Change Policy: What Do the Models Tell Us?," *Journal of Economic Literature* 51, no. 3 (September 2013): 860–872, https://doi.org/10.1257/jel.51.3.860.

75 Ibid.

76 "What Are Social Discount Rates?," Grantham Research Institute on climate change and the environment (The London School of Economics and Political Science , May 1, 2018), https://www.lse.ac.uk/granthaminstitute/explainers/what-are-social-discount-rates/.

77 Laurie T. Johnson and Chris Hope, "The Social Cost of Carbon in U.S. Regulatory Impact Analyses: An Introduction and Critique," *Journal of Environmental Studies and Sciences* 2, no. 3 (September 12, 2012): 205–221, https://doi.org/10.1007/s13412-012-0087-7.

78 Cass R. Sunstein, "The Catastrophic Harm Precautionary Principle."

79 Ibid.

80 "Global Risks Report 2022," World Economic Forum, January 11, 2022, https://www.weforum.org/reports/global-risks-report-2022.

81 David Deutsch, "TEDGlobal 2005 Transcript," David Deutsch, accessed June 6, 2022, https://www.daviddeutsch.org.uk/videos/tedglobal-2005-transcript/.
82 Terje Aven, "Risk Assessment and Risk Management."
83 Martin I. Boyer, "Der Mensch Tracht, UN Gott Lacht (Man Plans, and God Laughs)," *The Journal of Hand Surgery* 46, no. 2 (2021): 87–91, https://doi.org/10.1016/j.jhsa.2020.12.001.
84 Brian Walker et al., "Resilience, Adaptability and Transformability in Social-Ecological Systems," *Ecology and Society* 9, no. 2 (2004), https://doi.org/10.5751/es-00650-090205.
85 Terje Aven, "Risk Assessment and Risk Management."
86 Emmy Wassénius and Beatrice I. Crona, "Adapting Risk Assessments for a Complex Future," *One Earth* 5, no. 1 (January 21, 2022): 35–43, https://doi.org/10.1016/j.oneear.2021.12.004.
87 Johan Bergström et al., "On the Rationale of Resilience in the Domain of Safety: A Literature Review," *Reliability Engineering & System Safety* 141 (September 2015): 131–141, https://doi.org/10.1016/j.ress.2015.03.008.
88 Erik Hollnagel et al., *Resilience Engineering Concepts and Precepts* (Aldershot, UK: Ashgate, 2006).
89 Robert Lempert (@RobertLempert), "We live in a world with fundamental disagreements related to policy – not just choices, but about ends, means, and problem framings," Twitter, November 10, 2020, https://twitter.com/robertlempert/status/1326266225991667712.
90 Andreas Wieland and Carl Marcus Wallenburg, "Dealing with Supply Chain Risks," *International Journal of Physical Distribution & Logistics Management* 42, no. 10 (February 16, 2012): 887–905, https://doi.org/10.1108/09600031211281411.
91 Yakov Ben-Haim, "Info-Gap Decision Theory (IG)," *Decision Making under Deep Uncertainty*, 2019, 93–115, https://doi.org/10.1007/978-3-030-05252-2_5.
92 M. Obersteiner et al., "Managing Climate Risk," *Science* 294, no. 5543 (October 26, 2001): 786–787, https://doi.org/10.1126/science.294.5543.786b.
93 Terje Aven, "Climate Change Risk – What Is It and How Should It Be Expressed?," *Journal of Risk Research* 23, no. 11 (October 9, 2019): 1387–1404, https://doi.org/10.1080/13669877.2019.1687578.
94 Hanh Nhât and Mobi Warren, *Fragrant Palm Leaves: Journals 1962–1966* (Berkeley, CA: Parallax Press, 2020).
95 "United Nations Office for Disaster Risk Reduction (UNDRR)," United Nations Office for Disaster Risk Reduction (UNDRR), accessed June 6, 2022, https://www.undrr.org/.
96 Torben Juul Andersen and Luca Gatti, "Generating Solutions to Systemic Risks through on-Going Experimentation on Invested Space-Forms," in *GAR2022 Contributing Paper* (Geneva, CH: United Nations Office for Disaster Risk Reduction, 2022).
97 "United Nations Office for Disaster Risk Reduction (UNDRR)."
98 Ibid.
99 Scott E. Page, *The Model Thinker: What You Need to Know to Make Data Work for You* (New York, NY: Basic Books, 2018).
100 "Habits of a Systems Thinker Courses," Thinking tools studio, accessed June 6, 2022, https://thinkingtoolsstudio.waterscenterst.org/courses/habits.

Chapter Twelve

DECISION-MAKING UNDER DEEP UNCERTAINTY

"If, to take a somewhat jaundiced view of the situation, we view ourselves as looking for a lost object (the best climate policy) in a dark room, at least we now know the size and shape of the room. With luck, we might even find a flashlight."

— Legal scholar Daniel Farber[1]

Decision-making chooses among alternatives in order to change outcomes in a desired way.[2] Alternatives are problem-dependent and can include policies, strategies, plans, designs, or other actions.[3] Decision analysis supports risk management and policy making. Decision analysis frameworks and tools can provide systematic approaches for organizing the pros and cons of decision alternatives.

Classical decision analysis identifies an optimal choice among actions, based upon the probability of occurrence of possible outcomes and the decision maker's utility functions that measures preferences. Influence diagrams and decision trees are commonly used graphical representations of decision-analysis problems. Quantitative approaches such as Bayesian approaches and expected utility-based perspectives complement more qualitative decision tools. Uncertainties in the input parameters are propagated through a model to generate the expected utility of the different options. Decision rules are then applied (e.g., the maximum expected utility).[4] Classical decision analysis is useful for tame or simple problems.

However, nearly all decisions with high consequences are not simple—rather, they are characterized by complexity, uncertainty, and ambiguity. Examples include climate change, global pandemics, mega-scale infrastructure projects, selection of energy sources to rely on in the future, development of cities, and management of rare events such as a natural disaster, a financial crisis, or a terrorist attack.[5] For a complex decision with the potential for serious consequences, the classical analytical decision tools have severe limitations. But non-simple risks are commonly treated, assessed, and managed as if they were simple.[6] This maltreatment can result in social amplification of the risk, prolonged controversy, conflicts, stalemates, questions of legitimacy, expensive recovery procedures, and commitments that are difficult to adjust.[7]

The examples cited above are associated with "deep uncertainty." There can be many sources of deep uncertainty in a complex system. Fundamental indeterminacy or unpredictable system outcomes are common conditions of emergent, systemic risks (Section 10.1.2). Decision-making in the face of systemic complexity often faces disagreement about priorities, goals, and values with regards to the desirability of different outcomes. Further, there are irreducible uncertainties regarding the consequences of our actions in the presence of systemic risk. For long-term decision-making, deep uncertainties are in most cases a given.[8]

Is it possible to make good decisions in the face of deep uncertainty? In recent years, there has been a proliferation of ways to include systematic consideration of deep uncertainty in practical decision-making—referred to as Decision-Making Under Deep Uncertainty (DMDU). The main objective of DMDU approaches is to facilitate the development of policies that are robust and/ or adaptive, meaning that they perform satisfactorily under a wide variety of possible futures and can be adapted over time to unforeseen future conditions.

12.1 Classical Decision Analysis

"I take it we are all in complete agreement on the decision here? Then, I propose we postpone further discussion of this matter until our next meeting to give ourselves time to develop disagreement and perhaps gain some understanding of what this decision is about." (American executive Alfred P. Sloan)[9]

Consider the following example. Following the catastrophic impacts of Hurricane Katrina (2005) on New Orleans, economist Stephane Hallegatte conducted a comprehensive cost-benefit analysis on the New Orleans Flood protection system.[10]

Hurricane Katrina caused over 1,800 fatalities and $125 billion in damage in the city of New Orleans and the surrounding areas.[11] At landfall, Katrina was a Category 3 hurricane with a very large horizontal extent, resulting in a 27-foot storm surge.[12] Extensive flooding resulted in the inundation of 80 percent of New Orleans for weeks, destroying most of the city's transportation and communication facilities.[13]

In response to concerns about even stronger hurricanes in the future,[14] the cost/ benefit analysis (CBA) conducted by Hallegatte focused on the cost/benefits of a flood protection system for New Orleans to cope with a Category 5 hurricane.[15] CBA, as traditionally practiced, is an agree-on-assumptions process. Hallegatte evaluated the costs and benefits for a flood protection system able to cope with Category 5 hurricanes, compared to a system able to cope only with Category 3 hurricanes. The additional cost of Category 5 protection would be justified if the additional cost of the upgraded protection is lower than the expected benefits from reduced flood damages.[16]

The cost-benefit analysis conducted by Hallegatte depends on assumptions about construction costs, the probability of losses that would occur in the event of a Category 5 hurricane, and social discount rates. The construction costs are relatively simple to evaluate—engineers and state officials evaluated the cost of Category 5 protection at about US$32 billion, compared with a $1 billion cost to restore the original Category 3 protection.[17]

The benefits depend on the lifetime of the protection, the annual probability that a Category 5 hurricane hits New Orleans, the social discount rate, and an estimate of the annual direct cost of the New Orleans flooding. Based on probabilities derived from historical experience and direct cost estimates produced by insurance companies support assumptions of: a 1/500 annual chance of a Category 5 landfall, direct costs of flooding of $20 billion in damages, and $5 billion in human losses. Under these assumptions, the expected present benefit of the Category 5 flood protection system was calculated at $1.5 billion with a 7 percent discount rate and $6 billion with a 3 percent discount rate—substantially less than the $32 billion cost of the Category 5 flood protection system.[18]

However, there is no political or scientific consensus regarding the appropriate input assumptions from which to calculate the benefits. Almost all parameters utilized in a CBA for long-term investments are highly uncertain. The probability of occurrence of a Category 5 landfall may be larger than 1/500 in the presence of global warming. Further, flood damages from storm surge will also increase in the presence of global warming. A 1/100 chance of a Category 5 landfall increases the benefit to $23 billion with a 3 percent discount rate. Consideration of loss amplification effects produces damage estimates of $60 billion—well exceeding the costs of the Category 5 flood protection system.[19]

The end result was that by utilizing historically-based probabilities and insurance supplied cost estimates an inappropriately low-cost assessment was achieved, ruling out a more costly and robust flood protection system. However, this conclusion can be reversed by making less optimistic assumptions about possible increases in flood likelihoods while also considering estimates of second-order disaster costs, public risk-aversion, and damage heterogeneity.[20]

CBA provides useful information, but does not directly justify a particular decision with any confidence. Too often in such agree-on-assumptions analyses, individuals press for assumptions that will lead to the options that they favor, effectively defining the deep uncertainty out of the problem. This can lead to bias, gridlock, and brittle decisions. Brittle decisions are optimal for a particular set of assumptions but perform poorly or even disastrously under other assumptions.[21]

Sensitivity analyses using CBA models can explore the full range of plausible assumptions and future conditions, but this rarely provides an unambiguous basis for making a specific choice. Further, CBA assumes that society is risk neutral, weighing positive and negative risks equally. The horrors of complete devastation of a major city may outweigh any optimistic assumptions, with the precautionary principle associated with an intolerable risk then carrying the greatest weight in a decision to build category five protection.

A further consideration is countervailing risks and side effects. The implementation of a large-scale protection system is expected to attract more people to a risky location and increase exposure to floods if the defense fails. But the protection system can also improve infrastructure and create jobs and income, thus improving welfare more than is suggested by an analysis of direct costs. These factors could easily double (or halve) expected benefits.[22]

Hallegatte's analysis concluded with the following statement:[23]

"More generally, climate change creates an additional layer of uncertainty in infrastructure design that increases the probability of either under-adaptation (and increased disaster risk) or over-adaptation (and sunk costs in protection)."

As will be seen in the following sections, robust and adaptive decision-making strategies provide a richer framework for evaluating such decisions.

12.2 Decision-Making Under Deep Uncertainty (DMDU) Framework

"Decisions are being made today, even when they're not." (Physicist and statistician Leonard Smith)[24]

An excellent overview of decision-making under deep uncertainty is provided in the recent book entitled *Decision-Making Under Deep Uncertainty: From Theory to Practice*, edited by risk scientists from the Netherlands and The Rand Corporation.[25] The DMDU approach has been created to aid in iterative analysis of the most complex policy issues utilizing alternative assumptions and an evolving course of action.[26]

In contrast to strategies such as CBA, decision-making in the context of deep uncertainty is not based on predictions of the future—the "predict-then-act" or "agree-on-assumptions" paradigms. DMDU inverts the decision-making process via an "agree-upon-decision" approach that seeks a robust decision—one that performs well across a wide range of possible futures, preferences, and worldviews, although it may not be optimal for any particular one. The agree-upon-decision approach defers agreement until after the options have been evaluated under many different assumptions.[27]

A complementary element of DMDU is to prepare for uncertain events or conditions by monitoring how the future evolves and allowing adaptations over time as knowledge is gained. "Monitor-and-adapt" explicitly acknowledges the deep uncertainty encompassing decision-making over long time windows during which uncertain events may occur.[28]

The goal of DMDU is to seek out the alternatives within the possible pool of solutions that have lower risk for the given uncertainties over the short-term, while at the same time are expected to perform reasonably well as judged by the positive outcome measures.[29]

The agree-upon-decision approach bypasses provocative debates that try to discern whose unknowable facts are most accurate. Such debates lead to stalemates, as opposed to solution finding. By contrast, DMDU carries forward all sets of assumptions into the analysis stage and continues to explore courses of action that have the potential to satisfy all parties.[30]

DMDU approaches share several principles. The process starts with decision makers creating a shared strategic vision of future ambitions, exploring possible adaptive strategies in context of scenarios of the future, evaluating the strategies based on robustness metrics. Decision makers commit to short-term actions while keeping long-term options open, and prepare a framework that guides future actions.[31]

A major purpose of decision analysis is to provide a framework for discussion and a focus for subjective judgments. Decision-making on complex and uncertain systems generally involves multiple actors coming to agreement. This requires an iterative approach that facilitates learning across alternative framings of the problem, stakeholder preferences, and trade-offs, all in context of what is possible.[32] DMDU analyses regularly seek engagement with stakeholder groups, planners, and decision makers through the use of real-time analytical support, workshops, games, and Assumption-Based Planning pre-plan exercises.[33,34] Candidate policy decisions are evaluated by stress testing them over a wide range of uncertainties. Various decision support approaches and tools can be used to enable a constructive learning process among the stakeholders and analysts.[35] This approach moves away from an a priori agreement on assumptions, towards an a posteriori exploration of trade-offs among objectives and the robustness of the decisions across possible futures.[36]

12.3 Robust Decision-Making

"Sometimes, you make the right decision. Sometimes, you make the decision right." (Psychologist Phil McGraw)[37]

Robust Decision-Making (RDM) begins with the set of alternative decisions and aims to characterize the uncertainties in a way that usefully highlights the trade-offs among them. This inverted agree-upon-decisions process promotes consensus around robust decisions and helps manage uncertainty and disagreement.[38]

Making decisions with inverted processes that analyze options first and seek agreement second have the following advantages:[39]

- Create buy-in by incorporating diverse beliefs and transparent analyses
- Identify robust strategies likely to succeed under any future scenario
- Keep decision makers focused on tradeoffs between the available options
- Strive for agreement on the most critical matters: actions, not assumptions
- Ensures that decision-makers, rather than experts and analysts, make the decision

The agree-upon-decisions process starts by stress-testing the decision options under a wide range of plausible conditions. Decision options are evaluated repeatedly, under many different sets of assumptions—no agreement is needed on these assumptions. This analysis reveals which of the options are robust—meeting needs under a wide range of conditions, rather than performing well in only a few.[40]

Robust decision-making options include:[41]

- No- and low-regret decisions
- Reversible and flexible decisions
- Safety-margin decisions
- Decisions with reduced time horizons

No- and low-regret decisions are highly useful no matter what the future brings. Finding existing shortcomings in systems or processes can reveal no-regret or low-regret strategies for an uncertain future. For example, increasing energy efficiency, fixing water and methane leaks, and limiting development in flood prone areas are good investments, regardless of how the climate and other factors change. Such strategies are beneficial over the short term, easy to implement, and offer benefits under a wide range of future conditions.[42]

Reversible and flexible decisions are generally more robust than irreversible ones because they allow for adjusting decisions as new information becomes available, thus helping reduce regret. Annual adjustments of insurance are made in response to the new information on emerging risks, which is an example of flexibility. Delaying an infrastructure project as environmental and climate risks are further investigated is an example of a reversible decision—the decision not to build immediately is easily reversed following a favorable analysis that the location is suitable and safe. During the decision-making process, it is important to take into consideration the value of the reversibility of a strategy (the option value).[43]

Safety margin strategies are a relatively inexpensive way to reduce the risk of infrastructure decisions and increase robustness. Consider the following example of a failure to include a safety margin in a major infrastructure project. In 2013, the San Francisco-Oakland Bay Bridge ramp was completed in California (US), which carries 270,000 vehicles each day. Despite clear information about rising sea levels and storm surges, the ramp was rebuilt at an elevation that is at risk of flooding today under a 50-year storm surge, and is below the 100-year base flood elevation. The Bay Bridge ramp would be permanently inundated by three feet of sea level rise. As a result of failing to include a safety margin even for current conditions—not to mention expected sea level rise from global warming—a subsequent report by the Metropolitan Transportation Commission recommended a series of construction projects to protect the Bay Bridge, costing taxpayers additional millions of dollars.[44,45]

Safety margin strategies are justified if it is relatively inexpensive to design the infrastructure to cope with increased stresses such as sea level rise. Modifying the system after it has been built can be difficult and expensive.[46] Safety margins are especially important for adaptation measures that are not reversible or flexible, such as a bridge. Plausible worst-case scenarios (Chapter Nine) provide important input into designing safety margins. In the example of the California bridge, adding the safety margin in the original design and construction would have added a small incremental cost, which is much less than additional construction costs after the bridge has been built. The challenge is when the safety margin produces substantial additional cost, such as protecting New Orleans from a Category 5 hurricane (Section 12.1). In such cases, other adaptive and operational risk management strategies are needed.

Reducing uncertainty associated with a decision can also be achieved by shortening the lifetime of investments. If houses and other buildings are to be built in an area on the coast that is prone to hurricanes or in other flood prone locations, one strategy is to build cheaper buildings with shorter lifetimes, instead of high-quality buildings designed to last a century.[47]

12.4 Robustness Metrics

"One never knows, do one?" (American jazz pianist Fats Waller)[48]

Consider the following decision that farmers face annually. The farmer has a choice between two crops: Crop #1 produces a steady yield under all rainfall conditions, while Crop #2 provides a large yield only under conditions of high seasonal rainfall. Since we cannot reliably predict rainfall on seasonal time scales, Crop #1 is the safest choice, although yield will be suboptimal if rainfall is high. Robustness becomes an important decision criterion when the consequences of making a wrong decision are high. If crop insurance is available to protect against potentially poor yields, or if sufficient savings are available, Crop #2 may be the best strategy. If these tools and resources are not available and the consequences of a few years of low yields would be disastrous, then robustness becomes the priority. An alternative strategy is to hedge by planting mostly Crop #1, but devoting a fraction of the land to Crop #2.

The farmer's decision can be evaluated using robustness metrics. Robustness metrics are characterized by three features:[49,50]

(1) Preference for a system performance indicator (e.g., absolute performance, regret, or the satisfaction of constraints);
(2) The preference of a decision-maker toward a high or low level of risk aversion, which is addressed through scenario subset selection; and
(3) The decision-maker's preference toward maximizing average performance, minimizing variance, or some other higher-order moment over the range of selected scenarios.

Different robustness metrics can produce different rankings among decision options. Stability in the ranking is most sensitive to (3).[51]

A number of robustness metrics have been used to measure system performance under deep uncertainty. Classic decision theoretic metrics include:[52,53,54]

- Maximize expected utility—action with the most desirable outcomes on average compared to all other actions.
- Maximax—most optimistic (focuses on best case scenario)
- Maximin—most pessimistic (focuses on worst case scenario)
- Minimax regret—measures desirability; minimizing potential maximum regret by not selecting other options.
- Hurwicz optimism pessimism rule—blending of optimism and pessimism; percentage of the best-case and a percentage of the worst-case.

- LaPlace's principle of insufficient reason—principle of indifference; all outcomes are equally likely. A choice is made based on outcome with best potential payoff.

There is a well-established distinction between "regret" metrics and "satisficing" metrics.[55] Regret metrics are defined as the difference between the performance of the selected option for a particular scenario and the performance of the best possible option for that scenario.[56] Satisficing metrics aim at maximizing the number of scenarios in which the policy alternative under consideration meets minimum performance thresholds.[57] If there are hard constraints in the system, such as budget or the requirement to provide services, then satisficing metrics are most appropriate. By contrast, if optimizing system performance is the primary value, then regret-based metrics are appropriate.

The different metrics reflect varying levels of risk aversion, and differences about what is meant by robustness—ensuring insensitivity to future scenarios, versus reducing regret, versus avoiding very negative outcomes. Which robustness metric is most appropriate for a particular decision depends on a combination of the likely impact of system failure and the degree of risk aversion of the decision maker.[58]

For a complex policy issue, there is rarely agreement on a single metric sufficient to evaluate outcomes. Often, there are several assessment criteria that may even be in opposition to one another. Even if there is agreement about the values in play, there may be disagreement about priorities among the outcomes. Utilizing multiple risk metrics promotes a wider perspective on normative outcomes. Satisficing metrics are useful when there is disagreement about preferred outcomes.[59]

12.5 Dynamic Adaptive Decision-Making

"As for the future, your task is not to foresee it, but to enable it." (French writer Antoine de Saint-Exupéry)[60]

When there are many plausible scenarios for the future, it is often impossible to construct a single static plan that will perform well in all of them. Under conditions of deep uncertainty, static plans are likely to fail, become overly costly to protect against failure, or incapable of seizing opportunities. Alternatively, flexible plans can be designed that will adapt over time. In this way, a policy can be responsive to an evolving knowledge base and technologies.[61,62]

Political economist Charles Lindblom was an early proponent of various kinds of incrementalism in decision-making, which he called "muddling through." He brought into focus the prospect that we may not have knowable preferences. He realized that means and ends may not be separable and may be reshaped in the very process of making a decision. He supported the criteria of choosing incremental and corrigible changes in preference to large and irreversible ones. And finally, he observed that many decisions are embedded in institutional or social contexts that may be harnessed to enhance decision-making.[63]

More recently, structured stepwise approaches for dynamic adaptation have been developed. The main decision-making principle is that decision makers should take only those actions that are non-regret and time-urgent—postponing other actions to a later stage.[64,65]

Adaptive decision-making strategies are especially useful in complex coupled physical-technical-human systems, where the degree of uncertainty over the planning horizon is large, multiple management decisions need to be taken over long time periods, and the path dependency of these decisions is high. Adaptive strategies are particularly helpful when the implementation time is relatively short compared to the rate of change, there is flexibility in the solutions, and the functional lifetime of the decision is shorter than the planning horizon. Failing to consider the long term and path dependencies can result in maladaptive actions which can strand assets well in advance of reaching their anticipated end of life. Further, problems and opportunities may become more or less urgent than what is envisaged today, and alternative options to solve these issues may become available.[66]

Adaptive planning makes explicit the importance of monitoring and adapting to changes of time to prevent the initial policy from failing.[67]

Anticipatory actions can be taken immediately upon implementation of the plan to address vulnerabilities and opportunities, thus increasing the robustness of the initial plan. Anticipatory actions include:[68]

- Mitigating actions—reduce adverse impacts associated with likely vulnerabilities.
- Hedging actions—reduce adverse impacts on a plan or spread or reduce risks that stem from uncertain vulnerabilities.
- Seizing actions—take advantage of opportunities that may prove beneficial to the plan.
- Exploiting actions—take advantage of new developments that can make the plan more successful or succeed sooner.
- Shaping actions—proactively affect external events or conditions that could either reduce the plan's chance of failure or increase its chance of success.

Notes

1 Daniel A. Farber, "Coping with Uncertainty: Cost-Benefit Analysis, the Precautionary Principle, and Climate Change," *Washington Law Review* 90, no. 4 (December 2015).

2 Warren Walker, "Policy Analysis: A Systematic Approach to Supporting Policymaking in the Public Sector," *Journal of Multi-Criteria Decision Analysis* 9, no. 1–3 (2000): 11–27.

3 H. J. Miser and E. S. Quade, *Handbook of Systems Analysis* (Chichester, UK: Wiley, 1985).

4 Millet Granger Morgan et al., *Uncertainty: A Guide to Dealing with Uncertainty in Quantitative Risk and Policy Analysis* (Cambridge, UK: Cambridge University Press, 1990).

5 Muriel C. Stanton and Katy Roelich, "Decision Making under Deep Uncertainties: A Review of the Applicability of Methods in Practice," *Technological Forecasting and Social Change* 171 (October 2021): 120939, https://doi.org/10.1016/j.techfore.2021.120939.

6 Ortwin Renn et al., "Coping with Complexity, Uncertainty and Ambiguity in Risk Governance: A Synthesis," *AMBIO* 40, no. 2 (February 3, 2011): 231–246, https://doi.org/10.1007/s13280-010-0134-0.

7 Ibid.

8 Robert J Lempert, "Shaping the next One Hundred Years: New Methods for Quantitative, Long-Term Policy Analysis" (Santa Monica, CA: RAND, 2003).

9 David Burkus, "How Criticism Creates Innovative Teams," Harvard Business Review, July 22, 2013, https://hbr.org/2013/07/how-criticism-creates-innovati.

10 Stéphane Hallegatte, "A Cost-Benefit Analysis of the New Orleans Flood Protection System," HAL Open Science (HAL, 2007), https://hal.archives-ouvertes.fr/docs/00/16/46/28/PDF/Hallegatte_NewOrleans_CBA9.pdf.

11 Axel Graumann et al., "Hurricane Katrina, a Climatological Perspective," technical report; no. 2005-01 (National Climatic Data Center, 2006), https://repository.library.noaa.gov/view/noaa/13833/noaa_13833_DS1.pdf.

12 "Hurricane Katrina: A Nation Still Unprepared," Committee on Homeland Security and Governmental Affairs (United States Senate, 2006), https://www.govinfo.gov/content/pkg/CRPT-109srpt322/pdf/CRPT-109srpt322.pdf.

13 Allison Plyer, "Facts for Features: Katrina Impact," The Data Center, August 26, 2016, https://www.datacenterresearch.org/data-resources/katrina/facts-for-impact/.

14 P. J. Webster et al., "Changes in Tropical Cyclone Number, Duration, and Intensity in a Warming Environment," *Science* 309, no. 5742 (2005): 1844–1846, https://doi.org/10.1126/science.1116448.

15 Stéphane Hallegatte, "A Cost-Benefit Analysis."

16 Ibid.

17 Ibid.

18 Ibid.

19 Nidhi Kalra et al., "Agreeing on Robust Decisions: New Processes for Decision Making under Deep Uncertainty," *Policy Research Working Papers*, June 2014, https://doi.org/10.1596/1813-9450-6906.

20 Stéphane Hallegatte, "A Cost-Benefit Analysis."

21 Nidhi Kalra et al., "Agreeing on Robust Decisions."

22 Ibid.
23 Stéphane Hallegatte, "A Cost-Benefit Analysis."
24 Leonard Smith (@lynyrdsmyth), "Decisions are being made today, even when they're not," Twitter, September 5, 2021, https://twitter.com/lynyrdsmyth/status/1434504608634875904.
25 Vincent A. W. J. Marchau et al., eds., *Decision Making under Deep Uncertainty: From Theory to Practice* (Cham, CH: Springer International Publishing, 2019).
26 Steven W. Popper, "Reflections: DMDU and Public Policy for Uncertain Times," in *Decision Making under Deep Uncertainty: From Theory to Practice*, ed. Vincent A. W. J. Marchau et al. (Cham, CH: Springer International Publishing, 2019), 375–392.
27 R. J. Lempert, "Robust Decision Making (RDM)," in *Decision Making under Deep Uncertainty: From Theory to Practice*, ed. Vincent A. W. J. Marchau et al. (Cham, CH: Springer International Publishing, 2019), 23–51.
28 Vincent A. W. J. Marchau et al., "Introduction," in *Decision Making under Deep Uncertainty: From Theory to Practice*, ed. Vincent A. W. J. Marchau et al. (Cham, CH: Springer International Publishing, 2019), 1–20.
29 Steven W. Popper, "Reflections."
30 Ibid.
31 Vincent A. W. J. Marchau et al., "Conclusions and Outlook," in *Decision Making under Deep Uncertainty: From Theory to Practice*, ed. Vincent A. W. J. Marchau et al. (Cham, CH: Springer International Publishing, 2019), 393–400.
32 Jonathan D. Herman et al., "How Should Robustness Be Defined for Water Systems Planning under Change?," *Journal of Water Resources Planning and Management* 141, no. 10 (October 2015), https://doi.org/10.1061/(asce)wr.1943-5452.0000509.
33 Vincent A. W. J. Marchau et al., "Introduction."
34 Liisa Ecola et al., "The Road to Zero: A Vision for Achieving Zero Roadway Deaths by 2050," 2018, https://doi.org/10.7249/rr2333.
35 Jan H. Kwakkel and Marjolijin Haasnoot, "Supporting DMDU: A Taxonomy of Approaches and Tools," in *Decision Making under Deep Uncertainty: From Theory to Practice*, ed. Vincent A. W. J. Marchau et al. (Cham, CH: Springer International Publishing, 2019), 355–374.
36 Alexis Tsoukiàs, "From Decision Theory to Decision Aiding Methodology," *European Journal of Operational Research* 187, no. 1 (May 16, 2008): 138–161, https://doi.org/10.1016/j.ejor.2007.02.039.
37 Dr. Phil (@DrPhil), "Sometimes, you make the right decision. Sometimes, you make the decision right. #DrPhil," Twitter, May 31, 2014, https://twitter.com/drphil/status/472778786820198401.
38 Nidhi Kalra et al., "Agreeing on Robust Decisions."
39 Ibid.
40 Ibid.
41 Stéphane Hallegatte, "Strategies to Adapt to an Uncertain Climate Change," *Global Environmental Change* 19, no. 2 (May 2009): 240–247, https://doi.org/10.1016/j.gloenvcha.2008.12.003.
42 Nidhi Kalra et al., "Agreeing on Robust Decisions."
43 Ibid.

44 Juliet Christian-Smith, "A Bridge over Troubled Waters: How the Bay Bridge Was Rebuilt without Considering Climate Change," The Equation, February 6, 2015, https://blog.ucsusa.org/juliet-christian-smith/a-bridge-over-troubled-waters-how-the-bay-bridge-was-rebuilt-without-considering-climate-change/.

45 "Climate Change and Extreme Weather Adaptation Options for Transportation Assets in the Bay Area Pilot Project," 2014.

46 Stéphane Hallegatte, "Strategies to Adapt."

47 Nidhi Kalra et al., "Agreeing on Robust Decisions."

48 *"One Never Knows, Do One?" The Best of Fats Waller*, n.d.

49 Muriel C. Stanton and Katy Roelich, "Decision Making under Deep Uncertainties."

50 C. McPhail et al., "Robustness Metrics: How Are They Calculated, When Should They Be Used and Why Do They Give Different Results?," *Earth's Future* 6, no. 2 (January 8, 2018): 169–191, https://doi.org/10.1002/2017ef000649.

51 Ibid.

52 Abraham Wald, *Statistical Decision Functions* (New York, NY: John Wiley & Sons, 1950).

53 Leonid Hurwicz, "The Generalized Bayes Minimax Principle: a Criterion for Decision Making under Uncertainty," *Cowles Commission Discussion Paper: Statistics No. 355*, 1951.

54 Thierry Denœux, "Decision-Making with Belief Functions: A Review," *International Journal of Approximate Reasoning* 109 (December 13, 2019): 87–110, https://doi.org/10.1016/j.ijar.2019.03.009.

55 Robert J. Lempert and Myles T. Collins, "Managing the Risk of Uncertain Threshold Responses: Comparison of Robust, Optimum, and Precautionary Approaches," *Risk Analysis* 27, no. 4 (October 23, 2007): 1009–1026, https://doi.org/10.1111/j.1539-6924.2007.00940.x.

56 Leonard J. Savage, "The Foundations of Statistics," *Naval Research Logistics Quarterly* 1, no. 3 (1954): 294, https://doi.org/10.1002/nav.3800010316.

57 Herbert Alexander Simon, *The Sciences of the Artificial* (Cambridge, MA: The MIT Press, 1969).

58 C. McPhail et al., "Robustness Metrics."

59 Steven W. Popper, "Reflections."

60 Antoine de Saint-Exupéry and Stuart Gilbert, *The Wisdom of the Sands* (New York, NY: Harcourt, Brace and Company, 1950), 155.

61 Lawrence E. McCray et al., "Planned Adaptation in Risk Regulation: An Initial Survey of US Environmental, Health, and Safety Regulation," *Technological Forecasting and Social Change* 77, no. 6 (July 2010): 951–959, https://doi.org/10.1016/j.techfore.2009.12.001.

62 Jan H. Kwakkel and Marjolijin Haasnoot, "Supporting DMDU."

63 Charles E. Lindblom, "The Science of 'Muddling through,'" *Public Administration Review* 19, no. 2 (1959): 79–88, https://doi.org/10.2307/973677.

64 Warren E. Walker et al., "Adaptive Policies, Policy Analysis, and Policy-Making," *European Journal of Operational Research* 128, no. 2 (January 16, 2001): 282–289, https://doi.org/10.1016/s0377-2217(00)00071-0.

65 Jan H. Kwakkel and Marjolijin Haasnoot, "Supporting DMDU."

66 Marjolijn Haasnoot et al., "Dynamic Adaptive Policy Pathways (DAPP)," in *Decision Making under Deep Uncertainty: From Theory to Practice*, ed. Vincent A. W. J. Marchau et al. (Cham, CH: Springer International Publishing, 2019), 71–92.

67 Marjolijn Haasnoot et al., "Dynamic Adaptive Policy Pathways (DAPP)."

68 Warren E. Walker et al., "Uncertainty in the Framework of Policy Analysis," *International Series in Operations Research & Management Science*, October 5, 2012, 215–261, https://doi.org/10.1007/978-1-4614-4602-6_9.

Chapter Thirteen

ADAPTATION, RESILIENCE, AND DEVELOPMENT

"Adapt or perish, now as ever, is Nature's inexorable imperative."

—Writer H. G. Wells[1]

Adaptation is the process of discovering methods to remain flexible and thrive during periods of change. With climate change, the adaptation process seeks to minimize the negative impact of a changing climate, while leveraging any new opportunities that may appear. Climate adaptation can be reactive, occurring in response to extreme weather and climate impacts, or anticipatory, occurring before impacts of climate change are observed. It involves adjusting policies and actions because of observed or expected changes in climate.[2]

A "climate shock" is a shock to a place, impacting people who live and/or work, own assets, or use products grown or produced there.[3] Resilience of a community or country relates to how well the institutions and systems provide public goods during and after shocks. Resilience relates to the capacity to adapt. Building resilience through development is a key foundation for effective adaptation.

Science now provides a basis for communities to anticipate a range of potential climate conditions. Taking action before the worst impacts are incurred is different from how humans have adapted in the past. The challenge to pro-active adaptation is to avoid expensive measures that end up either being wasted when the expected conditions do not occur, or locked-in measures that are inadequate to protect from conditions that actually occur.

Unlike mitigation (reducing/eliminating CO_2 emissions) which operates on a global scale with potentially global benefits, the benefits of adaptation, resilience, and development are local to regional. Local manifestations and impacts of climate change are mediated by geography and local infrastructure, social, economic, and political environments.

At the local level, adaptation is a complex process that emerges as social systems reorganize through a series of responses to external stresses. The challenge is to enhance the capacity of communities to adapt by creating enabling environments through rising incomes, access to markets, and rising educational attainment.[4]

13.1 Context

"If you don't know where you are going, you will probably end up somewhere else." (Laurence J Peter, author of *The Peter Principle*)[5]

The Earth is a hostile place for humans, as evidenced by the struggles of our early ancestors, who adapted primarily through migration.[6] Slowly, humans learned to deal with their environment. Over the past century, adaptation has been substantially enabled by fossil fuels and machines. Modern mastery of our environment and climate through technology and infrastructure allows for humans to live anywhere on the planet, including the extremely cold temperatures of the Antarctic and the scorchingly hot temperatures of the Middle East.

While adaptation to weather and climate has been remarkably successful over the past century, adaptation has been less successful for underdeveloped countries lacking financial resources and good governance. Adaptive capacity is closely linked to social and economic development.

The pace of warming and sea level rise over the past century has allowed time for adaptation. Human societies will continue to adapt in response to potential adverse impacts of climate change as well. The concern is that more rapid rates of change, such as hypothesized with the more extreme projections of human-caused global warming, will stress our ability to adapt.

13.1.1 Adaptation Success Stories

"And how should a beautiful, ignorant stream of water know it heads for an early release — out across the desert, running toward the Gulf, below sea level, to murmur its lullaby, and see the Imperial Valley rise out of burning sand with cotton blossoms, wheat, watermelons, roses, how should it know?" (Poet Carl Sandburg)[7]

We are not passive victims in the face of climate variability and change. The success of human adaptation to weather and climate extremes is reflected by the fact that global deaths and economic damage from weather and climate disasters dropped 80–90 percent during the last four decades, when scaled for population and gross domestic product (GDP) changes.[8]

Agriculture is a major success story over the past century. Currently, we produce enough food for a population of 10 billion people—a 20 percent surplus. While land use for agriculture has increased by 8 percent since 1962, agricultural productivity has increased by 300 percent. Between 1900 and 2010, the intensification of agricultural productivity allowed France and Spain to reforest. Between 2000 and 2017, production of beef and cow's milk increased as total land used for pasture shrank.[9] Humankind's use of land for agriculture is likely near its peak and capable of declining in the near future. Food production and climate change does not seem to be a problem, although growing trends to use grains and seed oil in producing fuels is providing stresses on food supplies and prices (see Section 14.2.3). Nevertheless, there are local food shortages and famines. Indian economist and Nobel Laureate Amartya Sen showed that famines occur because of war, political oppression, and the collapse of food distribution—not a lack of food.[10]

Challenges are greatest in underdeveloped countries of the global South, but some of the poorest and most vulnerable countries have taken pro-active measures that are significantly reducing adverse impacts. Consider Bangladesh, which was hit by a devastating tropical cyclone in 1970 (Cyclone Bhola) that killed up to 500,000 people. This storm and its aftermath precipitated the separation of Bangladesh from Pakistan. More recently in 2007 Cyclone Sidr, of similar magnitude, struck Bangladesh and killed 3,000 people. This relative success was enabled by an extensive program of building cyclone shelters around the entire coast, along with developing early warning systems. A successful program to improve the resilience of coastal communities was designed to be accessible to women who are unable to travel far from their homes. The program's aim was to cultivate forests, fish, and fruit to improve mangrove forests restoration, which in turn serves as a buffer against storms and flooding while also increasing food production. Artificial dikes provide beds for fruits, vegetables, and trees, while the resulting ditches held aquaculture ponds. Since these structures are close to households, many more women gained the opportunity to generate food and income, while also promoting ecological co-benefits through the creation of natural sea barriers which reduce coastal erosion.[11] This is an example of a very positive synergy between development and adaptation.

By contrast, also in 2007, Cyclone Nargis hit the neighboring country of Myanmar (formerly Burma) and killed over 140,000 people—mainly due to the lack of any kind of warning or preparation by the government.[12]

Another success story involved my company, Climate Forecast Applications Network (CFAN). A project was undertaken to reduce adverse impacts of heatwaves in Ahmedabad—one of India's hottest cities. Extreme heat is common during the pre-monsoon months of March–May. To address this concern in Ahmedabad, a Heat Action Plan was developed in 2013 by an interdisciplinary consortium formed by the Natural Resources Defense Council that included the Ahmedabad Municipal Council, the Indian Institute of Public Health, Public

Health Foundation of India, Icahn School of Medicine at Mount Sinai, Rollins School of Public Health at Emory University and CFAN. The Plan developed an evidence-based heat preparedness plan and early warning system, based on CFAN's 7-day probabilistic forecasts of extreme heat.[13]

The Heat Action Plan offered ways to mitigate the health impacts of rising temperatures, including mapping high-risk heat areas, increasing access to drinking water stations and green spaces for shade, reducing urban heat island effects, ensuring new buildings are more heat-resilient, and developing transportation systems that help people avoid heat stress. The project educated people on how to protect themselves via campaigns on television, radio, and newspapers, as well as through messaging platforms such as WhatsApp. The government alerted residents to forecasts of very high temperatures through hospitals, community groups, media outlets, and government agencies. Cooling spaces were set up in temples, public buildings, and malls during the warmest months. Health workers were also trained to recognize the symptoms of heat stress and ensure emergency rooms and ambulances are stocked with ice packs.

Air conditioning is the ultimate solution to dealing with excess heat. However, the household ownership of air conditioners in India is estimated at 7 percent. Further, under extreme heat stress, the electricity supply becomes more unreliable, with brownouts occurring in response to excessive demand. And this is not to mention the impact of additional air conditioning on greenhouse gas emissions. The India Cooling Action Plan provides a 20-year roadmap to address India's future thermal comfort and the cooling needs of its people in a sustainable manner.[14] The government of India initiated the Global Cooling Prize, focusing on the single-room air conditioner similar to those commonly installed in apartment buildings in India. In November 2019, eight finalists were announced for the $1 million competition to design a room air conditioner reducing by five times associated greenhouse gas over the course of its lifetime than does a standard room unit.[15]

These examples show how synergy among technological innovation, infrastructure, and operational plans can reduce vulnerability to weather and climate extremes. However, apart from many adaptation success stories, most of the world—including some of the wealthiest countries—is poorly adapted even to the current climate, let alone to a changing climate.

13.1.2 Political Context

"That is a trick of human nature. We get used to things." (Author R. J. Palacio)[16]

Adaptation has long been viewed as problematic in climate policy and politics. By allowing for positive human outcomes as the climate changes, adaptation conflicts with the apocalyptic view of ever-worsening impacts as global temperatures increase.[17]

For the past 30 years, the dominant goal of the UNFCCC has been eliminating emissions from burning fossil fuels. Global warming activists have regarded efforts to adapt to climate extremes as capitulation and a distraction from the need to curb emissions from fossil fuel emissions—in essence, a copout that lets the fossil fuel companies off the hook.

While adaption has been mentioned in all of the IPCC assessment reports and UNFCCC reports and policies, there have only been token efforts in this direction until recently. Over the last decade, a shift toward accepting parallel efforts to reduce emissions and also adapt to climate change has been building. In his 2013 book *The Future*, former US Vice President Al Gore acknowledged that he made a mistake in 1992 in his book *Earth in the Balance* by regarding adaptation as a "kind of laziness."[18,19] Gore admitted that he was "wrong in not immediately grasping the moral imperative of pursuing both policies simultaneously."[20]

A number of adaptation funds exist under the UNFCCC financial mechanism: the Special Climate Change Fund, the Least Developed Countries Fund, the Adaptation Fund, and the Green Climate Fund.[21] The Paris Agreement of 2015 included several provisions for adaptation, including an agreement that developed countries should provide some financial support and technology transfer to promote adaptation in more vulnerable countries. Sustainable Development Goal 13 (Chapter 15), also of 2015, aims to strengthen countries' resilience and adaptive capacities to climate-related issues, and includes areas such as infrastructure and agriculture.[22] The most prominent sign of the rising profile of adaptation is the Global Commission on Adaptation launched in 2018, that has a commitment of US$200 billion in climate finance over a period of 5 years.[23]

However, the continued dominance of mitigation over adaption is apparent in the IPCC AR6 WGII Report on Impacts, Adaptation, and Vulnerability. While the IPCC AR5 WGII report addressed its mission credibly, the AR6 WGII strayed from its purpose and positioned itself more as an advocate for emissions reductions. The AR6 WGII report concludes the following:

"[I]mpacts will continue to increase if drastic cuts in greenhouse gas emissions are further delayed—affecting the lives of today's children tomorrow and those of their children much more than ours"[24] [...] "Any further delay in concerted global action will miss a brief and rapidly closing window to secure a liveable future."[25]

Part of the problem with the AR6 WGII assessment is their focus on the extreme emissions scenarios RCP8.5/SSP5–8.5, which have been assessed to be implausible (Section 7.1). More than half of the scenario mentions in the AR6 WGII are represented by RCP8.5.[26] If warming is extremely rapid, it will be difficult for adaptation to keep up. However, for more plausible rates of warming (Section 8.2.4), urgency is much less of an issue.

International climate politics are slowly realizing that even if emissions are stopped relatively soon, there is no guarantee of a quick benefit due to the inertia of the climate system. Further, countries and communities are vulnerable now to weather and climate extremes, supporting the argument for building adaptable, resilient communities.

At the UN Climate Summit in Copenhagen in 2009, wealthy nations promised to channel US$100 billion a year to less wealthy nations by 2020, to help them adapt to climate change and mitigate further rises in temperature. Compared with the trillions of dollars invested into emissions reduction, the US$100 billion pledge is minuscule. However, even that relatively small promise has not been kept. While there is much disagreement on accounting methods, a 2020 report concluded that the US$100 billion target is out of reach.[27]

13.1.3 Misplaced Blame

"Pointing the finger at natural causes creates a politically convenient crisis narrative that is used to justify reactive disaster laws and policies." (Disaster scientist Emmanuel Raju et al.)[28]

The impetus to blame human-caused climate change for every societal problem that is in some way related to weather and climate (Section 1.3) is not only driven by the activist strategy to use such issues to urge for action to reduce emissions. There are two additional incentives for local politicians to blame these problems on human caused climate change.

Pointing the finger at climate change creates a politically convenient crisis narrative that deflects from poor policies that have created and perpetuated social and physical vulnerability. These factors include: poor governance and corruption, pre-existing fragilities, unplanned urbanization, exploitation of environmental resources, systemic injustice, and marginalization due to religion, caste, class, ethnicity, gender, or age.[29] Climate-centric disaster framing for disaster and other social problems erases from view the very socioeconomic and political factors that actually cause vulnerability and suffering as a result of weather and climate extremes. Climate-centric disaster framing is politically useful to politicians desiring to divert attention from local, national, and international policy failures.[30]

In an article entitled "It's not just climate: Are we ignoring other causes of disasters?" science writer Fred Pearce describes numerous examples—two of which are described below—where disasters were incorrectly attributed to global warming.[31]

During the 2021 food crisis in Madagascar, more than a million people were starving after successive years of drought. Madagascar's President Andry Rajoelina said: "My countrymen are paying the price for a climate crisis that they did not create."[32] Subsequent research concluded that human-caused climate change played a minor role in the drought, which was caused by natural climate variability. The analysis demonstrated that the crisis was due to poverty and infrastructure issues like inadequate water supplies for crop irrigation. These issues had gone unaddressed by Rajoelina's government.[33]

Continued drying of Lake Chad in West Africa has huge security and humanitarian consequences. Over the past 50 years, the surface area of the Lake shrank by 95 percent. The lack of water has devastated local fishers, farmers, and herders. Deepening poverty has contributed to growing jihadism and the exodus of more than 2 million people.[34] Nigeria's president Muhammadu Buhari stated: "Climate change is largely responsible for the drying up of Lake Chad."[35] The initial water level decline was caused by long droughts in the 1970s and 1980s, but the lake has remained virtually empty over the past two decades as rainfall has recovered. During this time, rivers flowing into the lake from Cameroon, Chad, and Buhari's Nigeria have been diverted by government agencies to irrigate inefficient rice farms.[36] Robert Oakes of the UN Institute for Environment and Human Security in Bonn says that "the climate-change framing has prevented the identification and implementation of appropriate measures to address the challenges." The appropriate measures include restoration of water from rivers that once fed the lake.[37]

The other incentive to blame climate change for disasters relates to international aid for development, which is increasingly dominated by climate-relevant needs. The availability of climate adaptation aid motivated Bangladesh to seek funds to rehabilitate 17 polders to accommodate 65 cm sea level rise by 2080 from global warming. Funding of US$400 million was obtained from the World Bank.[38]

Coastal scientist John Pethick identified numerous problems with the World Bank's solution. Actual sea level rise along the Bangladesh coast is 5–10 times greater than the global average, owing largely to land subsidence from ground water withdrawal.[39] Pethick further argues that a major cause of the sea level rise problem is the polders themselves.[40] Pethick advocates abandoning some of the polders and setting embankments further back from estuaries to reduce the funneling effect.[41] The Bangladeshi government and the World Bank are aware of Pethick's work, along with other papers that document this issue.[42] Pethick stated "The reaction among Bangladeshi government officials has been to tell me that I must be wrong."[43]

So why this charade of spending all this money on something that won't work and may potentially make the problem worse? Well of course there is uncertainty in all this (including Pethick's estimates). The objective of reducing Bangladesh's vulnerability to storm surges and sea level rise seems to have gotten lost in the desire to spend and receive money and to play the politics of climate adaptation funds.

The unfortunate lesson of this example is that strategies to succeed at getting funds for loss and damage or climate change adaptation are based on torquing every problem to be caused by human caused global warming. This requires exaggerating or minimizing existing problems to fit into some preconceived global warming damage magnitude, and relatively ignoring other causes and potentially more serious problems. And the developed world pays the bills, often for things that do not help address the real problems. By tying a large fraction of global development aid to climate change, developing countries will not receive the help they need in dealing with their very real and urgent problems.

13.2 Adaptation Frameworks

"The bamboo that bends is stronger than the oak that resists." (Japanese proverb)[44]

This section considers a selection of topics that frame approaches to adaptation: the decision to resist or retreat; the microeconomics of adaptation that determine how the choices made by households, firms, and local governments interact; and the role of operational decision-making in dealing with disasters.

13.2.1 Resist or Retreat

"Retreat! Hell! We're just advancing in a different direction." (Statement made by US Marine Corps General Oliver P. Smith during the Korean War)[45]

One aspect of adaptation addresses the question of whether to resist or retreat in areas vulnerable to weather and climate change risks, including sea level rise, flooding, wildfires, heatwaves, hurricanes, and land instability or erosion.[46]

This subsection addresses this question in coastal regions that are subject to sea level rise, hurricanes, and erosion. Coastal adaptation strategies can be classified in terms of protect, accommodate, advance, and retreat. A comprehensive description of adaption issues in coastal regions is provided by the IPCC AR6 WGII.[47]

Hard engineering coastal protection measures include breakwaters, sea walls and dikes, large barriers, super-levees that enable construction on top of them, and pumps to drain excess water. The extent to which engineering measures can protect against storm surge and sea level rise is demonstrated by the Netherlands—more than 50 percent of the Netherlands is below sea level, with the lowest point about seven meters below sea level. Nevertheless, the Dutch have managed to fight the encroachment of the sea for centuries by constantly innovating, building new infrastructure, and investing in new technologies. It is estimated that almost 30 percent of the Dutch national GDP is spent on improving coastal infrastructure.[48,49]

The high costs of hard protection are not affordable or practical in many, if not most, coastal locations. Soft engineering and sediment-based measures include beach nourishment. However, there are limits to this strategy due to environmental impacts, costs, and the availability of sand.[50] Nature-based measures, such as retaining mangroves and marshes, have been successful in reducing deaths and damage due to storm surges in the United States, providing an estimated US$23 billion per year in storm protection services.[51] These measures also reduce inland propagation of high tides and storm surges and reduce shoreline erosion. Nature-based measures have the greatest potential in coastal deltas and estuaries.[52]

The most effective strategy is to avoid new development in coastal locations that are prone to storm surges or sea level rise impacts. For existing coastal developments, accommodations include elevation or flood-proofing of houses. Raising structures or land can deter flooding. This can be achieved via artificial or nature-based interventions such as diverting rivers or controlling estuary and delta locations. Due to economic constraints, land raising is typically cost beneficial only for small areas.[53]

A strategy to advance creates new land by building seaward. This is achieved via planting vegetation which encourages natural polderisation or land-filling based land reclamation.[54] Successful advance strategies have been used in open coasts (Singapore), small atolls (the Maldives), cities on estuaries (Rotterdam), deltas (Shanghai), and mountainous coasts (Hong Kong).[55,56]

Retreat is a strategy that moves people, assets, and activities out of hazard zones, such as flood plains, barrier islands, and mudflats.[57] Retreat includes adaptive migration, involuntary displacement, and planned relocation of population and assets from the coast.[58] Sea level rise is not just a problem that is related to warming; local subsidence of the land can also be a major factor, largely due to excessive withdrawal of ground water and the ensuing ground compaction.[59] The most rapid subsidence is occurring in coastal cities in Asia—Tianjin, Manila, Chittagong, Karachi, Jakarta, and Shanghai have subsidence rates as high as 10 times that of the rate of global sea level rise.[60] Subsidence has compelled Indonesia to shift its capital from Jakarta to inland Nusantara.[61]

Some native villages in Alaska have already relocated away from the coast, owing to the interacting effects of permafrost thaw and coastal erosion.[62] To date, relocation has mainly been reactive; however increased consideration is being given to preemptive resettlement and the associated governance, finance, and institutional arrangements that would be required.[63] The small island of Kiribati's government has purchased land in Fiji for a planned relocation and is working with both Australia and New Zealand in the development of workforce training for its people.[64]

Adaptation pathway approaches (Section 12.5) are increasingly being used to facilitate long-term thinking, foresee maladaptive consequences and lock-ins, and address dynamic risk in the face of sea level rise; and frame adaptation as a series of manageable steps over time.[65] A portfolio of hard, soft, and nature-based interventions can be used to implement strategies to protect, accommodate, retreat, and advance, individually or in combination. Phasing adaptation approaches over time can help to spread costs and minimize regret.

13.2.2 Microeconomics of Adaptation

"We are not passive victims as Mother Nature cranks up the heat." (Environmental and urban economist Matthew Kahn)[66]

The microeconomic approach for studying climate change adaptation explores how people, firms, places, and governments make new investments that together increase resilience, and how one builds this capacity.[67] Environmental and urban economist Matthew Kahn has conducted extensive research into the microeconomics of adaptation in the United States. Rather than putting government at the center of achieving adaptation progress, Kahn shows how market mechanisms work to increase local resilience.

Kahn posits that owing to free markets, we are growing increasingly resilient and suffering less from the "punch[es] that Mother Nature is throwing."[68] We face lower risks of death and risks to the capital stock, food supply, and worker productivity over time due to shifts in our economy. As a result, the climate damage function flattens over time due to adaptation efforts and the per-person Social Cost of Carbon declines over time.

As the spatial distribution of climate-related risks becomes more apparent, climate change will reorder the rankings of the desirability of various locations. Cities and geographic areas that develop an edge in being resilient in the face of climate risks will experience greater economic growth. This resilience edge can be built up either due to natural advantage (physical location and topography) or due to local collective action in figuring out how to protect themselves against threats ranging from extreme heat, to drought to sea level rise to wildfire risk. Property owners in locations that are particularly susceptible to climate-related risks have strong incentives to seek out solutions both through private markets and through local government policy. Climate resilience is an increasingly valued public good. Competition between cities in supply climate resilience helps provide options for urbanites. For example, if Miami fails to cope with sea level rise, other cities will gain population and jobs.[69]

Banks and insurers can nudge real estate buyers to reduce their demand in risk-prone locations and increase their demand for housing and real estate in safer places. Changes to zoning codes to up zone in safer places featuring less climate-related risk will result in increased housing supply in safer places. The end result is not to completely desert fire- and flood-prone areas, because these are often desirable and productive locations. Civil engineers can design productive real estate assets that are acclimated to the risks. For example, single-family dwellings in flood-prone areas can be replaced with wetlands and tall buildings that have empty lower floors. An alternative approach to building resilient real estate is to build less durable structures that are meant to have a lifetime of less than 20 years. The owner of the property would have less capital at risk and holds an option to rebuild in the future as climate science makes more progress concerning the spatially refined risks facing an area. If the area faces great risk, then less capital would be invested there.[70]

Kahn argues that the key to this smooth adaptation dynamic is for government to retreat. Government is currently taxing people on higher ground to subsidize people taking risks on lower ground—politically, this may not be a sustainable situation. When the government subsidizes insurance in flood zones and fire zones, this creates a moral hazard effect of reducing the likelihood that owners of at-risk property take appropriate precautions. As the federal government crowds out private insurance sector investment in addressing climate-related risk, adaptation efforts may be slowed.[71]

13.2.3 Planning to Fail Safely

"You've got to ask yourself one question: 'Do I feel lucky?' Well do you, punk?" (Actor Clint Eastwood in the movie Dirty Harry)[72]

Climate change and extreme weather events will continue to challenge the ability of infrastructure systems to manage resources and supply critical services. Critical infrastructure systems include electric power systems, oil and gas networks, water networks, transportation networks, telecommunications, and computer systems. These complex systems are increasingly interdependent on each other, at scales ranging from the local to global. For example, operation of water and telecommunications systems requires a steady supply of electricity. The generation and delivery of electric power requires the availability of fuel, water plus telecommunication, and computer services. These interdependencies can turn a local disturbance in a single system into a large-scale failure via cascading events, with catastrophic impacts.[73]

Infrastructures are largely designed to be fail-safe—they are not intended to fail. However, even with very substantial safety margins, a system can fail to operate or protect during an extreme weather event. When failure happens, the consequences can be severe, with cascading impacts. Such rigidity in design is misaligned with a non-stationary climate and the growing interconnectedness of infrastructure systems. Fail-safe is built on a risk management principle—how often does the failure happen, how potentially bad could it be, and what are the impacts?[74]

Some systems are designed to accommodate failures in their planning and construction phases, such as nuclear power plants. Safe-to-fail is an emerging paradigm that broadly describes adaptation strategies that allow infrastructure to fail but control or minimize the consequences of the failure.[75] Safe-to-fail approaches include technologies that internalize the consequences of infrastructure failure in the development process. Safe-to-fail infrastructure planning and design supports managing unpredictability and infrastructure to be more adaptable to a myriad of shocks, surprises, and environmental hazards, including changing climate conditions.[76]

The focus of this subsection is not so much on technological approaches to safe-to-fail, but rather on social and operational approaches to controlling and minimizing the consequences of the failure. Designing systems so they are safe-to-fail can be a cheaper and more efficient solution, provided that there is an operational plan to minimize the consequences of failure.[77] A straightforward example is building a dam designed to contain up to a 100-year flood, coupled with a comprehensive property protection and evacuation plan for the surrounding area in the event of a more severe flood. This strategy anticipates that the dam will not control extreme flooding, but adds other protective measures for safety.[78]

Power outages associated with extreme weather events are estimated to cost US businesses more than $27 billion per year, and the cost trend is increasing.[79] If outages occur without warning, US households have little adaptative capacity to cope without power, clean water, refrigeration, heating and air conditioning, and cell phone access. Extended power outages contribute to loss of life from extreme cold, [80] extreme heat,[81] carbon monoxide deaths,[82] accidents in the home due to lack of lighting, traffic accidents in the absence of signals, and compromised medical care in hospitals and nursing homes.[83]

Safe-to-fail strategies can minimize the damage from an extreme event even if the physical electric power infrastructure is inadequate. Operational resilience is enabled through: learning from past system failures and recoveries; system and situational awareness enabled by monitoring sensors, advanced communications, and analytics; advanced asset management and outage management systems; pre-positioning prior to the event of equipment, replacement components and crews for rapid system recovery; and the use of innovative technologies to aid consumers, communities, and institutions in continuing some level of normal function without complete access to their normal power sources. Stress testing of the entire system is accomplished using extreme event scenarios.[84]

The state of Florida has historically been impacted by 41 percent of all US landfalling hurricanes. The extremely high winds, lightning, and flooding invariably produce power outages from blown transformers, damaged substations, and downed utility poles and power lines. During Hurricane Wilma in 2005, approximately 2.5 million customers lost power in South Florida, with electrical outages lasting an average of 5.4 days but as long as 18 days.[85]

Florida Power & Light Company (FPL) is the third-largest electric utility in the United States, serving more than 5.6 million customers across nearly half of the state of Florida. Following Wilma, FPL embarked on a program called Storm Secure.[86] They have been making significant investments to harden the grid and have implemented strategies to make the grid smarter. They have also implemented protocols to rapidly restore power following a hurricane. Continued efforts to strengthen the energy grid, coupled with innovative smart grid technology and rapid restoration of power outages, have led to national recognition of FPL for best-in-nation service reliability.[87]

FPL makes extensive use of smart sensors, analytics and Artificial Intelligence to monitor the grid. Smart meters increase the accuracy of outage predictions and verify power restoration.[88] Intelligent distribution grid sensors equipped with advanced analytics and communications are used to detect defective equipment before it fails, and to predict the likelihood of future outages.[89] Drones are used to assess overhead power equipment, providing maintenance inspections to identify areas of concern before an outage can occur. During a disruption, distribution smart islanding uses adaptive load shedding. Following a severe weather event, drones help assess damage and support rapid restoration of power.[90]

To enable rapid response following a hurricane, up to a week in advance of a possible landfall, FPL uses specialized weather forecasts of landfalling winds to drive outage models that estimate the size of the repair crews that are needed and where to place the crews and other emergency management assets. For Hurricane Irma (2017), FPL employed a storm restoration workforce of 28,000, which included nearly 20,000 workers from 30 states and Canada to restore outages to 4.8 million customers. These workers, along with mobile command units, were pre-positioned several days prior to hurricane landfall.[91]

Each year, FPL tests the response of more than 3,000 employees to forecasts and simulated landfall of a hypothetical hurricane, drawn from an extensive catalog of historical storms and synthetic worst-case scenarios (Section 9.4.1). During the simulation, employees track outages, assess damage, communicate with customers and employees, and initiate service restoration. The drill also tests the company's storm plans and tactics, and applies lessons learned from previous hurricanes and other extreme weather events. The drill also involves local emergency responders and weather forecast providers and contractors and suppliers.[92]

Rapid restoration of electric power following a hurricane clearly has nothing to do with luck.

13.3 Adaptation Lessons and Challenges

"Out of life's school of war: what does not kill me, makes me stronger." (German philosopher Friedrich Nietzche)[93]

Social scientists have been studying climate adaptation for decades. A growing understanding is emerging of best practices, lessons learned, and challenges.

How to assess adaptation is not always straightforward. Outcomes are typically evaluated for effectiveness, adequacy, justice, and equity in both outcomes and processes, and also tradeoffs with mitigation, ecosystem functioning, and other societal goals. Adaptation interventions can have both positive and negative consequences, the latter referred to as maladaptation.

However, some general principles, guidelines, and recommendations have emerged.

13.3.1 Lessons

"Those who are easily shocked should be shocked more often." (American actress Mae West)[94]

The analysis presented here of lessons learned and general best practices are drawn from the World Resources Institute,[95] World Bank Report on Economics of Adaptation to Climate Change,[96] the IPCC AR6 WGII Chapter Seventeen,[97] and a Workshop that evaluated a comprehensive review of the adaptation literature.[98]

Adaptation is fundamentally local—tailored to specific locations so that communities can adopt flexible and incremental solutions. There is no universal quick-fix recipe that translates across countries and communities. When done effectively, the benefits of local adaptation outweigh the estimated costs. Avoiding loss and damage from weather and climate stressors unlocks economic potential for the region.

Adaptation effectiveness is evaluated through outcomes as well as by enabling processes (e.g., community participation) that prioritize and implement adaptation. With regard to outcomes, effective adaptation should improve well-being, reduce vulnerability or increase adaptive capacity, enhance resilience, provide benefits to equity (marginalized ethnic groups, gender, low-income populations), and have transformation potential. Adaptation should be economically, ecologically, and socially sustainable, with consideration of longer-term, cross-generational viability of the adaptation actions.[99]

There is growing appreciation for the importance of Community-based Adaptation, which focuses on increasing the participation and agency of communities in adaptation prioritization and implementation. Community-based Adaptation explicitly focuses on mainstreaming community priorities, needs, knowledge, and capacities into adaptation thereby aiming to empower people to adapt more effectively.[100]

And finally, the adaptation process should be transformative, changing human thinking and behaviors in the face of climate change, and overtly confronting the power structures that generate vulnerability.[101]

The World Bank *Report on Economics of Adaptation to Climate Change* summarized its findings with seven lessons for adaptation.[102] The key finding is that economic development is the best hope for climate change adaptation, provided that it is properly managed. Economic development enables an economy to diversify and generate resources needed for adaptation. Economic development provides opportunities to adapt to climate change at a relatively low cost by locating and designing new infrastructure in ways that accounts for the impacts of extreme weather and climate change. The Report recommends investing in human capital, developing competent and flexible institutions, and tackling the root causes of poverty. Eliminating poverty is central to both development and adaptation, since poverty exacerbates vulnerability to weather variability and climate change.[103]

Some additional lessons from the Report:[104]

- Avoid rushing into long-lived investments for adaptation unless these are robust to a wide-range of climate outcomes or until the degree of uncertainty regarding future weather variability and climate has narrowed. Start with low-regret policies that would be priorities for development even without climate change—especially in water supply and flood protection.
- The short-term priority should be to better prepare for the weather risks that countries are already facing, such as investment in water storage in drought-prone regions and protection against flooding in urban areas. Regions that learn to cope with existing climate variability will be more successful in adapting to future climate change.
- Steer clear of developing incentives that promote development in locations exposed to severe weather risks, such as flood plains, coastal zones, and regions that are prone to wildfires.
- A key challenge is to strike the right balance between hard (capital intensive) and soft (institutions and policies) adaptation. Good policies, planning, and institutions are vital in ensuring that capital-intensive actions produce the anticipated benefits.

There is no simple recipe for adaptation, and adherence to all of these recommendations and lessons may not be feasible for an individual adaptation project, and there are invariably tradeoffs to be made. However, blindly neglecting any of these without considering implications can lead to maladaptation. In summary, good adaptation outcomes require flexible and strong institutions, policy integration, dynamic risk management, broad participation from stakeholders, and account for long term goals.

13.3.2 Maladaptation

"Recall the face of the poorest and the weakest man whom you may have seen, and ask yourself, if the step you contemplate is going to be of any use to him. Will he gain anything by it? Will it restore him to a control over his own life and destiny?" (Indian leader Mahatma Gandhi)[105]

Apart from the political and financial challenges associated with adaptation projects, there are also potential problems associated with risk transference and maladaptation.

Maladaptation refers to adverse outcomes resulting from intentional adaptation measures.[106] Maladaptation relates to increasing vulnerability for the targeted or external sectors, eroding preconditions for sustainable development and future adaptation with increased risk of adverse climate-related results, increased susceptibility to climate change, or reduced welfare, now or in the future.[107] Maladaptation differs from failed or unsuccessful adaptation,[108] which describes a failed adaptation initiative that does not produce any significant detrimental effect.[109]

Given the economic, social, and biophysical connections between people and places, adaptation reactions meant to dampen the impacts of climate change can end up redistributing risks and vulnerability.[110]

Maladaptation does not necessarily occur in the geographic space or within the targeted group; it can extend social and geographic boundaries. In Vietnam, hydroelectric dam and forest protection policies to regulate floods in lowlands are beneficial for reducing local vulnerability to specific hazards. However, these policies undermined access to land and forest resources for mountain peoples upstream, resulting in their greater vulnerability to the impacts of weather and climate variability.[111]

Maladaptation also has a temporal element—adaptation actions taken today can be maladaptive in the future. For example, irrigation may bring short-term benefits by ensuring a harvest, but overutilization of ground water can produce long term water shortages if the water table is not replenished by heavier rainfall.[112]

There is no hard and fast delineation between successful adaptation versus maladaptation. Rather there is a continuum, and adaptation options can score high or low on different outcome criteria. Further, the location of a given adaption option along this continuum is dynamic, varying with changes in the characteristics of climate hazards and the effects of iterative risk management. In order to limit the risk of maladaptation, explicit effort is needed to avoid negative consequences of adaptation interventions, anticipate detrimental lock-ins and path dependence, and minimize spatiotemporal trade-offs.

A seminal paper was recently published entitled "Adaptation interventions and their effect on vulnerability in developing countries: Help, hindrance or irrelevance?"[113] The paper was authored by a group of 20 adaptation scientists and practitioners. The paper critically reviewed the outcomes of internationally funded interventions aimed at climate change adaptation and vulnerability reduction. The paper found that many internationally funded climate adaptation projects reinforce, redistribute or create new vulnerability in developing countries. People in developing countries are often worse off after climate change adaptation interventions. The crux of the problem is that the adaptation projects frequently focus on the impacts of climate change rather than on the root causes of vulnerability.[114]

Projects designed as part of the Green Climate Fund or the Adaptation Fund have their finances managed by an international agency like the United Nations. In turn, funds are sent to nongovernmental organizations or local research institutes where they implement the local project. The failures of these "top-down intervention" methods have been well documented for decades. At the core, a primary cause of maladaptation is the lack of engagement with local stakeholders.[115]

The development community continues to investigate the causes of maladaptation and how best to avoid it. Good intentions aside, replicating previous errors can result in just making the situation worse not better. Incomplete understanding with regard to vulnerability context and social inequities can result in the most marginalized groups being ignored in adaptation planning and evaluation.[116,117]

A project in Melbourne, Australia, illustrates different maladaptation pathways, highlighting specific challenges to be considered when developing and implementing adaptation actions. Melbourne experienced annual rainfall below the long-term average every year for the period 1996 through the late 2000s. The combination of drought and rapid population growth led the Victorian state Premier in 2007 to declare a water crisis, recognizing "that climate change and record low rainfall demands a dramatic new approach to how we plan for Victoria's water needs."[118] He announced plans for two schemes: a desalinization plant and a pipeline to transport the water.[119] The drought broke in 2010–11, and operations of the plant and pipeline began in 2012. However, it wasn't until 2017 that the first water was withdrawn from the facility, owing to favorable rainfall conditions. Substantial annual operating costs continue to be incurred even when no water is withdrawn. In recent years, flooding in eastern Australia has been a much greater problem than drought.

The desalinization/pipeline projects contributed to maladaptation in the following ways. The desalinization plants are placed on locations that are significant to the indigenous Aboriginal community, who vehemently opposed the development. The projects impact disproportionally on poorer households in the form of higher water costs, who pay a higher share of their incomes on water and power. The project reduced incentives to adapt—prior to the project, many households in Melbourne had taken actions to adjust to water scarcity: recycling grey water, capturing rainfall, and replacing non-indigenous plant species with native ones better acclimatized to long dry periods. The Melbourne desalinization and pipeline have high opportunity costs and created path dependency that reduces the portfolio of adaptation options in the future. And finally, desalinization and the cross-basin water transfer require substantial electric power, challenging the transition away from fossil fuels.[120]

The adaptation measures least likely to result in maladaptation are those that seek to improve adaptive capacity, while those that focus on exposure reduction carry the greatest maladaptation risk.[121]

13.3.3 Resilience Traps

"Blessed are the flexible, for they shall not be bent out of shape." (Chiropractor Michael McGriff)[122]

An additional concern about adaptation is the potential for creating "resilience traps."[123] An overemphasis on recovery without accounting for transformation entrenches resilience traps, where recovery acts to inhibit positive transformation and perpetuate maladaptive states.[124]

Distorted incentives and government policies can create resilience traps. In the United States, current policies distort incentives in a way that increases vulnerability to extreme weather events, resulting in public investment that protects unwise private investments. These policies include subsidized flood insurance and federal funding for reconstruction after a disaster, which encourages people to build in areas known to be vulnerable. Providing aid to rebuild the areas that were damaged reinforces the incentive to downplay risks.[125]

Consider the role that US federal policies have played in the growing vulnerability of Florida to hurricanes. The history of Florida is intimately connected with hurricanes. In the 1920s, Florida's new railroads spurred a land boom. Then the 1926 Miami Hurricane nearly destroyed the city. In 1928, the Okeechobee hurricane made landfall near Palm Beach, breaching a dike that killed over 2,000 people and destroyed two towns. The 1926 hurricane thrust Florida into an economic depression and the 1928 hurricane effectively ended Florida's land boom.[126]

From the 1920s to 1940, Florida's population increased by less than 1 million, and until the 1970s the Florida Keys were largely undeveloped. Between 1951 and until Hurricane Andrew in 1992, only four major hurricanes struck the state of Florida, and the population increased by 10 million. A lull in hurricane landfalls during the 1970s and 1980s and rapid real estate development encouraged insurers to continue driving down the overall cost of the homeowner's insurance.

The most politically important hurricane that you have probably never heard of is Hurricane Frederic, a Category 3 hurricane that struck Alabama and Mississippi in 1979. This landfall occurred shortly after the US Federal Emergency Management Administration (FEMA) was established, and was the focal point for nearly $250 million in federal aid for recovery. In 1992, following the catastrophic damage to Miami from Hurricane Andrew, Robert Sheets (then

Director of the National Hurricane Center) stated in Congressional testimony that he credited the aid for Frederic's recovery with spurring development in hurricane prone regions.[127]

The landfall of Hurricane Andrew caused the largest catastrophic loss that the insurance industry had ever experienced up to that time and emphasized the increased exposure along Florida's coastline. Following Hurricane Andrew, Florida implemented strict building codes to withstand hurricane winds.[128] However, catastrophic losses from hurricanes have continued, particularly during 2004–05, and population and property development have continued to increase, with Florida's current population of more than 21 million people (in 2021) making it the third most populous state in the United States.

Also of concern is FEMA's National Flood Insurance Program, which provides coverage at subsidized rates for homes deemed too risky for commercial insurers. The Program's artificially low cost encourages residents to repeatedly rebuild their homes rather than move away. Efforts to reform the program have stalled because the resulting rate hikes would disadvantage low-income policyholders. As of 2012, repetitive loss properties that have been rebuilt multiple times using federal flood insurance payouts had cost US taxpayers more than $12 billion.[129]

The emphasis on restoring the status quo engenders norms and policies that inhibit the ability of communities to transform. Changing the motivations to locate in places at risk is far more cost-effective than fostering and defending unwise investments.[130]

13.4 Development and Resilience

"Be formless, shapeless, like water. Now, you put water into a cup, it becomes the cup. Put it into a teapot, it becomes the teapot. Now, water can flow or creep or drip or crash. Be water, my friend." (Hong Kong martial artist and actor Bruce Lee)[131]

There is growing recognition that climate change is as much a development problem as it is an environmental one. Development deals with the alleviation (or eradication) of poverty. More recent notions of development include sustainable development and climate resilient development, which emphasizes economic development without depletion of natural resources. Climate resilient development is the focus of the IPCC AR6 WGII (Chapter Eighteen). Tensions arise when the "sustainable" part—which includes no new fossil fuel-based energy systems—conflicts with poverty eradication.

An investigation of climate change vulnerability from a development perspective has recommended first assessing present adaptation to existing climate variability. Steps to improve present levels and types of adaptation to reduce present vulnerability are regarded as essential to reducing vulnerability climate change in the future, and so support economic development.[132]

Disaster reduction efforts have been central to international development strategies. Vulnerability to disasters and challenges to international development programs are bound up with deep-rooted, structural drivers related to development ideologies, cultural factors, ingrained habits, social inequality, and other processes that all have a role in the creation of poverty and disasters.

Development and resilience are potentially being slowed down by a growing emphasis on linking international development funds to reducing emissions. This emphasis comes at the expense of development funds that have historically been targeted for poverty reduction.

13.4.1 Adaptive Capacity

"The gods had condemned Sisyphus to ceaselessly rolling a rock to the top of a mountain, whence the stone would fall back of its own weight. They had thought with some reason that there is no more dreadful punishment than futile and hopeless labor." (French philosopher Albert Camus)[133]

Adaptive capacity is the property of a human system to adjust and expand its coping range to climate variability and change. A system may be defined as a region, a community, a household, an economic sector, a business, a population group, or ecological system. In practical terms, adaptive capacity is the ability to design and implement effective adaptation strategies to moderate potential damages, take advantage of opportunities, and cope with the consequences. Resources for adaptation may be natural, financial, institutional, or human, and include access to ecosystems, information, expertise, and social networks.[134]

Adaptive capacity is closely linked to social and economic development. Adaptive capacity is commonly assessed using income indicators, education statistics, emergency response protocols, business continuity schemes, and plans for adaptation. More specific determinants of adaptive capacity include: the availability of financial resources; the availability of technology and people trained to utilize it effectively; access to information; the existence of legal, social, and organizational arrangements; and effective leadership and governance.

Higher levels of economic development are generally associated with higher adaptive capacity, but some development locks people into certain patterns or behaviors. And the most developed areas may have low adaptation capacity to new hazards. For example, a region that is well adapted to floods may have a catastrophic outcome to an extreme heat wave.

Whereas the problem for wealthy countries is adapting sensibly and cost-effectively, lower income countries have difficulties in adapting at all. Common factors hampering the effective adaptation of poor countries include: violence and insecurity that make investments questionable; corrupt rulers who leave their populace in poverty while appropriating foreign aid and economic surpluses or for their personal benefit; and lack of clear property and land rights that are essential for investment incentives. Further impediments include closed political systems that exclude most of their population from meaningful involvement, and public works projects that benefit a narrow political support base but not the entire country.[135]

Capacity development refers to the process of expanding the coping range and strengthening the coping capacity of a priority system. Capacity development is viewed as a central goal of most adaptation strategies.[136] As such, capacity development is a primary objective of international development agencies and programs. Human development adaptation solutions include poverty reduction and risk-spreading through income diversification.[137]

In poor countries with a low human development index, in-country capacity building can be a substantial challenge.[138]

13.4.2 Disaster Reduction

"Antifragility is beyond resilience or robustness. The resilient resists shocks and stays the same; the antifragile gets better." (Mathematician Nassim Nicholas Taleb)[139]

Disasters can arise from small-scale recurrent hazards such as landslides or floods, relatively infrequent hazards such as a major hurricane, or slow onset crises such as drought and sea level rise. Climate change is one contributor to disaster risk amongst many, and not necessarily the most prominent or fundamental contributor. However, the political importance of climate change provides an opportunity to tackle the deep-rooted sources of vulnerability.[140]

Disasters occur from the intersection of hazards, vulnerability, and exposure of people and assets. Disasters are the effects of human choices, and are created through a combination of human interference with natural processes, the social production of vulnerability, and the neglect of response capacities.[141]

Poverty contributes to increased disaster vulnerability and reduced capacity to cope with and recover from disasters. Disasters, in turn significantly increase poverty by shrinking or eliminating coping capacity. Post disaster relief can create a resilience trap, if no efforts are made to mitigate the underlying poverty.

Good disaster risk governance aims to avoid the creation of vulnerability and exposure by tackling drivers and root causes of risk. Poverty is both the most persistent cause and the most pernicious effect of human disaster. The same loss in dollars affects the poor far more because they have fewer assets, their consumption is at subsistence level, they cannot rely on savings, their health and education are at greater risk, and they need more time and resources to recover and rebuild. The poor are also less likely to be adequately covered by social assistance or insurance programs that can reimburse at least part of their losses.[142]

While almost every country in Asia has made substantial progress in recent decades in saving the lives of victims of slow-onset flood disasters, these floods remain relentlessly impoverishing. In India alone, on average six million hectares of land is inundated each year, affecting 35–40 million people. The flooding occurs in the fertile flood plains of major rivers, causing the loss of agricultural inputs (seed, fertilizer, and pesticides) and livestock, with costs in excess of US$1 billion for an average flood or drought event. Smallholders nearly always purchase agricultural inputs on credit against repayment from the expected harvest. The loss of crops, agricultural inputs, and livestock typically place a farming family in debt for several years, by which time the disaster cycle is repeated, condemning successive generations to the treadmill of poverty.[143]

The UN *Sendai Framework on Disaster Risk Reduction 2015–2030* focuses on the adoption of measures that address all dimensions of disaster risk—hazard, exposure, vulnerability, and coping capacity.[144] The Sendai Framework includes seven targets intended to define and measure progress towards its overall goal to increase resilience by reducing risk. The first four targets are to substantially reduce disaster impacts: mortality, number of people affected, economic loss, and damage to critical infrastructure, and disruption of basic services. The other three targets are to substantially increase the adoption of national and local disaster risk reduction strategies, international cooperation to developing countries, and access to multi-hazard early warning systems.[145]

Now, halfway through the period for the Sendai Framework (2015–30), none of the Sendai Framework's "substantially reduce" targets are on track to be achieved by 2030. Instead, direct economic loss and damage to critical infrastructure having increased substantially over the past decade. Adoption of multi-hazard early warning systems may be the most successful element of the Framework, although these efforts need to be scaled up and made more effective. While disaster-related financing has increased since 2010, most of the resources have supported activities to respond to and recover from disasters.[146]

Disaster risk and impacts can be reduced by tackling fundamental issues that cause vulnerability, no matter what the evolution of weather and climate. Risk management, risk sharing, and warning strategies are key tasks for adapting to climate change. Rather than "bouncing back" from a disaster, resilience could be enhanced by "build forward better"[147] or "bouncing forward"[148] approaches. Re-establishing beneficial practices, ensuring that everyday needs are fulfilled, and changing social processes and structures that sustain vulnerability would increase resilience and diminish the impacts of hazards.[149]

13.4.3 Conflicts with Mitigation

"Working in global energy and development, I often hear people say, 'Because of climate, we just can't afford for everyone to live our lifestyles.' That viewpoint is worse than patronizing. It's a form of racism, and it's creating a two-tier global energy system, with energy abundance for the rich and tiny solar lamps for Africans." (Kenyan activist and materials scientist Rose Mutiso)[150]

Simply put, people are considerably less exposed to weather and climate shocks if they aren't poor. Creating more resiliency in poor countries will require energy-intensive investments in housing, transportation infrastructure, and agricultural technology. Economic development requires the availability of cheap, reliable, and abundant energy.[151]

International funds for development are being redirected away from reducing poverty and increasing resilience, and towards reducing carbon emissions. By limiting development of electric power, this redirection is exacerbating the harms of weather hazards and climate change for the world's poor. Development and poverty reduction requires abundant and cheap energy, and natural gas is regarded as the best near-term solution for most countries. Working against this need in developing countries, UN Secretary-General António Guterres has called on countries to end all new fossil fuel exploration and production. The United Kingdom, the United States, and the European Union are aggressively limiting fossil fuel investments; the World Bank, the International Monetary Fund (IMF), and other development banks are being pressured to do the same. The African Development Bank is increasingly unable to support large natural gas projects in the face of European shareholder pressure.[152]

Limiting the development of fossil fuel projects is profoundly hampering development in Africa. Africa is starved for energy; sub-Saharan Africa's one billion people have the power generation capacity that is less than the United Kingdom with 67 million people. Natural gas is needed not only for power, but also for industry and fertilizer and for cleaner cooking to avoid loss of life from indoor air pollution. One cannot overemphasize the significance of natural gas as a transition fuel in developing countries, especially in Africa. The irony is that even tripling energy use and emissions in sub-Saharan Africa driven by natural gas would add a meager 0.6 percent to overall global emissions.[153] The IEA projects that Africa won't exceed 4 percent of global CO_2 emissions by 2050 regardless of energy scenario.[154]

The World Bank and other development banks provide loans and grants for development projects, and the IMF helps poor countries overcome currency crises and keep their finances stable. However, these institutions are under pressure from their donor governments to focus on climate change—specifically, to reduce emissions.[155] The IMF recently proposed the creation of a US$50 billion Resilience and Sustainability Trust to help countries tackle climate change, where support would be contingent on recipient countries' plans to reduce emissions.[156] Similarly, the World Bank has unveiled a climate action plan promising to align all future projects with the Paris Agreement to slash emissions.[157]

Development experts Vijaya Ramachandran and Arthur Baker argue that the poorest countries should not be compelled to pivot to emissions reduction as a prerequisite for qualification for aid and loans. They argue that the shift of focus from poverty to climate is ineffective and unjust with potential disastrous outcomes for the world's poor. This is happening because the World Bank and IMF are deprioritizing poverty reduction, despite the vital need to protect developing countries from climate shocks caused by historical emissions from wealthy countries. Given that poor countries make up only a tiny fraction of global emissions and their share will remain small even with a marked increase in fossil fuel usage, this shift will prove ineffective. Finally, the shift will be disastrous for the 3 billion people struggling to escape misery because every dollar spent on reducing carbon emissions can have a significantly greater impact if directed into education, medical services, food security, and critical infrastructure. To promote human well-being and thrivability, climate action in the poorest countries should concentrate on reducing poverty, increasing energy access, and building resilience via investments in housing, transportation, infrastructure, and agricultural technology.[158]

Leaders from developing countries have been outspoken in criticizing these changes in international funding practices. Ugandan President Yoweri Museveni warns that by pushing climate mitigation on African countries, the West will "forestall Africa's attempts to rise out of poverty."[159] A widely viewed Technology, Entertainment, Design (TED) Talk by Kenyan energy expert Rose Mutiso characterized forcing emissions mitigation on the world's poor that is widening economic inequality as equivalent to "energy apartheid."[160]

Africa's fragile progress in recent decades could be undone by international efforts to curb investments in all fossil fuels. However, these same countries that are working to restrict fossil fuel investments in Africa include natural gas in their own multidecade plans to transition to clean energy. The greatest hypocrisy is that some of the largest private European and US firms are developing natural gas in Ghana, Mozambique, Nigeria, and Senegal to export to Asia and Europe, since it cannot be used in the countries of origin for lack of infrastructure.[161]

13.4.4 Bangladesh

"Time and time again I have seen [non-governmental organization] NGOs and politicians in rich countries advocate that the poor follow a path that they, the rich, never have followed nor are willing to follow." (John Briscoe, environmental engineer and 2014 recipient of the Stockholm Water Prize)[162]

In many ways, Bangladesh is a remarkable success story for development and for building resilience to weather and climate extremes, with numerous lessons to be gleaned from their success. Bangladesh's low elevation and high population density make it particularly vulnerable to rising sea levels, flooding, and tropical cyclones.

In 1970, the devastating Bhola tropical cyclone struck East Pakistan, causing an estimated 500,000 deaths. The woefully insufficient response to the disaster from Pakistan's government triggered a war for independence by East Pakistan. In 1971, the newly formed Bangladesh was the poster child for poverty and weather/climate disasters. The dire situation in Bangladesh even penetrated my own teenaged awareness, with Beatle George Harrison's Concerts for Bangladesh.[163]

In 1974, Bangladesh's government nationalized energy resources. The natural gas resources were concentrated into power generation supporting critical industrial growth, development of a fertilizer sector, cement production, and irrigation water supply. These investments provided the foundation for Bangladesh's modernization. In 2015, the World Bank upgraded Bangladesh to a lower-middle income country, with a life expectancy within seven years of the United States. By 2019, Bangladesh had become self-sufficient in food as well as a significant exporter of textiles, apparel, and leather products. While its economy is now the fastest growing in South Asia, Bangladesh is greener than many neighboring countries (like India) which still depend on coal for electricity.[164]

Bangladesh has adapted to extreme weather events and climate with improved forecasting, community-wide training and education campaigns, and infrastructure investments. My company CFAN has played a role in this adaptation, under the leadership of CFAN's Chief Scientist and co-owner, Peter Webster.

Every few years, major floods engorge the Brahmaputra and Ganges Rivers for periods ranging from a few days to a month or more, often displacing tens of millions of people and devastating agricultural production. CFAN overcame numerous technical hurdles to develop an extended-range probabilistic flood forecasting system for the Ganges and Brahmaputra (on time scales from days to 6 months) to predict the probability that river water level heights will exceed critical levels. The flood forecast system became operational in 2003 for the Ganges and Brahmaputra Rivers. In 2007, a new experimental dissemination program brought warnings directly via a cell phone network to more than 100,000 residents in five rural provinces in Bangladesh. On the basis of these forecasts, entire areas were evacuated ahead of the flooding. Early harvesting of some crops occurred and livestock and belongings were saved.[165]

"We disseminate the forecast information and how to read the flag and flood pillar to understand the risk during the prayer time. In my field, T. Aman (a rice variety) was at seedling and transplanting stage, I used the flood forecast information for harvesting crops and making decision for seedling and transplantation of T. Aman. Also, we saved household assets." (The Imam from the Mosquein Koijuri Union of Sirajgong District in Bangladesh 2008)[166]

Bangladesh's development approach has often run contrary to recommendations from international-development institutions, nongovernmental organizations, and global environmental groups. Instead, by using its own natural gas resources, Bangladesh has increased its prosperity and reduced its vulnerability, buying time to make an energy transition while not compromising the needs and aspirations of its people.[167]

To continue lifting its population out of poverty and to raise living standards across the country, Bangladesh needs to generate and consume increasing levels of energy. The challenge now should not be focused on how to phase out fossil fuels quickly, but instead how best to make the cleanest use of them. Prioritizing productive domestic use of the country's remaining natural-gas reserves will support efforts to reduce vulnerability, increase prosperity, and to eventually transition to cleaner energy sources.[168]

Notes

1 H. G. Wells, *Mind at the End of Its Tether* (London, UK: William Heinemann Limited, 1945), 19.
2 "An Introduction to Climate Change Adaptation," Government of Canada (Government of Canada, November 13, 2015), https://www.nrcan.gc.ca/changements-climatiques/impacts-adaptation/chapter-1-introduction-climate-change-adaptation/10081.

3 Matthew E. Kahn, *Adapting to Climate Change: Markets and the Management of an Uncertain Future* (New Haven, CT: Yale University Press, 2021).
4 Ibid.
5 Laurence J. Peter, *Peter's Quotations: Ideas for Our Time* (New York, NY: William Morrow and Company, Inc., 1977), 125.
6 David Graeber and David Wengrow, *The Dawn of Everything: A New History of Humanity* (London, UK: Allen Lane, 2021).
7 Carl Sandburg, *The Complete Poems of Carl Sandburg* (London, UK: Harcourt, 2003), 433.
8 Michael Shellenberger, *Apocalypse Never: Why Environmental Alarmism Hurts Us All* (New York, NY: HarperCollins, 2020).
9 Ibid.
10 Amartya Sen, *Poverty and Famines: An Essay on Entitlement and Deprivation* (Oxford, UK: Clarendon Press, 1981).
11 Emma Illick-Frank, "5 Benefits to Local Action on Climate Resilience," World Resources Institute, June 23, 2020, https://www.wri.org/insights/5-benefits-local-action-climate-resilience.
12 Peter J. Webster, "Myanmar's Deadly Daffodil," *Nature Geoscience* 1, no. 8 (July 20, 2008): 488–490, https://doi.org/10.1038/ngeo257.
13 Kim Knowlton et al., "Development and Implementation of South Asia's First Heat-Health Action Plan in Ahmedabad (Gujarat, India)," *International Journal of Environmental Research and Public Health* 11, no. 4 (March 25, 2014): 3473–3492, https://doi.org/10.3390/ijerph110403473.
14 PIB Delhi, "India Cooling Action Plan Launched," Press Information Bureau, March 8, 2019, https://pib.gov.in/PressReleaseIframePage.aspx?PRID=1568328.
15 Emily Underwood, "How to Prevent Air Conditioners from Heating the Planet," Anthropocene Magazine (Scientific American, June 23, 2021), https://www.scientificamerican.com/article/how-to-prevent-air-conditioners-from-heating-the-planet/.
16 R. J. Palacio, *White Bird: A Wonder Story (a Graphic Novel)* (New York, NY: Random House USA, 2022), 126.
17 Roger Pielke Jr., "A Rapidly Closing Window to Secure a Liveable Future," A Rapidly Closing Window to Secure a Liveable Future (The Honest Broker by Roger Pielke Jr., March 2, 2022), https://rogerpielkejr.substack.com/p/a-rapidly-closing-window-to-secure?s=r.
18 Al Gore, *The Future: Six Drivers of Global Change* (New York, NY: Random House, 2013).
19 Al Gore, *Earth in the Balance: Ecology and the Human Spirit* (Boston, MA: Houghton Mifflin, 1992).
20 Al Gore, *The Future*.
21 Ibid.
22 United Nations, "Resolution Adopted by the General Assembly on 25 September 2015," United Nations, September 25, 2015, https://www.un.org/en/development/desa/population/migration/generalassembly/docs/globalcompact/A_RES_70_1_E.pdf.
23 "The Global Commission on Adaptation," Global Center on Adaptation, accessed July 6, 2022, https://gca.org/about-us/the-global-commission-on-adaptation/.

24 "FAQ 3: How Will Climate Change Affect the Lives of Today's Children Tomorrow, If No Immediate Action Is Taken?," Climate Change 2022: Impacts, Adaptation and Vulnerability, accessed July 6, 2022, https://www.ipcc.ch/report/ar6/wg2/about/frequently-asked-questions/keyfaq3/.

25 "Press Release," Climate Change 2022: Impacts, Adaptation and Vulnerability, February 28, 2022, https://www.ipcc.ch/report/ar6/wg2/resources/press/press-release/.

26 Roger Pielke Jr. (@RogerPielkeJr), "The first observation is that the report is more heavily weighted to implausible scenarios than any previous IPCC assessment report—In particular, RCP8.5 represents ~57% of scenario mentions—This alone accounts for the apocalyptic tone and conclusions throughout the report," Twitter, February 28, 2022, https://twitter.com/RogerPielkeJr/status/1498318212248735750.

27 Amar Bhattacharya et al., "G-24 Technical Session on: Delivering on Climate Finance to Support Better Recovery and Climate Goals," Intergovernmental Group of 24, March 11, 2021, https://www.g24.org/wp-content/uploads/2021/03/Richard-Calland-and-Amar-Bhattacharya_Independent-Expert-Group-on-CF-MARCH-2021.pdf.

28 Emmanuel Raju et al., "Stop Blaming the Climate for Disasters," *Communications Earth &Amp; Environment* 3, no. 1 (January 10, 2022), https://doi.org/10.1038/s43247-021-00332-2.

29 Ibid.

30 Myanna Lahsen and Jesse Ribot, "Politics of Attributing Extreme Events and Disasters to Climate Change," *WIREs Climate Change* 13, no. 1 (December 8, 2021), https://doi.org/10.1002/wcc.750.

31 Fred Pearce, "It's Not Just Climate: Are We Ignoring Other Causes of Disasters?," Grist, February 19, 2022, https://grist.org/climate/its-not-just-climate-are-we-ignoring-other-causes-of-disasters/.

32 Ibid.

33 Ibid.

34 Ibid.

35 Ripples Nigeria, "Fact Check: Is Lake Chad Drying up?," Latest Nigeria News | Top Stories from Ripples Nigeria, November 10, 2021, https://www.ripplesnigeria.com/fact-check-is-lake-chad-drying-up/.

36 Fred Pearce, "It's Not Just Climate."

37 Fred Pearce, "It's Not Just Climate: Are We Ignoring Other Causes of Disasters?," Yale E360, February 8, 2022, https://e360.yale.edu/features/its-not-just-climate-are-we-ignoring-other-causes-of-disasters.

38 World Bank Group, "Bangladesh Receives $400 Million to Rehabilitate 600km Embankments in Coastal Districts," World Bank (World Bank Group, October 1, 2013), https://www.worldbank.org/en/news/press-release/2013/10/01/bangladesh-receives-400-million-to-rehabilitate-600km-embankments-in-coastal-districts.

39 John Pethick and Julian D. Orford, "Rapid Rise in Effective Sea-Level in Southwest Bangladesh: Its Causes and Contemporary Rates," Global and Planetary Change 111 (October 7, 2013): 237–245, https://doi.org/10.1016/j.gloplacha.2013.09.019.

40 Fred Pearce, "Bangladesh's Sea Walls May Make Floods Worse," New Scientist (New Scientist, June 25, 2014), https://www.newscientist.com/article/mg22229752-700-bangladeshs-sea-walls-may-make-floods-worse/.

41 Ibid.

42 World Bank, "Building Resilience for Sustainable Development of the Sundarbans : Strategy Report," Open Knowledge Repository (World Bank Group, 2014), https://openknowledge.worldbank.org/handle/10986/20116.

43 Gardiner Harris, "Borrowed Time on Disappearing Land," The New York Times (The New York Times, March 28, 2014), https://www.nytimes.com/2014/03/29/world/asia/facing-rising-seas-bangladesh-confronts-the-consequences-of-climate-change.html.

44 Erin Niimi Longhurst and Ryo Takemasa, *A Little Book of Japanese Contentments: Ikigai, Forest Bathing, Wabi-Sabi, and More* (San Francisco, CA: Chronicle Books, 2018).

45 Clifton La Bree, *The Gentle Warrior: General Oliver Prince Smith, USMC* (Kent, OH: Kent State Univ. Press, 2001).

46 Mark Scott et al., "Climate Disruption and Planning: Resistance or Retreat?," *Planning Theory & Practice* 21, no. 1 (January 14, 2020): 125–154, https://doi.org/10.1080/14649357.2020.1704130.

47 B Glavovic et al., "Cities and Settlements by the Sea," in *Climate Change 2022: Impacts, Adaptation, and Vulnerability. Contribution of Working Group II to the Sixth Assessment Report of the Intergovernmental Panel on Climate Change* (Geneva, CH: Intergovernmental Panel on Climate Change, 2022).

48 Ignacio Juan Vázquez Carneiro, "The Netherlands and Its Biggest Challenge: Stopping Sea Level Rise," european student think tank, April 6, 2021, https://esthinktank.com/2021/04/06/the-netherlands-and-its-biggest-challenge-stopping-sea-level-rise/.

49 "Netherland's Flood Management Is a Climate Adaption Model for the World," PreventionWeb (UNDRR, September 28, 2021), https://www.preventionweb.net/news/netherlands-flood-management-climate-adaption-model-world.

50 B Glavovic et al., "Cities and Settlements by the Sea."

51 Firas Saleh and Michael P. Weinstein, "The Role of Nature-Based Infrastructure (NBI) in Coastal Resiliency Planning: A Literature Review," *Journal of Environmental Management* 183 (December 2016): 1088–1098, https://doi.org/10.1016/j.jenvman.2016.09.077.

52 B Glavovic et al., "Cities and Settlements by the Sea."

53 Ibid.

54 Wei Wang et al., "Development and Management of Land Reclamation in China," *Ocean & Coastal Management* 102 (December 2014): 415–425, https://doi.org/10.1016/j.ocecoaman.2014.03.009.

55 Jochen Hinkel et al., "The Ability of Societies to Adapt to Twenty-First-Century Sea-Level Rise," *Nature Climate Change* 8, no. 7 (June 25, 2018): 570–578, https://doi.org/10.1038/s41558-018-0176-z.

56 Dhritiraj Sengupta et al., "Gaining or Losing Ground? Tracking Asia's Hunger for 'New' Coastal Land in the Era of Sea Level Rise," *Science of The Total Environment* 732 (August 2020): 139290, https://doi.org/10.1016/j.scitotenv.2020.139290.

57 M Oppenheimer et al., "Sea Level Rise and Implications for Low Lying Islands, Coasts and Communities," in *IPCC Special Report on the Ocean and Cryosphere in a Changing Climate* (Geneva, CH: Intergovernmental Panel on Climate Change, 2019).

58 C Cissé et al., "Health, Wellbeing, and the Changing Structure of Communities," in *Climate Change 2022: Impacts, Adaptation, and Vulnerability. Contribution of Working*

Group II to the Sixth Assessment Report of the Intergovernmental Panel on Climate Change (Geneva, CH: Intergovernmental Panel on Climate Change, 2022).

59 Pei-Chin Wu et al., "Subsidence in Coastal Cities throughout the World Observed by Insar," *Geophysical Research Letters* 49, no. 7 (March 24, 2022), https://doi.org/10.1029/2022gl098477.

60 Ibid.

61 "Indonesia Picks Borneo Island as Site of New Capital," BBC News (BBC, August 26, 2019), https://www.bbc.com/news/world-asia-49470258.

62 https://qz.com/994459/the-us-is-relocating-an-entire-town-because-of-climate-change-and-this-is-just-the-beginning/

63 B Glavovic et al., "Cities and Settlements by the Sea."

64 Silja Klepp, "Framing Climate Change Adaptation from a Pacific Island Perspective *–the Anthropology of Emerging Legal Orders*," *Sociologus* 68, no. 2 (2018): 149–170, https://doi.org/10.3790/soc.68.2.149.

65 Robert E. Kopp et al., "Usable Science for Managing the Risks of Sea-Level Rise," *Earth's Future* 7, no. 12 (October 16, 2019): 1235–1269, https://doi.org/10.1029/2018ef001145.

66 Matthew E. Kahn, "Some Microeconomics of Extreme Heat Exposure in the United States," Environmental and Urban Economics, September 20, 2021, http://greeneconomics.blogspot.com/2021/09/some-microeconomics-of-extreme-heat.html.

67 Matthew E. Kahn, *Adapting to Climate Change*.

68 Matthew E. Kahn (@mattkahn1966), "Our cities have demonstrated great resilience in the face of these shocks and we will learn from this shock so that the next Ida causes less damage. The adaptation hypothesis posits that each punch that Mother Nature throws causes less damage.," Twitter, September 5, 2021, https://twitter.com/mattkahn1966/status/1434507734443909130.

69 Matthew E. Kahn, *Adapting to Climate Change*.

70 Ibid.

71 Ibid.

72 Terrence Malick et al., *Dirty Harry: Screenplay* (Hollywood, CA, 1970).

73 L. Mili et al., "Risk Assessment of Catastrophic Failures in Electric Power Systems," *International Journal of Critical Infrastructures* 1, no. 1 (February 4, 2004): 38–63, https://doi.org/10.1504/ijcis.2004.003795.

74 Heidi Waterhouse, "If You're Going to Fail, Fail Safely," LaunchDarkly (LaunchDarkly, January 15, 2018), https://launchdarkly.com/blog/if-youre-going-to-fail-fail-safely/.

75 Yeowon Kim et al., "The Infrastructure Trolley Problem: Positioning Safe-to-Fail Infrastructure for Climate Change Adaptation," *Earth's Future* 7, no. 7 (April 30, 2019): 704–717, https://doi.org/10.1029/2019ef001208.

76 Yeowon Kim et al., "Leveraging Sets Resilience Capabilities for Safe-to-Fail Infrastructure under Climate Change," *Current Opinion in Environmental Sustainability* 54 (February 2022): 101153, https://doi.org/10.1016/j.cosust.2022.101153.

77 Jennifer Weeks, "Failure Becomes an Option for Infrastructure Engineers Facing Climate Change," Scientific American (Scientific American, March 20, 2013), https://www.scientificamerican.com/article/failure-becomes-an-option-for-infrastructure-engineers-facing-climate-change/.

78 Yeowon Kim et al., "Fail-Safe and Safe-to-Fail Adaptation: Decision-Making for Urban Flooding under Climate Change," *Climatic Change* 145, no. 3–4 (October 26, 2017): 397–412, https://doi.org/10.1007/s10584-017-2090-1.

79 Kym Wootton, "E Source Market Research Reveals That Power Outages Cost Businesses over $27 Billion Annually, Winter Storm Jonas Makes It Worse," E source, January 27, 2016, https://www.esource.com/ES-PR-Outages-2016-01/Press-Release/Outages.

80 Peter Aldhous et al., "The Texas Winter Storm and Power Outages Killed Hundreds More People than the State Says," BuzzFeed News (BuzzFeed News, May 26, 2021), https://www.buzzfeednews.com/article/peteraldhous/texas-winter-storm-power-outage-death-toll.

81 Christopher Flavelle, "A New, Deadly Risk for Cities in Summer: Power Failures during Heat Waves," The New York Times (The New York Times, May 3, 2021), https://www.nytimes.com/2021/05/03/climate/heat-climate-health-risks.html.

82 Amy Norton, "Gun Deaths Continue to Rise in America's Cities—US News & World Report," US News & World Report, January 17, 2022, https://www.usnews.com/news/health-news/articles/2022-01-10/gun-deaths-continue-to-rise-in-americas-cities.

83 Matt Daniel, "Who Died during Hurricane Sandy, and Why? Find out on Earthsky.: Earth," EarthSky, November 27, 2012, https://earthsky.org/earth/who-died-during-hurricane-sandy-and-why/.

84 Clark W. Gellings, *Smart Grid Planning and Implementation* (Atlanta, GA: The Fairmont Press, 2015).

85 Manny Miranda, "Hurricane Wilma 15 Years Later," FPL, accessed July 11, 2022, https://www.fpl.com/blog/hurricane-wilma.html.

86 Ibid.

87 "Florida Power & Light Wins ReliabilityOne National Reliability Award for Fourth Time in Five Years," UtilityProducts, November 26, 2019, https://www.utilityproducts.com/line-construction-maintenance/article/14072736/florida-power-light-wins-reliabilityone-national-reliability-award-for-fourth-time-in-five-years.

88 "FPL Showcases Category 5-Rated Control Center during Hurricane Drill," T&D World, July 18, 2019, https://www.tdworld.com/smart-utility/outage-management/article/20972872/fpl-showcases-category-5rated-control-center-during-hurricane-drill.

89 "Sentient Energy Signs Deal with Florida Power & Light to Deliver 20,000 Distribution Line Sensors," Business Wire, December 8, 2015, https://www.businesswire.com/news/home/20151208005085/en/Sentient-Energy-Signs-Deal-with-Florida-Power-Light-to-Deliver-20000-Distribution-Line-Sensors.

90 "You Depend on Reliable Energy," FPL, accessed July 11, 2022, https://www.fpl.com/reliability/drones.html.

91 "FPL Showcases Category 5-Rated Control Center during Hurricane Drill."

92 Ibid.

93 Friedrich Willhelm Nietzsche, *Twillight of the Idols*, 1888.

94 Erica Brown, *Take Your Soul to Work: 365 Meditations on Every Day Leadership* (New York, NY: Simon & Schuster, 2015).

95 Emma Illick-Frank, "5 Benefits to Local Action on Climate Resilience."

96 World Bank, "Economics of Adaptation to Climate Change."

97 M New et al., "Decision Making Options for Managing Risk," in *Climate Change 2022: Impacts, Adaptation, and Vulnerability. Contribution of Working Group II to the Sixth Assessment Report of the Intergovernmental Panel on Climate Change* (Geneva, CH: Intergovernmental Panel on Climate Change, 2022).

98 Chandni Singh et al., "Interrogating 'Effectiveness' in Climate Change Adaptation: 11 Guiding Principles for Adaptation Research and Practice," *Climate and Development*, August 24, 2021, 1–15, https://doi.org/10.1080/175655 29.2021.1964937.

99 Ibid.

100 Karen Elizabeth McNamara and Lisa Buggy, "Community-Based Climate Change Adaptation: A Review of Academic Literature," *Local Environment* 22, no. 4 (August 5, 2016): 443–460, https://doi.org/10.1080/13549839.2016.1216954.

101 Chandni Singh et al., "Interrogating 'Effectiveness' in Climate Change Adaptation."

102 World Bank, "Economics of Adaptation to Climate Change."

103 Ibid.

104 Ibid.

105 Mahatma Gandhi and Pyarelal, *Mahatma Gandhi, The Last Phase, Vol II* (Ahmedabad, IN: Navajivan Publishing House, 1958).

106 J D Scheraga and A E Grambsch, "Risks, Opportunities and Adaptation to Climate Change," *Climate Research* 11 (December 10, 1998): 85–95, https://doi.org/10.3354/cr011085.

107 Philip Antwi-Agyei et al., "Adaptation Opportunities and Maladaptive Outcomes in Climate Vulnerability Hotspots of Northern Ghana," *Climate Risk Management* 19 (2018): 83–93, https://doi.org/10.1016/j.crm.2017.11.003.

108 E.L.F. Schipper et al., "The Debate: Is Global Development Adapting to Climate Change?," *World Development Perspectives* 18 (June 2020): 100205, https://doi.org/10.1016/j.wdp.2020.100205.

109 Alexandre Magnan and Gaëll Mainguy, "Avoiding Maladaptation to Climate Change: towards Guiding Principles," *S.A.P.I.EN.S* 7.1 (2014).

110 Aaron Atteridge and Elise Remling, "Is Adaptation Reducing Vulnerability or Redistributing It?," *WIREs Climate Change* 9, no. 1 (October 26, 2017), https://doi.org/10.1002/wcc.500.

111 Siri Eriksen et al., "Adaptation Interventions and Their Effect on Vulnerability in Developing Countries: Help, Hindrance or Irrelevance?," *World Development* 141 (May 2021): 105383, https://doi.org/10.1016/j.worlddev.2020.105383.

112 Ibid.

113 Ibid.

114 Rishika Pardikar, "When Climate Adaptation Intervention Risks Further Marginalization," Eos, February 22, 2021, https://eos.org/articles/when-climate-adaptation-intervention-risks-further-marginalization.

115 Ibid.

116 Lisa Schipper et al., "Why Avoiding Climate Change 'Maladaptation' Is Vital," Carbon Brief, February 10, 2021, https://www.carbonbrief.org/guest-post-why-avoiding-climate-change-maladaptation-is-vital/.

117 Siri Eriksen et al., "Adaptation Interventions."

118 Jon Barnett and Saffron O'Neill, "Maladaptation," *Global Environmental Change* 20, no. 2 (May 2010): 211–213, https://doi.org/10.1016/j.gloenvcha.2009.11.004.

119 Ibid.

120 Ibid.

121 Antje Lang, "Maladaptation: An Introduction," weADAPT, January 8, 2019, https://www.weadapt.org/knowledge-base/vulnerability/maladaptation-an-introduction.

122 John C. Maxwell, *The 17 Essential Qualities of a Team Player: Becoming the Kind of Person Every Team Wants* (Nashville, TN: Thomas Nelson, 2002), 1.

123 Benjamin Rachunok and Roshanak Nateghi, "Overemphasis on Recovery Inhibits Community Transformation and Creates Resilience Traps," *Nature Communications* 12, no. 1 (December 17, 2021), https://doi.org/10.1038/s41467-021-27359-5.

124 Ibid.

125 W. David Montgomery, "The Costs of Inaction: The Economic and Budgetary Consequences of Climate Change," US Senate Budget Committee, July 29, 2014, https://www.budget.senate.gov/imo/media/doc/Montgomery%20July%2029%20Testimony.pdf.

126 "The Great Florida Land Boom," Florida in the 1920's (Florida History Internet Center), accessed July 13, 2022, http://floridahistory.org/landboom.htm.

127 "Hurricane Frederic," Wikipedia (Wikimedia Foundation, September 6, 2004), https://en.wikipedia.org/wiki/Hurricane_Frederic.

128 "Setting New Standards for Safety—The Florida Building Code," Florida Building Commission—Florida Department of Community Affairs, accessed July 13, 2022, https://www.floridabuilding.org/fbc/publications/fbc.pdf.

129 Rawle O. King, "National Flood Insurance Program: Background, Challenges, and Financial Status," Washington Pst, June 12, 2012, https://www.washingtonpost.com/wp-srv/business/documents/health-science-NFIP-123110.pdf.

130 W. David Montgomery, "The Costs of Inaction."

131 *Bruce Lee: A Warrior's Journey* directed by John Little (Warner Home Video, 2000).

132 Ian Burton, "Come Hell or High Water—Integrating Climate Change Vulnerability and Adaption into Bank Work," World Bank, October 31, 1999, https://documentos.bancomundial.org/es/publication/documents-reports/documentdetail/212171468756566936/come-hell-or-high-water-integrating-climate-change-vulnerability-and-adaption-into-bank-work.

133 Albert Camus, *The Myth of Sisyphus*, trans. Justin O'Brien (New York, NY: Vintage International, 2018).

134 Nick Brooks et al., "Assessing and Enhancing Adaptive Capacity," UNFCCC, 2004, https://www4.unfccc.int/sites/NAPC/Country%20Documents/General/apf%20technical%20paper07.pdf.

135 W. David Montgomery, "The Costs of Inaction."

136 Nick Brooks et al., "Assessing and Enhancing Adaptive Capacity."

137 P. M. Kelly and W. N. Adger, "Theory and Practice in Assessing Vulnerability to Climate Change and Facilitating Adaptation," *Climatic Change* 47, no. 4 (December 2000): 325–352, https://doi.org/10.1023/a:1005627828199.

138 Francisco R. Sagasti, *Knowledge and Innovation for Development the Sisyphus Challenge of the 21st Century* (Cheltenham, UK: E. Elgar, 2004).

139 Nassim Nicholas Taleb, *Antifragile: Things That Gain from Disorder* (New York, NY: Random House, 2014).

140 I. Kelman et al., "Learning from the History of Disaster Vulnerability and Resilience Research and Practice for Climate Change," *Natural Hazards* 82, no. S1 (March 21, 2016): 129–143, https://doi.org/10.1007/s11069-016-2294-0.

141 Mami Mizutori et al., *Global Assessment Report on Disaster Risk Reduction 2022: Our World At Risk* (Geneva, CH: United Nations Office for Disaster Risk Reduction, 2022).

142 Ibid.

143 Peter J. Webster et al., "Extended-Range Probabilistic Forecasts of Ganges and Brahmaputra Floods in Bangladesh," *Bulletin of the American Meteorological Society* 91, no. 11 (November 1, 2010): 1493–1514, https://doi.org/10.1175/2010bams2911.1.

144 "Sendai Framework for Disaster Risk Reduction 2015–2030," United Nations Office for Disaster Risk Reduction, March 18, 2015, https://www.undrr.org/publication/sendai-framework-disaster-risk-reduction-2015-2030.

145 Ibid.

146 Ibid.

147 Michael P. Murphy and Alan Ricks, "Beyond Shelter: Architecture and Human Dignity," *The Journal of Architecture* 18, no. 1 (March 8, 2013): 111–114, https://doi.org/10.1080/13602365.2013.767054.

148 S. B. Manyena, "Disaster Resilience: A Bounce Back or Bounce Forward Ability?," *Local Environment* 16, no. 5 (May 2011): 417–424, https://doi.org/10.1080/13549839.2011.583049.

149 I. Kelman et al., "Learning from the History of Disaster."

150 Rose M. Mutiso, "The Energy Africa Needs to Develop -- and Fight Climate Change," Rose M. Mutiso: The energy Africa needs to develop -- and fight climate change | TED Talk, November 9, 2020, https://www.ted.com/talks/rose_m_mutiso_the_energy_africa_needs_to_develop_and_fight_climate_change?language=en.

151 Vijaya Ramachandran and Arthur Baker, "The World Bank and IMF Are Getting It Wrong on Climate Change," Foreign Policy, April 11, 2022, https://foreignpolicy.com/2022/04/11/the-world-bank-and-imf-are-getting-it-wrong-on-climate-change/.

152 Yemi Osinbajo, "The Divestment Delusion," Foreign Affairs, August 31, 2021, https://www.foreignaffairs.com/articles/africa/2021-08-31/divestment-delusion.

153 Ibid.

154 IEA, "Key Findings – Africa Energy Outlook 2022," IEA, accessed August 5, 2022, https://www.iea.org/reports/africa-energy-outlook-2022/key-findings.

155 Vijaya Ramachandran and Arthur Baker, "The World Bank and IMF."

156 Ceyla Pazarbasioglu and Uma Ramakrishnan, "A New Trust to Help Countries Build Resilience and Sustainability," IMF Blog, January 20, 2022, https://blogs.imf.org/2022/01/20/a-new-trust-to-help-countries-build-resilience-and-sustainability/.

157 "World Bank Group Climate Change Action Plan (2021–2025) Infographic," World Bank, June 22, 2021, https://www.worldbank.org/en/news/infographic/2021/06/22/climate-change-action-plan-2021-2025.

158 Vijaya Ramachandran and Arthur Baker, "The World Bank and IMF."

159 Yoweri K. Museveni, "Opinion | Solar and Wind Force Poverty on Africa," The Wall Street Journal (Dow Jones & Company, October 24, 2021), https://www.wsj.com/articles/solar-wind-force-poverty-on-africa-climate-change-uganda-11635092219.

160 Rose M. Mutiso, "The Energy Africa Needs to Develop."
161 Yemi Osinbajo, "The Divestment Delusion."
162 John Briscoe, "Hydropower for Me but Not for Thee--with Two Postscripts," Center for Global Development | Ideas to Action, March 6, 2014, https://www.cgdev.org/blog/hydropower-me-not-thee.
163 "The Concert for Bangladesh," Wikipedia (Wikimedia Foundation), accessed July 13, 2022, https://en.wikipedia.org/wiki/The_Concert_for_Bangladesh.
164 Joyashree Roy, "Bangladesh Really Is a Climate Success Story," The Atlantic (Atlantic Media Company, October 19, 2021), https://www.theatlantic.com/ideas/archive/2021/10/bangladesh-climate-change/620224/.
165 Peter J. Webster et al., "Extended-Range Probabilistic Forecasts."
166 Ibid.
167 Joyashree Roy, "Bangladesh Really Is a Climate Success Story."
168 Ibid.

Chapter Fourteen

MITIGATION

"Make me chaste and celibate—but not yet!"
—Fourth-century philosopher Saint Augustine[1]

Climate change mitigation refers to efforts to control the concentration of atmospheric CO_2 and other greenhouse gases by reducing or preventing emission of greenhouse gases and enhancing sinks of carbon.[2] By contrast with adaptation, which manages the local impacts of a hazard, mitigation seeks to prevent the hazard from occurring.

14.1 Carbon Mitigation and Management

"Shutting down investment in fossil fuels before you have a plan to replace their role in the energy system is neither resilient nor just." (Energy finance expert Michael Liebreich)[3]

Carbon management is targeted at limiting the atmospheric concentration of CO_2 and additional compounds that comprise carbon-based air pollution (e.g., methane and soot). Several comprehensive schemes have been formulated for atmospheric CO_2 emissions and carbon-based air pollution, including the IPCC AR6 WGIII Report,[4] Climate Stabilization Wedges,[5] and Project Drawdown.[6] While the emphasis has been on atmospheric CO_2, management of other carbon pollutants is important in efforts to reduce warming. These mitigation schemes focus primarily on the sources of emissions: energy systems, transportation, industry, agriculture and forests, buildings, and urban systems.

Reducing CO_2 emissions has become an end in itself, with the implicit assumption that reducing CO_2 emissions will rapidly decrease atmospheric CO_2 and improve the climate. However, rigorous detection and attribution of the climate impacts of even very strong emission mitigation efforts will be very challenging. Emergence of a climate mitigation signal beyond natural variability can never be proven, as we would be comparing it to an unknown, counterfactual world. The challenge is to understand how atmospheric carbon will evolve in response to emissions reductions, and how the fast and slow components of the climate system will respond.

14.1.1 Global Carbon Cycle, Feedbacks and Budget

"There is room for words on subjects other than last words." (Philosopher Robert Nozick, author of *Anarchy, State, and Utopia*)[7]

Reservoirs of carbon in the earth system include the atmosphere, land ecosystems, the ocean, sediments, and the Earth's interior. Excluding rocks, by far the largest reservoir of carbon is the ocean. Exchanges of carbon between reservoirs occurs via various chemical, physical, geological, and biological processes. These exchanges establish a dynamical equilibrium over time in the absence of large external perturbations to the system. Since the advent of agriculture, humans have gradually influenced the carbon cycle by modifying the vegetation in land ecosystems. Since the industrial revolution, humans have modified the carbon cycle via production and burning of fossil fuels that emit CO_2 and other carbon-based pollutants into the atmosphere.[8]

Some of the carbon fluxes between reservoirs are sensitive to the state of the climate, and their resulting responses to changes in climate are referred to as "carbon cycle-climate feedbacks."[9] An example of a positive feedback is an increase in temperature increases the flux of carbon from the ocean surface into the atmosphere—more carbon in the atmosphere leads to further climate warming, further increasing carbon fluxes to the atmosphere.[10]

Removal of carbon from the atmosphere is achieved via photosynthesis into leaves, roots, stems, and woody biomass. Carbon is returned to the atmosphere through the respiration of plants and the decomposition of plant detritus or dead organic material and also via soil carbon. Disturbances such as fire, insect outbreaks, and timber harvesting provide substantial local perturbations to the land-atmosphere carbon exchange. In oceanic regions of ocean upwelling with nutrient-rich warm waters, carbon is outgassed into the atmosphere. By contrast, where there is cold, sinking water, such as in the North Atlantic and Southern Ocean, carbon is removed from the atmosphere. Carbon is exchanged between land and ocean reservoirs via river transport from interior regions to the coastal ocean.[11]

The El Niño Southern Oscillation (ENSO) is a significant driver of tropical carbon flux variability for both the ocean and terrestrial ecosystems. During the warm phase of ENSO (El Niño), the ocean takes up more carbon because of reduced upwelling and outgassing from the eastern Tropical Pacific. On land, El Niño is associated with drought and warmer temperatures, resulting in outgassing from land ecosystems. There is also substantial decadal variability in the land and ocean carbon sinks, driven by atmospheric and oceanic circulation patterns in the extratropics.[12]

The annual estimation of global CO_2 emissions and sinks is a major effort by the international carbon cycle research community that includes compilation and synthesis of measurements, statistical estimates, and model results. The ocean CO_2 sink showed rapid growth during the past decade, after little to no growth during 1991–2002. The land CO_2 sink demonstrated a continued increase during the last decade (2011–20), driven predominantly by increased atmospheric CO_2.[13]

While there is good qualitative understanding of the relevant processes involved in the global carbon cycle, our quantitative understanding is relatively weak.[14] There is persistent large uncertainty in the estimate of CO_2 fluxes from land-use changes. There is low agreement between the varied methods with respect to the magnitude of the land CO_2 flux in the northern extra-tropics. There is also substantial discrepancy between the different methods on the strength of the ocean sink over the last decade, particularly in the Southern Ocean.[15] There is also concern about permafrost feedback loops, which could release more carbon into the atmosphere with warming.

New research continues to provide surprises. Phytoplankton, which live on the warm, light-filled ocean surface, use atmospheric CO_2 for food. Phytoplankton sequester carbon and once they die and sink, carry it to the ocean depths. A recent study found evidence that phytoplankton may become more efficient at sequestering carbon as the ocean warms. Another recent article reported the discovery of a widely circulated microbe variety that demonstrates the potential to sequester carbon.[16]

The storage of CO_2 on land and in the ocean has effectively subsidized the emission of CO_2 in the atmosphere, buying time in the fight against climate change. The big question is to what extent land and ocean sinks will continue to sequester carbon from the atmosphere.[17]

The vagaries of the carbon cycle, in combination with natural climate variability, make it difficult to identify a measurable change in the evolution of global warming in response to emissions reduction. Inertia in the ocean and ice sheets along with natural internal variability of the climate system will delay the emergence of a discernible response even to strong mitigation.[18]

Even with large reductions in carbon emissions, a corresponding significant shift in surface temperature evolution is not anticipated until decades later.[19] It is unclear how the climate will evolve after net-zero emissions is achieved, and how much warming there is in the pipeline. To address this issue, the Zero Emissions Commitment Model Intercomparison Project (ZECMIP) used multiple Earth System Models to investigate how the climate system including the carbon cycle will respond to zero emissions.[20] Fifty years after an immediate cessation of CO_2 emissions, the models showed a temperature change of -0.36–$0.29°C$. The models exhibit a wide variety of behaviors, with some models continuing to warm for decades to millennia while others cool. Carbon uptake by both the ocean and the terrestrial biosphere is shown to be important in counteracting the warming effect created by reduction in ocean heat uptake anticipated decades after emissions cease. This response is difficult to constrain primarily given the high uncertainty in the effectiveness of ocean heat uptake.[21]

The bottom line is that there is substantial inertia in the global carbon cycle and the climate system. Even if emissions are successfully reduced/eliminated, it takes time for the CO_2 concentration in the atmosphere to respond to the emissions reduction and it takes time for the climate to respond to the change in atmospheric CO_2 concentration. There is substantial uncertainty regarding how much time this will take—we may not see much of a beneficial change to the climate before the twenty-second century even if emissions are successfully eliminated.

14.1.2 Carbon Sequestration

"Just then they came in sight of thirty or forty windmills that rise from that plain. And no sooner did Don Quixote see them that he said to his squire, 'Do you see over yonder, friend Sancho, thirty or forty hulking giants? I intend to do battle with them and slay them. With their spoils we shall begin to be rich for this is a righteous war and the removal of so foul a brood from off the face of the earth is a service God will bless." (Don Quixote)[22]

Carbon management techniques can be divided into two broad categories: natural and active carbon management. Natural carbon management seeks to increase the amount of carbon that is stored in a semipermanent state in natural sinks. Biological carbon management techniques include afforestation and reforestation, soil and agricultural sequestration, biochar, and ocean fertilization. While natural carbon management is an important component of climate mitigation strategy that also has numerous ancillary benefits, its ability to reduce the increase in atmospheric CO_2 is limited. Further, natural carbon sinks are not permanent, with disruptions such as wildfires and drought resulting in re-release of captured carbon.[23]

Active carbon management utilizes engineering methods to capture CO_2 emissions from power plants and factories, or by directly withdrawing it from the atmosphere. The captured CO_2 is then trapped deep underground, or alternatively employed in industrial and commercial activities. Active carbon management is implemented more quickly than planting forests and insuring their protection until maturity. Further, underground reservoirs are considerably less likely to re-release carbon than soils and forests.[24]

The goal of carbon capture and sequestration (CCS) is to prevent emissions from ever entering the atmosphere, thus decreasing the flow of carbon. By contrast, direct air capture (DAC) of CO_2 directly lowers levels of atmospheric CO_2. Small-scale DAC technology has been created for utilization in specialized applications such as making air breathable in submarines and spacecraft. Bioenergy with carbon capture and storage (BECCS) is a hybrid active/natural carbon management method that uses the natural process of capturing carbon via growing fuel crops while concurrently capturing emissions when the fuel is utilized to generate electricity and heat.[25]

Once captured, CO_2 can be sequestered permanently underground. Unless it is stored at the same site where it is produced or extracted, the purified CO_2 must be compressed and transported. The CO_2 is then injected in a supercritical, liquid-like state into deep saline reservoirs or coal seams. In the United States, most of the potentially suitable sites for sequestration are located off the coasts and in the interior of the country.[26] Once pumped underground, the sequestration sites must be monitored to ensure that CO_2 does not leak into the atmosphere or surrounding water tables.[27]

Use of captured CO_2 is an alternative to sequestration. The most common use of captured CO_2 is for increasing oilfield production. CO_2 is also used in food processing. Research is ongoing into new uses of CO_2 as an input for the production of plastics and polymers, fertilizer, synthetic aviation fuels, and building materials. Commercializing these products would reduce the costs for active carbon management.[28]

From the perspectives of risk management and decision-making under deep uncertainty, how should we evaluate the various options for carbon management? Natural carbon management is relatively straightforward and low cost, with co-benefits for forestry, agricultural, and land restoration. As such, these strategies are associated with no regrets and should be pursued, but with regards to competing uses for land. Active carbon management expands our options for dealing with climate change. However, the different proposed strategies are evaluated differently from a risk management perspective. Carbon capture from electric utilities and industry that can be then used or upcycled in industrial processes or in products makes economic sense, even apart from climate change concerns.

In the context of the political imperative of net-zero emissions by 2050 and expected emissions trajectories, CCS, BECCS, and DAC may become cost effective strategies in the future that are preferred over attempting to eliminate emissions from certain sectors. The big problem is this: CCS, DAC, and BECCS technologies are in relatively early stages of development, and at the time of this writing there is only one large-scale operational CCS plant, in Canada.[29] Further, these technologies also have large land use requirements.

Prudent risk management would continue to develop these carbon removal options, although their development costs compete with funds to develop improved technologies for energy production and transmission and industrial processes. It is important to assess not only the feasibility of these technologies, but also their expected impacts on land and water use, food security, and biodiversity. If warming in the coming decades turns out to be lower than expected, then the value of CCS technologies will be much lower.

In spite of these technical uncertainties, CCS, DAC, and BECCS have become central to climate mitigation strategies. The IPCC AR6 WGIII states that:

"Carbon Dioxide Removal (CDR) is necessary to achieve net zero CO_2 and [greenhouse gas] emissions both globally and nationally, counterbalancing 'hard-to-abate' residual emissions. CDR is also an essential element of scenarios that limit warming to 1.5°C or likely below 2°C by 2100, regardless of whether global emissions reach near zero, net zero, or net negative levels."[30]

Risk management that relies on un- and under-developed technologies is not a robust strategy and has a high chance of failure.

14.2 Short-Lived Carbon Pollutants

"We have to win the sprint to slow warming in the near term by tackling the short-lived climate pollutants, so that we can stay in the race to win the marathon against CO_2." (Gabrielle Dreyfus, chief scientist for the Institute for Governance & Sustainable Development)[31]

Short-lived climate pollutants include methane, black carbon particulates, and hydrofluorocarbons (HFCs). These short-lived climate pollutants are responsible for up to 45 percent of radiative forcing from human emissions. Unlike CO_2, these pollutants have short atmospheric lifetimes. Hence their mitigation could potentially reduce global mean warming by about 0.5°C by 2050, with improvements to air quality, human health, and agricultural productivity as co-benefits.[32]

Methane (CH_4) is a powerful greenhouse gas with both natural and human sources. Although methane is much less abundant than CO_2 in the atmosphere, on a per mass basis and over 100 years methane is about 25 times more effective at warming the surface than CO_2. Recent methane emissions are roughly

40 percent from natural sources, primarily from waterlogged wetlands and geological emissions. Human sources of atmospheric methane are fossil fuel production, livestock and rice paddies, sewage and landfills, and burning of biomass and biofuels. Chemical reactions in the atmosphere result in a relatively short atmospheric lifetime of methane of 9 to 12 years. Atmospheric methane increased rapidly during the 1980s and early 1990s before leveling off between the mid-1990s and early 2000s. Atmospheric methane began increasing around 2006, with the increase accelerating since 2014. Causes of the changing growth rate of atmospheric methane are the subject of much debate.[33]

In addition to its impact on climate through the greenhouse effect, methane has adverse impacts on human health, crop yields, and the quality and productivity of vegetation through its role as an important precursor to the formation of ozone (itself a greenhouse gas). Ground level ozone aggravates asthma and respiratory diseases and also leads to reduced agricultural crop and commercial forest yields and increased susceptibility to diseases and pests.

Various strategies have been proposed for reducing methane emissions or enhancing the atmospheric chemical sink for methane. These strategies include reducing methane associated with oil and natural gas production, improved agricultural practices, and novel waste management strategies. A recent report by the UN Environmental Program showed that human-caused methane emissions can be reduced by up to 45 percent this decade. Such reductions would avoid nearly 0.3°C of global warming by 2045.[34] In 2020, the European Commission adopted the European Union Methane Strategy that outlines measures to cut methane emissions in Europe and internationally.[35]

Fine particles of black carbon, commonly referred to as soot, form from the incomplete combustion of fossil fuels, wood, and other fuels. Black carbon is a component of fine particulate matter that when inhaled at high concentrations can cause lung disease and severe cardiovascular and neurological problems. Black carbon particulates remain in the atmosphere on a timescale of days to weeks, so effects are strongly regional. While not a gas, black carbon particles have an overall warming effect on local surface temperature. More significantly, the deposition of black carbon darkens snow and ice, increasing the absorption of sunlight and promoting melting. Apart from increasing the melt of ice sheets that contribute to sea level rise, black carbon is contributing to the melting of glaciers that are a major source of fresh water for millions. This is of particular concern in the Himalayas, which is the head source of major rivers in Asia that support half of the world's population.[36]

Decreasing black carbon emissions should reduce warming impacts and slow glacier melting. Black carbon reductions would also result in considerable health benefits, primarily in economically challenged regions where biomass burning is a significant source of black carbon. Measures to reduce black carbon emissions include eliminating high-emitting vehicles, introducing clean burning cook stoves in low-income countries, and transitioning away from coal power plants.[37]

HFCs are industrial chemicals predominantly used in refrigeration, but also in other applications such as foam blowing aerosol propellants and as solvents. HFCs were developed as a result of the 1987 Montreal Protocol to eliminate use of substances that deplete the Ozone Layer.[38] Many HFCs are powerful greenhouse gases, with relatively short atmospheric lifespans ranging from 15 to 29 years.[39] Despite HFCs representing around 1 percent of total greenhouse gases, they are hundreds to thousands of times more potent than CO_2. HFC utilization was predicated to double between 2010 and 2020, which could contribute substantially to radiative forcing by the mid-century.[40] The 2019 Kigali Amendment to the Montreal Protocol commits countries to cut the manufacture and use of HFCs by more than 80 percent over the next 30 years.[41] There are already replacements for many HFCs and more are developed each year to help reduce HFC emissions.

The math is simple. Global mean surface temperature has warmed by about 1.1°C since the late nineteenth century. With the expectation of passing 1.5°C warming within several decades, efforts to reduce the short-lived climate pollutants could have a critical policy-relevant impact by preventing up to 0.5°C of global warming on the time scale of a few decades. This would reduce the urgency and extend the timeline for the energy transition so that better solutions can be deployed.

Reducing the short-lived climate pollutants is much simpler and less expensive than eliminating CO_2 emissions. Further, there are economic, health, and agricultural co-benefits to mitigation of the short-lived climate pollutions. And unlike CO_2 mitigation, there is no transition risk associated with mitigating the short-lived climate pollutions (Section 14.3).

14.3 Energy Transitions

"We are drowning in promises." (Ugandan climate activist Vanessa Nakate)[42]

For the past two centuries, fossil fuels have fueled the progress of humanity, improved standards of living, and increased the life span for billions of people.[43] In the twenty-first century, a rapid transition away from fossil fuels has become an international imperative for climate change mitigation under the auspices of the UNFCCC Paris Agreement.

Currently, there is rapid technological innovation across all domains of the global energy sector. Innovation is transforming every part of the modern energy system: long-distance transmission and power grid control, energy storage, residential heating, electric vehicles, and remarkable progress in advanced nuclear designs.

We do not know how the twenty-first century energy transition will play out, but there is much we can learn from the history of previous energy transitions, the status of the current transition, a more holistic envisioning of future energy systems, and the principles of risk management and decision-making under deep uncertainty.

14.3.1 History of Previous Energy Transitions

"I am looking for a lot of people who have an infinite capacity to not know what can't be done." (American inventor Henry Ford)[44]

To understand the current state of the transition away from fossil fuels, it is instructive to examine the history of previous energy transitions.[45] A fascinating analysis of previous energy transitions is provided by Australian entrepreneur Tsung Xu.[46]

Through history, key energy transitions have all proceeded through similar phases. Following discovery, the energy source (or carrier, in the case of electricity) finds a foothold in niche markets where it outperforms existing technologies. As the technology matures, it enters larger markets. Then, in the most transformative phase, new infrastructure emerges for the energy source, new materials are enabled, and new industries emerge. In the transformative phase, transitions become mutually reinforcing, with better distribution technologies and materials driving further adoption and investment into the energy system.[47]

The rapidity at which key energy transitions have occurred in the past is rather astonishing:[48]

- In the 1740s, coked coal was first used to smelt iron. Less than 60 years later, iron had replaced wood in most manufacturing and construction uses in Britain.
- In the 1880s, the first city-wide electricity grids were installed in the United States. By the end of the 1920s, about 70 percent of US homes had access to electricity.
- In 1900 in New York, there were few gasoline-powered cars on horse-dominated roads. Within two decades, almost no horses were found on the streets of New York City.
- Generation of cheap electricity enabled materials like aluminum to be produced cost-effectively. Global aluminum production increased by 15 times

between 1900 and 1916 as electrolysis costs fell and the metal found more uses. Longer range electricity transmission lines were enabled by inexpensive aluminum, further increasing the availability and use of electricity.

These previous transitions were sparked by technical innovations in the United States and UK. The transformative stage was key to accelerating the transitions.[49] How does the putative twenty-first century energy transition differ from these previous transitions?

- The current transition is motivated by top-down international imperatives to reduce/eliminate CO_2 emissions from burning fossil fuels.
- The transition is decreed to be urgent, so that existing technologies are to be deployed. The favored technologies are wind and solar.
- The scale and scope of the transition is much greater than previous transitions owing to much larger populations and inertia in existing electric power infrastructure.
- The entrance of wind and solar power into electricity production has been heavily subsidized by governments; this has not been a natural market evolution.
- Wind and solar are viewed as replacing the capabilities now provided by fossil fuels, rather than enabling new capabilities.

The following subsections discuss how this transition is evolving, and presents an alternative vision for a twenty-first century electricity system.

14.3.2 Current State of the Energy Transition

"Everyone has a plan until they get punched in the mouth." (American boxer Mike Tyson)[50]

The electricity system began transitioning two decades ago. The old system was characterized by a relatively small number of large generators that were connected to a transmission grid. There were baseload and peak generators to accommodate variations in weather-driven demand. Coal reserves guaranteed an inexpensive supply of fuel if demand was high or there were supply or cost issues with natural gas.

Over the past two decades, the electricity system has evolved away from coal and has connected enormous numbers of smaller generators from wind and solar to the grid. Weather-driven variations now occur in both supply and demand, which are managed by demand response, storage, overcapacity, and interconnections with neighboring systems.

In 2020, 82 percent of new electricity generation globally was from wind and solar.[51] Although deployment is increasing rapidly, renewable energy is not displacing fossil fuel generation overall, owing to the continuous increase in electricity demand. Globally, population growth and Gross Domestic Product (GDP) per capita continue to be the strongest drivers of CO_2 emissions from fossil fuel combustion.[52] In 2021, fossil fuels accounted for 83 percent of the overall global energy consumption.[53]

Intermittency of wind and solar is a major problem. In the third quarter of 2021, Europe's largest wind energy generating countries (Britain, Germany, and France) produced only 14 percent of installed capacity from wind, compared with a 20–26 percent average in previous years. The shortfall was unexpected and not properly planned for. In 2021, China experienced a prolonged period of low-wind weather impacting generation, resulting in rolling blackouts to deal with the situation.[54]

Brownouts and electricity curtailments have become more frequent during both cold and heat extremes. The worst problems are associated with continental-scale high pressure systems during winter, which produce very cold temperature and still winds—demand is exceptionally high and supply from renewables is very low. Continental-scale high pressure systems during summer produce very hot temperatures under still conditions, but solar power production is high. Even a continental-scale macrogrid transmission system won't help much if nearly all of the continent is suffering under the same weather conditions.

Mistaken ideas about carbon accounting, political pressures, and short-sighted economics are perpetuating the use of biofuels. Biofuels have played a major role in global food crises in 2008, 2011, and 2022.[55,56] In 2022, global food insecurity hit record highs.[57] Nevertheless, approximately 10 percent of the world's grains are being turned into biofuels.[58] Palm and soy oil from Indonesia and South America are also being burned for fuel and it is estimated that 58 percent of rapeseed oil in Europe is burned for fuel, despite soaring prices for cooking oil.[59] The European Union plans to allocate one-fifth of Europe's cropland to producing fuels for bioenergy and also plans a four-fold increase of wood imports to burn for energy equivalent to approximately 40 percent of Canada's (the world's largest exporter) annual wood harvest.[60]

The United States uses about 40 percent of its annual corn crop grown on tens of millions of acres of cropland to produce ethanol that comprises only about 10 percent of US transportation fuel.[61,62] It has been estimated that the life-cycle greenhouse gas emissions of ethanol are no less than those of gasoline, and likely greater. Corn ethanol has exacerbated environmental problems such as soil erosion and poor water quality, contributing to the degradation of agricultural land that would be more importantly used for food production.[63]

If the crops for biofuels are irrigated, this can exacerbate water supply issues during droughts.[64] The net effect of biofuels is lifecycle emissions that can be worse than the displaced fossil fuels, exacerbated food shortages, and degraded farmland.[65]

The realization is growing that countries face substantial economic and geopolitical risks if they reduce the production of fossil fuels under the assumption that renewables can quickly replace them. Premature retirements of baseload generating units, such as coal and nuclear plants, combined with the intermittency of wind and solar as power sources, have seriously impaired grid resiliency and reliability in some regions and countries. The success of renewables depends on investment in transmission, synchronous backup power for times when renewable generation is low, and improved energy storage.

In 2022, the energy transition has been disrupted by Russia's war on Ukraine, with the ensuing gas and oil shortages and price spikes leading to political pressures to abandon green energy pledges and return to coal. The transition has been further disrupted by supply-chain problems and an affordability crisis for materials needed for wind, solar, and batteries, and declining government subsidies in some countries. The IEA has found that global solar photovoltaic supplies are over-concentrated in China, representing a considerable vulnerability to global supply chains.[66] In combination, these factors are slowing down the build-out of wind and solar.

In the United States, the onshore wind industry is facing challenges from rural landowners who don't want wind turbines nearby—they are concerned about noise, declining property values, and destruction of views. Over the past eight years, 328 wind farm proposals have been rejected in the United States.[67]

There are substantial institutional and structural barriers in the United States that are slowing down or preventing wind and solar generating capacity from being integrated into transmission grids. The US transmission grid has been growing very slowly in recent decades, at a pace that is a fraction of that required for net-zero emissions plans. Transmission and renewable energy projects are being blocked across the country by landowners, consumer, and environmental groups. Even when all relevant parties agree to proceed with new transmission lines, the cost allocation process can take years.[68] A further challenge is that utilities and grid operators need to analyze the impacts of new generating projects when added to the grid.[69]

In the United States, electric vehicles (EVs) are rapidly growing in popularity, but it is becoming increasingly difficult to actually purchase an EV. Tesla CEO Elon Musk said his electric-car factories are "losing billions of dollars" as global supply-chain disruptions and challenges in battery manufacturing constrain the company's ability to scale up production.[70] According to the CEO of Rivian, a manufacturer of electric adventure vehicles, EV battery supply chain, pressures could surpass the current semiconductor shortage. He stated, "All the world's cell production combined represents well under 10% of what we will need in 10 years … meaning 90–95% of the battery supply chain does not exist."[71]

The costs of household solar are unnecessarily high in the United States—almost three times more expensive than in Australia.[72] Permitting costs, city-level red tape, arcane code restrictions, and very high customer acquisition costs are making rooftop solar in the United States far less affordable than it could be.

The net outcome of the energy transition to date is that in 2022, very few of the world's countries are on track to meet their emissions reductions commitment. Further, the shortages and price spikes in the global natural gas and oil supply caused by Russia's war on Ukraine and the supply chain have demonstrated the current fragility of the transition.

14.3.3 Vision—2100

"Tell me how this ends." (US General David Petraeus)[73]

People are less inclined to act on the energy transition if they can't imagine the endpoint, and can't see themselves better off in the future.

The overall vision as per the IPCC AR6 WGIII Report is predicated around net-zero emissions, with energy systems having the following characteristics: (1) electricity systems that produce no net CO_2 or remove CO_2 from the atmosphere; (2) widespread electrification of end uses; (3) substantially lower use of fossil fuels; (4) use of hydrogen, bioenergy, and ammonia in sectors less amenable to electrification; (5) more efficient use of energy; (6) greater energy system integration across regions and components; and (7) use of CO_2 removal technologies.[74]

A more holistic vision for energy systems considers a broader range of values plus potential dangers and risks associated with the transition. Table 14.1 provides a list of relevant values and the associated risks or dangers to be considered while envisioning electric power systems humans will want and need to thrive on the threshold of the twenty-second century.

The starting point for this more holistic vision is to acknowledge that the world will need much more energy than it is currently consuming. Electricity underpins a wide range of products and services that enhance our quality of life and stimulate economic productivity. Even if global population remains around

Table 14.1 Values and risks/dangers associated with electric power systems.

Values	Risks/Dangers
Abundant	Structural inadequacies to meet energy needs
Reliable	Catastrophic power cuts in the face of weather extremes
Secure	Subject to supply shocks (availability, cost); cyberattacks
Clean	Pollution from emissions, mining; ecosystem and human health concerns
Food & Water	High cost and/or lower food supply; competition for scarce water resources
Local control	Loss of autonomy; loss of economic opportunity
Minimal land use	Interference with other land use priorities and ecosystems
Minimal material use	Scarcity of rare minerals; scope and scale of mining; supply chain issues
No CO_2 emissions	Long-term concerns about adverse impacts of climate change

8 billion, people in developing countries will use substantially greater amounts of electricity in the future.

Apart from supporting human development, more electricity can help reduce our vulnerability to the weather and climate: air conditioners, water desalination plants, irrigation, vertical farming operations, water pumps, coastal defenses, and environmental monitoring systems. And this is not to mention the large amounts of electricity needed for Direct Air Capture of CO_2. A substantial increase in computational power will be needed for many things, including weather and climate models and artificial intelligence to keep the energy supply secure in the face of weather disruptions. Further, abundant electricity is key to innovations in advanced materials, advanced manufacturing, artificial intelligence, blockchain, robotics, photonics, electronics, quantum computing, and others that are currently unforeseen or unimagined.

The energy choices that are laid out are fossil fuels (with carbon capture and removal as needed), renewable energy, and nuclear energy. Of these three choices, nuclear has the greatest potential to provide the very large amounts of energy that we will need heading into the twenty-second century. Different countries and locales will use different combinations of these energy sources based upon their climate, local resources, power needs, and sociopolitical preferences.

While fossil fuels have propelled human progress in the twentieth and early twenty-first centuries, there is a strong rationale for reducing our reliance on fossil fuels for energy—independent of their impact on atmospheric CO_2 concentrations and air and water quality. Mining for fossil fuels has large, continuing economic and environmental costs. Supplies of fossil fuels are finite, and will become increasingly expensive to extract by the twenty-second century. The Russian war on Ukraine highlights the vulnerability of fossil fuel supply chains and price spikes to geopolitical instability. That said, it is also easy to imagine the continued burning of fossil fuels into the twenty-second century for certain niche applications and in countries that have abundant fossil fuel resources. However, burning may not be the best use petroleum—petrochemicals are used to manufacture thousands of different products, including plastics, fertilizers, synthetic fibers, cosmetics, paint, medicines and detergents.

Does our future hold a plethora of wind turbines, solar farms, and transmission lines covering an ever-growing fraction of the planet's surface as energy demand increases? The output of farmland and forests being burned to provide power? The amount of land required for renewable energy is an issue of growing concern that has received surprisingly little attention. The current global energy system exists on a relatively small land footprint (only 0.4 percent of ice-free land), which is two orders of magnitude less than the area utilized by agriculture.[75] Plans for an entirely renewable energy system for the globe will substantially increase the amount of land use for energy production, particularly with growing energy consumption and the widespread electrification of heating, transportation, and industry.

The land footprint of energy systems displaces natural ecosystems, leads to land degradation, and creates trade-offs for food production, urban development, and conservation. In densely populated countries such as Japan, Bangladesh, Lebanon, South Korea, India, Netherlands, Belgium, Bahrain, and Israel, there simply is not sufficient land to support a majority of the energy supply coming from renewables.[76] Emerging economies face larger challenges with more dynamic land use in the face of urbanization, industrialization, and agriculture.

A recent article calculated the land-use intensity of energy (LUIE) from actual data for all major sources of electricity.[77] The LUIE values include both land used for actual energy generation as well as land used for fuel sourcing. The study found a range of LUIE values that spanned 4 orders of magnitude, from the lowest value for nuclear to the highest value for dedicated biomass. Relative to the LUIE value for natural gas, the sources with lower values of LUIE are nuclear, geothermal, rooftop solar, and residue biomass. Evaluating the LUIE for wind power is complicated by the fact that the land footprint of an individual wind turbine is relatively small; however, the overall footprint of a wind farm is not just the sum of the land footprints of individual turbines, but rather the area within the perimeter of the wind farm. Wind turbines that are built on degraded, contaminated, otherwise unusable land or on top of agricultural land aren't competing with other uses for that land and don't unduly interfere with surface ecosystems. The problem is that such locations are far from population centers where the bulk of the energy is needed. Offshore wind helps with the land use issue (although there are competing uses of coastal waters) and also places wind farms in closer proximity to coastal population centers. The land-use implications of carbon capture and storage (CCS) for fossil- or biomass-fueled power plants have been estimated to increase the land use footprint by 40 percent compared to a plant without CCS.[78,79]

Another issue of growing importance is that households, companies, communities, and countries want as much autonomy, reliability, and security as possible in their energy supply. This implies mostly local/regional production of energy, rather than relying primarily on fuel or transmission from another country. A hierarchical transmission grid that includes microgrids connected to the high-voltage central utility grid can increase reliability, efficiency, security, and sustainability of the electricity supply, and provide the desired autonomy.

Microgrids are comparatively small power systems composed of one or more generation units connected to neighboring users. They can be operated in conjunction with or independently from central transmission systems. Microgrids support a flexible and efficient electric grid by enabling integration of distributed energy resources such as solar. Microgrids serving local loads helps reduce energy losses in transmission, increasing efficiency of the electric delivery system. Microgrids also can help ease strain on the central transmission grid during periods of peak demand. Microgrids can supply power to critical facilities after weather or security related outages impact the broader grid. Microgrids can also support central grid recovery by maintaining services needed by restoration crews and by aiding to re-energize the central grid.[80]

Envision microgrids powered by small modular nuclear reactors and localized renewable energy sources that are able to work seamlessly in conjunction with the central transmission grid, prioritizing use of the lowest carbon resources to the extent possible. When automated with smart technology and intelligent controls, microgrids can foresee complications and failures allowing them to reconfigure in real-time to minimize impacts. A smart microgrid can make use of excess renewable energy that would otherwise be curtailed to produce synthetic and hydrogen fuels.[81]

So, what will our energy systems look like circa 2100? It is of course difficult to predict the future, especially with regards to technological advances, but some rough constraints are becoming apparent. Energy systems producing abundant amounts of electricity driven solely by renewable energy would be prohibitive in most regions in terms of land and coastal ocean use. It is difficult to imagine electric power systems circa 2100 in most regions without a backbone of nuclear power. Windfarms are a viable solution where land and coastal use considerations permit. Rooftop solar is a good solution and supports some level of local autonomy. Advanced geothermal is desirable, but it is not clear at present how the new technologies will evolve. I don't see a role for biofuels in the future, where other power sources are available. A combination of smart continental-scale macrogrids with local microgrids will enable both local autonomy and the advantage of importing and exporting energy to meet supply surpluses and deficits from demand.

14.4 Managing Transition Risk: Electric Power Systems

"Science does not say you shouldn't pee on high-voltage lines. It says urine is an excellent conductor." (Physicist Sabine Hossenfelder, author of *Existential Physics: A Scientist's Guide to Life's Biggest Questions*)[82]

Electric power systems face numerous risks: supply and demand shocks, severe weather events, cyberbreaches, solar storms, cascading voltage collapse. The tightly integrated system of systems that provides the backbone for advanced economies—power, transport, telecommunications, health services, logistics, payments, emergency services, public information —all depend on electricity.

Germany's Office of Technology Assessment published a report in 2011 on the risk associated with a prolonged power outage.[83] Communications would fail. All rail and air transportation systems would halt, and fuel would become unavailable. Food processing would end due to lack of refrigeration and distribution would cease without viable transportation. Fresh water cannot be delivered, nor can sewage be processed. Financial, healthcare, and virtually all commercial activities begin to fail. Emergency services and military are severely hampered by the lack of transportation and communications networks. Civil unrest breaks out and if it happens during winter, people starve. The report concluded that it would be almost impossible to prevent a collapse of society as a whole.[84]

The rapid transition of electric power systems away from fossil fuels to meet net-zero emissions targets is introducing substantial new risks to electric power systems. A transition of the electric power system that produces reduced amounts of electricity, less reliable electricity, and/or more expensive electricity to achieve net-zero goals would be a tourniquet that restricts the lifeblood of modern society and hampers development and may thwart sustainability efforts.

14.4.1 Relevant Risk Management Principles

"What all the wise men promised has not happened, and what all the damned fools said would happen has come to pass." (Eighteenth-century UK Prime Minister Lord Melbourne)[85]

Transition towards the general vision for energy systems in 2100 outlined in Section 14.3.3 involves systems with high levels of complexity, uncertainty with regards to future technologies and weather/climate risk, and ambiguity with regards to competing societal values. Because of the role of electricity as the backbone of societal support systems, risks to the electric power system are systemic risks. Risk and uncertainty sciences provides structure and logic on how to think and act in relation to risk and uncertainties (Chapters Eleven and Twelve).

The risk management principle of proportionality provides a framework to guide action when there are competing demands on public policy decisions. Focusing on one set of risks—CO_2 emissions—can create other, potentially more dangerous risks. If efforts to reduce CO_2 emissions result in electric power that is less abundant, less reliable, and less secure, then the transition will make people worse off now and very possibly in the future.

The Russian war on Ukraine provides a stark conflict between net-zero emissions goals versus immediate needs for abundant, reliable, and secure energy. The dangers from inadequate, unreliable, and insecure electricity supply are well known and becoming increasingly apparent as European and other countries struggle with inadequate natural gas supplies that they had been receiving from Russia. By contrast, the dangers from CO_2 emissions are much more uncertain, with a long time horizon and a far weaker knowledge base. The debate is then between imposition of certain, intolerable risks from the rapid transition away from fossil fuels, versus the highly uncertain future impacts from climate change.

This conflict can be resolved by relaxing the time horizon for reducing CO_2 emissions while maintaining energy abundance, reliability, and security through the energy transition. Yes, CO_2 emissions are a problem and should be reduced, but not as an urgent problem that trumps the need for abundant, reliable, and secure sources of energy for the global population. As discussed previously in this book, the urgency for reducing CO_2 emissions has been diminished by acknowledgment of the implausibility of extreme emissions scenarios (Section 7.1) and postulated extremely high values of climate sensitivity to CO_2 (Section 7.2). Plausible scenarios of natural climate variability in the mid-twenty-first century point to a slowdown in the rate of global warming (Section 8.2). Growing clarification of the role of natural weather climate variability, land use, and governance issues in causing societal problems is diminishing their attribution to

CO_2 emissions (Section 13.1.3). And mitigation of short-lived carbon pollutants (Section 14.2) buys additional time for reducing CO_2 emissions.

In evaluating risk management strategies for electric power systems, an assessment is needed of the feasibility and costs of controlling the various risks. A practical difficulty is being forced to retain a risk that is recognized as being beyond the risk appetite, or even the risk capacity. The low feasibility and high costs of reaching net-zero emissions targets by 2050 are at the heart of the debate over allowing near-term net-zero targets to dominate future energy systems.

Attempts to speed up the transition away from fossil fuels by restricting the production of fossil fuels and new generating plants has backfired by making many countries reliant on Russia's fossil fuels. The geopolitical instability associated with Russia's war against Ukraine highlights the importance of having multiple options and safety margins—key characteristics of robustness. Safety margin strategies in electric power systems include redundancy, a range of different power sources, and reserve power. Overcapacity should be a feature of future energy systems, not a bug.[86]

The long time horizons of the transition and uncertainties about both the available technologies and future climate impacts are best handled by adaptive risk management, which includes learning from trial and error and incorporating changes in the technologies and knowledge base over time. Given the uncertainties of the physical and socioeconomic systems, the concept of robustness argues for building a broad portfolio of technologies and other measures that might be useful/effective.[87]

Risk assessment should be comprehensive, so that initiatives address all the aspects of the system at risk. In evaluating different sources of electric power, a holistic evaluation is needed of all relevant factors: lifecycle costs of power production, transmission, and the required infrastructure; land use requirements; costs of unreliability of electric power; and lifecycle environmental impacts including CO_2 emissions. Too often, only the cost of power production and CO_2 emissions are considered.

14.4.2 Nuclear Power

"I'm so pro-nuclear now that I would be in favor of it even if climate change and greenhouse gasses were not an issue." (Environmentalist Steward Brand, author of the *Whole Earth Catalog*.)[88]

The role of nuclear power in twenty-first century energy systems is subject to intense debate: one perspective is that nuclear power has an unrealized potential for abundant and clean energy, versus a perspective that views nuclear power plants as ticking time bombs that produce unmanageable radioactive waste.[89]

The fear of nuclear power (section 10.1.1) is not consistent with the actual risks. Around 10 percent of the world's electricity is generated by large nuclear fission plants, from a high of over 17 percent in the mid-1990s.[90] The decline is partly because of public safety concerns over meltdowns like Fukushima, Chernobyl, and Three Mile Island.[91] Permitting and regulatory delays along with cost overruns have made large nuclear plants uncompetitive with other sources of energy.

The current light-water nuclear reactors are associated with risks from uranium mining, reactor safety, nuclear waste, and nuclear proliferation. Modern designs developed over the last four decades virtually eliminate meltdown risk.[92] Storing spent nuclear fuel in large, thick concrete dry casks has an excellent safety record over decades of operation.[93] Nuclear waste can be even more safely deposited deep in geologically stable bedrock, as is being done in Finland.[94] Neither plant safety nor waste disposal is considered a serious risk by most energy experts, and nuclear power is regarded as far safer than burning fossil fuels.[95]

A climate-relevant risk associated with the current generation of nuclear reactors that rely on water cooling is that proximity to the coast or rivers is required. River-based cooling can result in shutdown of a nuclear power plant during droughts when river levels are too low or during periods of extreme heat when the river water temperatures are too warm to serve the cooling function. Coastal plants relying on ocean water for cooling face concerns over sea level rise and storm surge. Nuclear power plants located on the coast have been shut down numerous times when jellyfish clogged the sea water-cooling intake pipes at the plant.[96]

There are many possible designs for nuclear reactors, and the current light-water reactors are far from the best design. Many of these options were on the table in the United States prior to the 1950s. However, the actual development of nuclear power for civilian use was tied to the development of nuclear-powered submarines, which used the light-water reactor. The light-water reactor requires enriched uranium as a fuel, which locked in nuclear power with the technologies used in producing nuclear weapons.[97]

China currently has the fastest-growing nuclear energy sector in the world.[98] Russia is the world's top nuclear technology exporter.[99] The United States is making a very large investment in the next phase of the nuclear energy development. The US effort includes a range of technologies for advanced nuclear power plants, including small modular nuclear reactors that can solve the cost problem with modular design and mass production. Further, some new designs are impossible to melt down and are not cooled by water; hence they can be located anywhere and would be a step toward more local community control of power sources. Additional innovations include new fuels, more complete utilization of nuclear fuel, lower generation of waste, and enhanced resistance to nuclear proliferation.[100] Accelerated ramping of power generation can balance fluctuations in renewable energy supply. Advanced reactor designs with high-temperature characteristics can be used for industrial applications, including production of hydrogen, petrochemicals, ammonia, steel, and cement manufacturing.[101]

Many options for nuclear power plants are being developed, that can support a range of particular needs and applications. In the United States, estimated project completion dates for advanced nuclear plants typically lie in the late 2020s and early 2030s, with plans to expand deployment further upon successful demonstration. Initial construction costs of advanced reactors are high but are projected to fall with lessons learned from expanded commercial deployments, similar to the solar energy industry's trajectory. Commercially successful designs will be those with fast learning curves and falling costs.

14.5 Mid Transition

"Pious, performative, broad-brush bans on fossil fuels help no one. A more intelligent, data-led approach is needed to better protect the climate alongside vulnerable people in developing nations." (Economist Vijaya Ramachandran)[102]

How do we get from where we are now with the energy transition (Section 14.3.2) to where we would like to be circa 2100 (Section 14.3.3)?

The priority under the auspices of the UNFCCC Paris Agreement is net-zero CO_2 emissions. The UNFCCC starts from goals of zero net emissions by 2050 and works backwards by assuming actions that meet these goals, with actual socioeconomic needs and technical capabilities receiving little consideration.[103] The imposed urgency of making such a colossal transformation can lead to poor decisions that introduce a path dependency that not only harms overall human well-being, but also slows down progress on reducing carbon emissions.

The energy transition can be facilitated with minimal regrets by:

- Accepting that the world will continue to need and desire much more energy—energy austerity such as during the 1970s is off the table.
- Accepting that we will need more fossil fuels in the near term to maintain energy security and reliability and to facilitate the transition in terms of

developing and implementing new, cleaner technologies.

- Continuing to develop and test a range of options for energy production, transmission, and other technologies that address goals of lessening the environmental impact of energy production, CO_2 emissions, and other societal values (Table 14.1)
- Using the next two to three decades as a learning period with new technologies, experimentation, and intelligent trial and error, without the restrictions of near-term targets for CO_2 emissions.

In the near term, laying the foundation for zero-carbon electricity is substantially more important than trying to immediately stamp out fossil fuel use. Africa can develop its own natural gas resources. The transition focuses on developing and deploying new sources of clean energy. The transition does not focus on eliminating electricity from fossil fuels, since we will need much more energy to support the materials required for renewable energy and battery storage and building nuclear power plants, as well as to support growing numbers of electric vehicles and heat pumps.

To maintain the reliability of electric power under rapidly increasing demand for electricity will require all available generators of nuclear, natural gas, solar, wind, and hydro. While coal is an exceptionally dirty fuel, even coal power plants could be needed in the face of extreme demand shocks or natural gas supply shocks (such as occurring with Russia's war on Ukraine). Coal has two advantages over natural gas: relatively stable prices and storage which supports an adequate reserve margin.

Wind and solar power have developed synergistically with natural gas power plants, since it is easy to turn gas power plants off and on to balance the intermittent energy supplies from wind and solar.[104] Wind turbines, solar panels, and natural gas power plants all have average lifetimes around 20 years, so new natural gas power plants in the 2020s can be built with the expectation that they will not become stranded assets or promote lock-in.[105] The combination of wind, solar and natural gas is reducing greenhouse gas emissions and air pollution, at least on the current margin.

Natural gas price shocks and constraints on natural gas associated with Russia's war on Ukraine raise obvious concerns over geopolitical instability of natural gas as a fuel. A recent article in *Foreign Affairs* outlines a strategy for a shock-proof energy economy that would stabilize oil and gas supplies and prices. The strategy includes heavy production of natural gas and oil in North America.[106]

The build out of wind, solar and natural gas can fuel the transition, but this combination probably will not survive competition from new and better technologies that become available in the coming decades. It may be viable and affordable to take wind and solar to about 50 percent of a power system under very favorable climate conditions.[107] But unless there is hydropower backup, energy storage, or remote transmission capability, the cost profile for additional wind and solar becomes increasingly unfavorable and there are increasingly adverse consequences for electric power system reliability and performance.[108]

The push for weather-based renewable energy (wind, solar, hydro) seems somewhat ironic. One of the main motivations for transitioning away from fossil fuels is to avoid the extreme weather that is alleged to be associated with increasing CO_2 levels. So why subject our energy supply to the vagaries of water droughts and wind droughts, snow and icing, and forest fires? In any event, the growth of renewable energy has been a substantial boon to private sector weather forecasting companies that support the energy sector.

One work around for wind and solar intermittency is energy storage. The amount of electricity storage that is currently available from batteries can be measured in minutes or hours, used to buffer wind intermittency or the rapid transition created when the sun sets. While battery technologies are rapidly advancing with applications towards electric vehicles and back up batteries for household solar, feasible technologies for multi-month utility-scale battery storage is probably decades away. Longer duration storage has possible solutions like pumped hydro, mechanical and various electrochemical solutions, but it is too early to tell whether any of these technologies will scale. Load following nuclear power may be a more viable solution.

Transmission upgrades are an important factor in the transition. Modernization of the transmission grid is needed to enhance reliability and resiliency, improve cybersecurity and prevent outages due to extreme weather. Smart grids can allow advanced control of load supply and demand. There is a debate in the electric utilities community as to whether it makes sense to plan and build a macrogrid that would link far away regions to take advantage of variations in wind and solar power.[109] However, evolving technology and unpredictable markets suggest an incremental build-out incorporating smart microgrids would provide greater flexibility for an evolving power system. Early investment in a continental-scale macrogrid needs to be evaluated against the possibility of a post 2050 energy system that is mostly nuclear with relatively little wind and solar, which could introduce regrets and path dependence associated with a macrogrid. Developing and evaluating microgrids would substantially support the learning curve for incorporating distributed energy resources to increase the flexibility of transmission.

Transformation of the electric power sector will require considerable inputs of raw materials, including rare earth and semi-precious metals and structural materials such as cement, steel, and fiberglass. Apart from the amount of energy required to mine, process, and refine these materials, some critical materials are concentrated in a few countries, which will change the geopolitical dynamics. Building stable supply chains for critical materials is critical for the growth of wind, solar and battery storage. A circular economy with reuse and recycling will help minimize environmental impacts and supply chain shocks. Technologies that use low cost, readily available materials will have an advantage in being adopted.

Multiple options will allow the energy system transition will evolve in many different trajectories based on evolving values and choices made by different regions, countries, and utility companies. Each of these trajectories can be viewed as an experiment, with learning from the successes and failures of each of these experiments.

One question is the speed at which the transition can occur. China has developed its energy systems at an astonishing speed.[110] An autocratic government with a top-down energy strategy can rapidly implement changes. However, there are disadvantages to such an autocratic approach relative to the more chaotic, bottom-up approach of developing energy systems in the United States. Yes, making changes to an electric utility system in the United States must confront permitting, regulations, public approval, litigation, delays, cost overruns, and an archaic financial model for electric utilities. However, the more bottom-up approach in the United States (individual states and electric utility companies) provides many different opportunities to experiment and learn, thus producing in the end electric power systems that may be more anti-fragile with a broader array of options.

Learning curves are an essential feature of the transition. The basic idea is that the more you build of a technology—solar panels, lithium-ion batteries, small nuclear reactors—the cheaper it gets.[111] Economies of scale, learning-by-doing, and ongoing innovations are part of the so-called "green vortex," a self-accelerating process that occurs via a positive feedback loop.[112]

Notes

1 Saint Augustine, *Confessions*, trans. Henry Chadwick (Oxford, UK: Oxford University Press, 2008).
2 "Introduction to Mitigation," United Nations Climate Change (United Nations Framework Convention on Climate Change), accessed July 25, 2022, https://unfccc.int/topics/mitigation/the-big-picture/introduction-to-mitigation.
3 Michael Liebreich, "Liebreich: The Quest for Resilience—What Could Possibly Go Wrong?," BloombergNEF, March 1, 2022, https://about.bnef.com/blog/liebreich-the-quest-for-resilience-what-could-possibly-go-wrong/.
4 S Priyadarshi et al., "Summary for Policymakers," in *Climate Change 2022: Mitigation of Climate Change. Contribution of Working Group III to the Sixth Assessment Report of the*

Intergovernmental Panel on Climate Change (Geneva, CH: Intergovernmental Panel on Climate Change, 2022).

5 S. Pacala and R. Socolow, "Stabilization Wedges: Solving the Climate Problem for the next 50 Years with Current Technologies," *Science* 305, no. 5686 (August 13, 2004): 968–972, https://doi.org/10.1126/science.1100103.

6 "Project Drawdown," Project Drawdown, accessed July 25, 2022, https://drawdown.org/.

7 Robert Nozick, *Anarchy, State, and Utopia* (Oxford, UK: Basil Blackwell, 1974).

8 "Chapter 1: Overview of the Global Carbon Cycle," Second State of the Carbon Cycle Report (U.S. Global Change Research Program), accessed July 25, 2022, https://carbon2018.globalchange.gov/chapter/1/.

9 Ibid.

10 Ibid.

11 Ibid.

12 Peter Landschützer et al., "Decadal Variations and Trends of the Global Ocean Carbon Sink," *Global Biogeochemical Cycles* 30, no. 10 (October 5, 2016): 1396–1417, https://doi.org/10.1002/2015gb005359.

13 Pierre Friedlingstein et al., "Global Carbon Budget 2021," *Earth System Science Data* 14, no. 4 (April 26, 2022): 1917–2005, https://doi.org/10.5194/essd-14-1917-2022.

14 Ibid.

15 Ibid.

16 Nancy Averett, "The Ocean Is Still Sucking up Carbon-Maybe More than We Think," Eos, May 3, 2022, https://eos.org/articles/the-ocean-is-still-sucking-up-carbon-maybe-more-than-we-think.

17 Anthony P. Walker et al., "Integrating the Evidence for a Terrestrial Carbon Sink Caused by Increasing Atmospheric CO_2," *New Phytologist* 229, no. 5 (August 12, 2020): 2413–2445, https://doi.org/10.1111/nph.16866.

18 B. H. Samset et al., "Delayed Emergence of a Global Temperature Response after Emission Mitigation," *Nature Communications* 11, no. 1 (July 7, 2020), https://doi.org/10.1038/s41467-020-17001-1.

19 Ibid.

20 Andrew H. MacDougall et al., "Is There Warming in the Pipeline? A Multi-Model Analysis of the Zero Emissions Commitment from CO_2," *Biogeosciences* 17, no. 11 (June 15, 2020): 2987–3016, https://doi.org/10.5194/bg-17-2987-2020.

21 Ibid.

22 Miguel de Cervantes Saavedra, *Don Quixote*, trans. Walter Starkie (New York, NY: Signet Classic, 2001).

23 Stefan Koester and David M Hart, "Active Carbon Management: Critical Tools in the Climate Toolbox," Information Technology and Innovation Foundation, April 8, 2022, https://itif.org/publications/2022/04/18/active-carbon-management-critical-tools.

24 Ibid.

25 Ibid.

26 Raimund Malischek and Samantha McCulloch, "The World Has Vast Capacity to Store CO_2: Net Zero Means We'll Need It ," IEA, April 1, 2021, https://www.iea.org/commentaries/the-world-has-vast-capacity-to-store-co2-net-zero-means-we-ll-need-it.

27 "Supply, Underground Injection, and Geologic Sequestration of Carbon Dioxide," EPA (Environmental Protection Agency), accessed July 26, 2022, https://www.epa.gov/ghgreporting/supply-underground-injection-and-geologic-sequestration-carbon-dioxide.

28 Stefan Koester and David M Hart, "Active Carbon Management."

29 Ibid.

30 M Pathak et al., "Technical Summary," in *Climate Change 2022: Mitigation of Climate Change. Contribution of Working Group III to the Sixth Assessment Report of the Intergovernmental Panel on Climate Change* (Geneva, CH: Intergovernmental Panel on Climate Change, 2022).

31 Gabrielle B. Dreyfus et al., "Mitigating Climate Disruption in Time: A Self-Consistent Approach for Avoiding Both near-Term and Long-Term Global Warming," *Proceedings of the National Academy of Sciences* 119, no. 22 (May 23, 2022), https://doi.org/10.1073/pnas.2123536119.

32 Phil McKenna, "New Study Says World Must Cut Short-Lived Climate Pollutants as Well as Carbon Dioxide to Meet Paris Agreement Goals," Inside Climate News, May 23, 2022, https://insideclimatenews.org/news/23052022/short-lived-super-climate-pollutants-impact/.

33 S Szopa et al., "Short-Lived Climate Forcers," *In Climate Change 2021: The Physical Science Basis.*
Contribution of Working Group I to the Sixth Assessment Report of the Intergovernmental Panel on Climate Change (Geneva, CH: Intergovernmental Panel on Climate Change, 2021).

34 "Global Methane Assessment: Benefits and Costs of Mitigating Methane Emissions," UNEP, May 6, 2021, https://www.unep.org/resources/report/global-methane-assessment-benefits-and-costs-mitigating-methane-emissions.

35 Press Release, "Reducing Greenhouse Gas Emissions: Commission Adopts EU Methane Strategy as Part of European Green Deal," European Commission, October 14, 2020, https://ec.europa.eu/commission/presscorner/detail/en/ip_20_1833.

36 Muthukumara Mani, "Glaciers of the Himalayas," Open Knowledge Repository (Washington, DC: World Bank, June 3, 2021), https://openknowledge.worldbank.org/handle/10986/35600.

37 United Nations Environment Programme, "Near-Term Climate Protection and Clean Air Benefits: Actions for Controlling Short-Lived Climate Forcers," UN Environment Document Repository Home (UNEP, 2011), https://wedocs.unep.org/handle/20.500.11822/8048.

38 "Recent International Developments under the Montreal Protocol," EPA (Environmental Protection Agency), accessed July 26, 2022, https://www.epa.gov/ozone-layer-protection/recent-international-developments-under-montreal-protocol.

39 "What Are Hydrofluorocarbons?," EIA US, accessed July 26, 2022, https://us.eia.org/campaigns/climate/what-are-hydrofluorocarbons/.

40 United Nations Environment Programme, "Integrated Assessment of Short-Lived Climate Pollutants in Latin America and the Caribbean: Summary for Decision Makers," UN Environment Document Repository Home (United Nations Environment Programme, 2016), https://wedocs.unep.org/handle/20.500.11822/8972.

41 "Recent International Developments under the Montreal Protocol."

42 "COP26 Ends with Agreement but Falls Short on Climate Action," UNEP, November 15, 2021, https://www.unep.org/news-and-stories/story/cop26-ends-agreement-falls-short-climate-action.

43 Alex Epstein, *The Moral Case for Fossil Fuels* (New York, NY: Portfolio/Penguin, 2014).

44 "Henry Ford 150 Years," Home | Ford Media Center, accessed July 27, 2022, https://media.ford.com/content/fordmedia/fna/us/en.html.

45 Vaclav Smil, *How the World Really Works: The Science behind How We Got Here and Where We're Going* (New York, NY: Viking, 2022).

46 Tsung Xu, "The Transition to Clean Energy Is Accelerating," Tsung Xu (Tsung Xu, March 1, 2022), https://www.tsungxu.com/clean-energy-transition-guide/.

47 Ibid.

48 Ibid.

49 Norbert Edomah et al., "Sociotechnical Typologies for National Energy Transitions," *Environmental Research Letters* 15, no. 11 (October 16, 2020): 111001, https://doi.org/10.1088/1748-9326/abba54.

50 Paul Campbell, "When Mike Tyson Bit Evander Holyfield: 20 Years On," The Irish Times (The Irish Times, June 28, 2017), https://www.irishtimes.com/sport/other-sports/when-mike-tyson-bit-evander-holyfield-20-years-on-1.3136759.

51 "Renewable Capacity Highlights," International Renewable Energy Association, March 31, 2021, https://www.irena.org/-/media/Files/IRENA/Agency/Publication/2021/Apr/IRENA_-RE_Capacity_Highlights_2021.pdf.

52 M Pathak et al., "Technical Summary."

53 "Statistical Review of World Energy: Energy Economics: Home," BP Global, accessed July 27, 2022, https://www.bp.com/en/global/corporate/energy-economics/statistical-review-of-world-energy.html.

54 Gail Tverberg, "Limits to Green Energy Are Becoming Much Clearer," Our Finite World, February 9, 2022, https://ourfiniteworld.com/2022/02/09/limits-to-green-energy-are-becoming-much-clearer/.

55 "Biofuels Policies Drive up Food Prices, Say over 100 Studies," Transport & Environment, October 3, 2017, https://www.transportenvironment.org/discover/biofuels-policies-drive-food-prices-say-over-100-studies/.

56 "Fueling the Food Crisis: The Cost to Developing Countries of US Corn Ethanol Expansion," ReliefWeb, October 11, 2012, https://reliefweb.int/report/world/fueling-food-crisis-cost-developing-countries-us-corn-ethanol-expansion.

57 "Global Report on Food Crises: Acute Food Insecurity Hits New Highs," The Food and Agriculture Organization, May 4, 2022, https://www.fao.org/newsroom/detail/global-report-on-food-crises-acute-food-insecurity-hits-new-highs/en.

58 Henry Boucher, "Guest View: Global Hunger Fight Means No Biofuel," Reuters (Thomson Reuters, June 6, 2022), https://www.reuters.com/breakingviews/guest-view-global-hunger-fight-means-no-biofuel-2022-06-06/.

59 Arthur Neslen, "Europe and UK Pour 17,000 Tons of Cooking Oil into Vehicles a Day," The Guardian (Guardian News and Media, June 21, 2022), https://www.theguardian.com/environment/2022/jun/22/europe-and-uk-pour-17000-tons-of-cooking-oil-into-vehicles-a-day.

60 "'The EU's Proposed Bioenergy Policies Risk Additional Deforestation and Biodiversity Loss'," Le Monde, July 5, 2022, https://www.lemonde.fr/en/opinion/article/2022/07/05/the-eu-s-proposed-bioenergy-policies-risk-additional-deforestation-and-biodiversity-loss_5989103_23.html.

61 Carson Vaughan, "Ethanol Market Is Disturbing to American Farmers. and Now There's Covid-19.," Successful Farming (Successful Farming, March 30, 2020), https://www.agriculture.com/news/business/ethanol-market-is-disturbing-as-hell-to-american-farmers-and-now-there-s-covid-19.

62 "Biofuels Factsheet," Center for Sustainable Systems, accessed July 27, 2022, https://css.umich.edu/publications/factsheets/energy/biofuels-factsheet.

63 Jason Hill, "The Sobering Truth about Corn Ethanol," *Proceedings of the National Academy of Sciences* 119, no. 11 (March 9, 2022), https://doi.org/10.1073/pnas.2200997119.

64 "The Nexus of Biofuels, Climate Change, and Human Health: Workshop Summary," National Center for Biotechnology Information (U.S. National Library of Medicine, April 2, 2014), https://www.ncbi.nlm.nih.gov/books.

65 Damian Carrington, "Biofuels Needed but Some More Polluting than Fossil Fuels, Report Warns," The Guardian (Guardian News and Media, July 13, 2017), https://www.theguardian.com/environment/2017/jul/14/biofuels-need-to-be-improved-for-battle-against-climate-change.

66 Isla Binnie, "IEA Warns Global Solar Supply Chains Are Too Concentrated in China," Reuters (Thomson Reuters, July 7, 2022), https://www.reuters.com/business/energy/iea-warns-global-solar-supply-chains-are-too-concentrated-china-2022-07-07/.

67 Robert Bryce, "Wind Projects Rejected in Nebraska and Ohio, Wind Rejections across U.S. Now Total 328 since 2015," Forbes (Forbes Magazine, April 29, 2022), https://www.forbes.com/sites/robertbryce/2022/04/29/wind-projects-rejected-in-nebraska-and-ohio-wind-rejections-across-us-now-total-328-since-2015/?sh=37607b513bab.

68 "2021 Annual Energy Paper—J.P. Morgan," J.P. Morgan, May 2022, https://am.jpmorgan.com/content/dam/jpm-am-aem/global/en/insights/eye-on-the-market/future-shock-amv.pdf.

69 Jeff St. John, "The US Has More Clean Energy Projects Planned than the Grid Can Handle," Canary Media, April 20, 2022, https://www.canarymedia.com/articles/transmission/the-us-has-more-clean-energy-projects-planned-than-the-grid-can-handle.

70 Annabelle Timsit, "Elon Musk Says Tesla's Car Factories Are 'Gigantic Money Furnaces'," The Washington Post (WP Company, June 23, 2022), https://www.washingtonpost.com/technology/2022/06/23/elon-musk-tesla-factories-gigantic-money-furnaces/.

71 Sean McLain and Scott Patterson, "Rivian CEO Warns of Looming Electric-Vehicle Battery Shortage," The Wall Street Journal (Dow Jones & Company, April 18, 2022), https://www.wsj.com/articles/rivian-ceo-warns-of-looming-electric-vehicle-battery-shortage-11650276000.

72 Andrew Birch, "How to Halve the Cost of Residential Solar in the US," Greentech Media (Greentech Media, January 5, 2018), https://www.greentechmedia.com/articles/read/how-to-halve-the-cost-of-residential-solar-in-the-us.

73 Rick Atkinson, *In the Company of Soldiers: A Chronicle of Combat* (New York, NY: Henry Holt, 2004).

74 S Priyadarshi et al., "Summary for Policymakers."

75 Jessica Lovering et al., "Land-Use Intensity of Electricity Production and Tomorrow's Energy Landscape," *PLOS ONE* 17, no. 7 (July 6, 2022), https://doi.org/10.1371/journal.pone.0270155.

76 Katinka M. Brouwer, ed., "Road to EU Climate Neutrality by 2050," EGR Group / renew europe, January 2021, https://roadtoclimateneutrality.eu/Energy_Study_Full.pdf.
77 Jessica Lovering et al., "Land-Use Intensity of Electricity Production."
78 Edgar G. Hertwich et al., "Integrated Life-Cycle Assessment of Electricity-Supply Scenarios Confirms Global Environmental Benefit of Low-Carbon Technologies," *Proceedings of the National Academy of Sciences* 112, no. 20 (October 6, 2014): 6277–6282, https://doi.org/10.1073/pnas.1312753111.
79 Jessica Lovering et al., "Land-Use Intensity of Electricity Production."
80 Ibid.
81 "Microgrids—the next Evolution of the Grid," sol-up, November 2, 2018, https://solup.com/microgrids-next-evolution-grid/.
82 Sabine Hossenfelder, "Follow the Science? Nonsense, I Say.," YouTube (YouTube, September 23, 2020), https://www.youtube.com/watch?v=nGVIJSW0Y3k.
83 T. Petermann et al., "What Happens during a Blackout: Consequences of a Prolonged and Wide-Ranging Power Outage," The Office of Technology Assessment at the German Bundestag, 2011, https://publikationen.bibliothek.kit.edu/1000103292.
84 T. Petermann et al., "What Happens during a Blackout."
85 Henry Dunckley, *Lord Melbourne* (New York, NY: Harper & Brothers, 1890).
86 Michael Liebreich, "Liebreich: The Quest for Resilience."
87 Terje Aven, "Climate Change Risk—What Is It and How Should It Be Expressed?," *Journal of Risk Research* 23, no. 11 (November 9, 2019): 1387–1404, https://doi.org/10.1080/13669877.2019.1687578.
88 " 'Whole Earth's' Stewart Brand Backs Nuclear Power," NPR (NPR, February 20, 2010), https://www.npr.org/templates/story/story.php?storyId=123919458.
89 Taylor Dotson and Michael Bouchey, "Democracy and the Nuclear Stalemate," The New Atlantis, December 22, 2020, https://www.thenewatlantis.com/publications/democracy-and-the-nuclear-stalemate.
90 Mycle Schneider et al., "World Nuclear Industry Status Report 2017," World Nuclear Industry Status Report, September 2017, https://www.worldnuclearreport.org/IMG/pdf/20170912wnisr2017-en-lr.pdf.
91 David Warmflash, "Perspective: Why Nuclear Power Is Good for Public Health," NEO.LIFE, accessed July 29, 2022, https://neo.life/2022/06/perspective-why-nuclear-power-is-good-for-public-health/.
92 Ibid.
93 "Storage of Spent Nuclear Fuel," nrc.gov, accessed July 29, 2022, https://www.nrc.gov/waste/spent-fuel-storage.html.
94 Sedeer El-Showk, "Finland Built This Tomb to Store Nuclear Waste. Can It Survive for 100,000 Years?," Science, February 24, 2022, https://www.science.org/content/article/finland-built-tomb-store-nuclear-waste-can-it-survive-100000-years.
95 Kerry Emanuel, "Nuclear Fear: The Irrational Obstacle to Real Climate Action," Bulletin of Atomic Scientists, November 15, 2021, https://thebulletin.org/premium/2021-11/nuclear-fear-the-irrational-obstacle-to-real-climate-action/.
96 Susan D'Agostino, "Jellyfish Attack Nuclear Power Plant. Again.," Bulletin of the Atomic Scientists, October 28, 2021, https://thebulletin.org/2021/10/jellyfish-attack-nuclear-power-plant-again/.

97 Robin Cowan, "Nuclear Power Reactors: A Study in Technological Lock-In," *The Journal of Economic History* 50, no. 3 (September 1990): 541–567, https://doi.org/10.1017/s0022050700037153.

98 "Plans For New Reactors Worldwide," World Nuclear Association, accessed July 29, 2022, https://world-nuclear.org/information-library/current-and-future-generation/plans-for-new-reactors-worldwide.aspx.

99 Hartmut Winkler, "Russia's Nuclear Power Exports: Will They Stand the Strain of the War in Ukraine?," The Conversation, March 6, 2022, https://theconversation.com/russias-nuclear-power-exports-will-they-stand-the-strain-of-the-war-in-ukraine-178250#:~:text=Unlike%20China%2C%20the%20country%20most,of%20Fukushima%20incident%20in%202011.

100 Darrell Proctor, "Climate Ripe for Nuclear Advancements," POWER Magazine, January 3, 2022, https://www.powermag.com/climate-ripe-for-nuclear-advancements/.

101 Adam Stein et al., "Advancing Nuclear Energy."

102 Vijaya Ramachandran, "Blanket Bans on Fossil Fuels Hurt Women and Lower-Income Countries," Nature News (Nature Publishing Group, July 5, 2022), https://www.nature.com/articles/d41586-022-01821-w.

103 "Vaclav Smil, *How the World Really Works*."

104 Emily Grubert and Sara Hastings-Simon, "Designing the Mid-Transition: A Review of Medium-Term Challenges for Coordinated Decarbonization in the United States," *WIREs Climate Change* 13, no. 3 (February 8, 2022), https://doi.org/10.1002/wcc.768.

105 "Natural Gas Generators Make up the Largest Share of Overall U.S. Generation Capacity," U.S. Energy Information Administration (EIA), December 18, 2017, https://www.eia.gov/todayinenergy/detail.php?id=34172.

106 Gregory Brew, "A Shock-Proof Energy Economy," Foreign Affairs, July 13, 2022, https://www.foreignaffairs.com/articles/united-states/2022-07-13/shock-proof-energy-economy.

107 Aaron Bloom, "Four Reasons 30% Wind and Solar Is Technically No Big Deal," Utility Dive, December 5, 2016, https://www.utilitydive.com/news/four-reasons-30-wind-and-solar-is-technically-no-big-deal/431686/.

108 Semich Impram et al., "Challenges of Renewable Energy Penetration on Power System Flexibility: A Survey," *Energy Strategy Reviews* 31 (September 2020): p. 100539, https://doi.org/10.1016/j.esr.2020.100539.

109 Steve Hanley, "Macrogrids or Microgrids: Which Is the Key to the Renewable Energy Revolution?," CleanTechnica, June 19, 2020, https://cleantechnica.com/2020/06/19/macrogrids-or-microgrids-which-is-the-key-to-the-renewable-energy-revolution/.

110 Arendse Huld, "China's Energy Transition—How Far Has the Country Come?," China Briefing News, April 22, 2022, https://www.china-briefing.com/news/earth-day-2022-whats-the-state-of-chinas-energy-transition/.

111 Goksin Kavlak et al., "Evaluating the Causes of Cost Reduction in Photovoltaic Modules," *Energy Policy* 123 (October 11, 2018): 700–710, https://doi.org/10.1016/j.enpol.2018.08.015.

112 Robinson Meyer, "How the U.S. Made Progress on Climate Change without Ever Passing a Bill," The Atlantic (Atlantic Media Company, June 16, 2021), https://www.theatlantic.com/science/archive/2021/06/climate-change-green-vortex-america/619228/.

Chapter Fifteen

CLIMATE RISK AND THE POLICY DISCOURSE

"The best way to predict your future is to invent it."
—Computer scientist Alan Kay[1]

As I write this final chapter in mid-summer of 2022, the world is embroiled in geopolitical and financial instability from the COVID-19 pandemic and Russia's war on Ukraine. In the midst of this instability, we are seeing the inevitable clash between alarming proclamations about the climate crisis, the priorities of food and energy and poverty reduction, and the costs and difficulties of transitioning to net-zero CO_2 emissions.

Recent headlines include:

- "German cities impose cold showers and turn off lights amid Russian gas crisis"[2]
- "Hungary declares state of emergency over threat of energy shortages"[3]
- "Almost half of UK adults fear falling into fuel poverty before the years end"[4]
- "The West's Green Delusions Empowered Putin"[5]
- "Russia's war is the end of climate policy as we know it"[6]
- "Trudeau moves forward with fertilizer reduction climate policy"[7]
- "Why Dutch farmers are protesting over emissions cuts"[8]
- "Ireland debates a 30% emissions cap on farmers"[9]
- "Green dogma behind fall of Sri Lanka"[10]
- "Rich countries' climate policies are colonialism in green"[11]
- "Barbados Resists Climate Colonialism in an Effort to Survive the Costs of Global Warming"[12]
- "African nations expected to make case for big rise in fossil fuel output"[13]
- "UN climate talks end in stalemate and 'hypocrisy' allegation"[14]

How to respond to the climate "crisis" in the midst of genuine crises associated with food and energy shortages and the humanitarian crisis in Ukraine is best reflected by the response of the New Zealand government. In defending its decision to issue fossil fuel prospecting permits in spite of declaring a climate emergency, the New Zealand government stated that the climate crisis was "insufficient" to halt oil and gas exploration.[15] Climate change is indeed a crisis of insufficient weight that is now being all but ignored by many countries as they grapple with the basic human needs for energy and food.

In 2015, the world's nations agreed on a set of 17 interlinked Sustainable Development Goals to support future global development.[16] These goals include, in ranked order:[17]

1. No poverty
2. Zero hunger
7. Affordable and clean energy
13. Climate action

Should one element of Goal 13, related to net-zero emissions, trump the higher priority goals of poverty and hunger and the availability of energy? Not if human well-being, flourishing, and thriving are the objectives.

Climate change and its interactions with humans and their societies are exceedingly complex issues. The misidentification of climate change as a "crisis" and the ensuing precautionary mandate to rapidly eliminate the use of fossil fuels is creating new risks, while failing to address the current emergency risks associated with extreme weather events. Simply put, the basic risk management principles of **P**roportionate, **A**ligned, **C**omprehensive, **E**mbedded, and **D**ynamic (PACED; Section 11.1) are being ignored with regards to climate change policy.

Striking a balance between the security of basic food, water, energy, and material supplies and the least possible impact on the environment including CO_2 emissions is arguably the greatest challenge of the twenty-first century. This final chapter argues for pushing the reset button by opening up the climate policy envelope to new possibilities that support human wellbeing and thrivability in the twenty-first century.[18]

15.1 Moral Dilemmas and the Fallacy of Control

"God, give us grace to accept with serenity the things that cannot be changed, courage to change the things that should be changed, and the wisdom to distinguish the one from the other." (American theologian Reinhold Niebuhr)[19]

At the heart of the current debate over climate change is different views of the relationship between humans and the environment and our moral obligations. A common environmental debate is whether humans have an obligation to the planet, beyond the human values of ecosystem services and enjoyment of nature. Traditional environmentalists view people as surviving by virtue of nature's benevolence, and thus focus on conservation. By contrast, ecomodernism embraces substituting natural ecological services with energy, technology, and synthetic solutions, which improves human wellbeing through eco-economic decoupling while incidentally protecting nature.[20] The climate change issue provides cognitive dissonance to traditional conservation-minded environmentalists who protest against wind and solar farms, hydropower, nuclear power plants, and new transmission lines, while at the same time claiming to be ardent supporters of decarbonization.[21]

Another overarching moral dilemma is whether there is a moral obligation for humans to preserve the climate in its present, or otherwise specified, state.[22] Implicit in the various temperature and emissions targets under the UNFCCC is that the world does have such an obligation. If the world's climate were warming at the current rate solely owing to natural causes, would humanity feel obligated to slow down future warming (perhaps by Direct Air Capture of CO_2 or solar radiation engineering)? Unlikely; such control of the climate would rightfully be regarded as futile and/or dangerous. People would adapt to the changing climate as they always have, perhaps with some imaginative engineering to shield Greenland from summertime melt to slowdown sea level rise, which would be perceived as the main future impact of warming. There would be no motivation to attribute every extreme weather event to the warming, because there would be no political gain to be obtained from such attribution.

The current alarm is about the possibility of much faster rates of warming in the future, which are associated with implausible emissions scenarios and high values of climate sensitivity to CO_2. Even with a continued rate of warming of 0.18°C/decade—an additional 1.4°C of warming by 2100 (totaling 2.5°C since the late nineteenth century)—all of the IPCC's socioeconomic scenarios expect humans to be substantially better off than at present.[23] The current dread of such a magnitude of warming is not so much the magnitude per se, but the cause of the warming (human versus natural).

Humanity has a moral compass that points to caring about all people currently alive as well as the collective future of our species. With regards to the future of our species, there are many different perspectives and imagined futures.[24]

Prior to imagining the future circa 2100—78 years hence relative to 2022—it is useful to consider the world of 78 years ago, circa 1944. Near the end of World War II, it would have been difficult to imagine: the existence of the European Union and the rise of China as a superpower, a global population of eight billion that is far less impacted by hunger and poverty, and globalization. It would be equally difficult to imagine a myriad of technological developments including satellite communications, computers, the internet, advanced medical diagnostics and treatments, advanced military capabilities, and sophisticated household appliances. Also, the major social changes including advancement of women and some fundamental changes in values within cultures. It would also have been difficult to imagine in 1944 that in 2022 the world's governments would be preoccupied with human-caused climate change, a priority that is effectively superseding poverty reduction and energy access for developing countries (Section 13.5).

The UNFCCC has presupposed a moral obligation to control climate change by eliminating fossil fuels and achieving net-zero emissions. They have bypassed the moral dilemma of preventing future harm from climate change versus fulfilling our duties to currently living humans. Given our inability to predict what the world of 2100 will be like and the world population's preferences and values (not to mention our inability to predict the climate in 2100), what kind of legacy can we hope to leave our descendants circa 2100? Beyond a culture of freedom and prosperity and an environment that can sustain, does it make sense to torque current policies around imagining what kind of climate the world's citizens would prefer in 2100? Given our inability to understand the world circa 2100 including likely losses from human-caused climate change, sacrificing the wellbeing of the current population by restricting energy access seems neither moral nor just.

Apart from the moral dilemmas, the UNFCCC strategy to control the climate or even emissions needs to be challenged. Even beyond the technical issues, greater realism is needed about the uncertainties and politics underpinning the pursuit of control.[25] The pandemic illustrates that our tools for acting on a complex global problem—experts, precise scientific metrics, computer models, enforced restrictions—have resulted in much less than the desired quality of control.

The global energy transition and worldwide transformations to sustainability are far more challenging than the global COVID-19 pandemic. Our hubristic aspirations for control fail to acknowledge the wickedness and systemic aspects of the climate change problem and its proposed solutions. Yes, we can seek to lower emissions As Low As Reasonably Practical (ALARP, Section 11.1.1), ideally while minimizing our regrets and maximizing our opportunities through the energy transition. But we should not pretend that we are controlling the climate. The modernist paradigm of mastery, planning, and optimization is not appropriate for the wicked problems of the twenty-first century.[26]

15.2 Towards Post-Apocalyptic Climate Politics

"Climate change' is just a mental tattoo—a phrase we invoke with an air of scientific sophistication to give some sense of knowledgeability about the unknowable." (Anthropologist Peter Wood)[27]

The climate catastrophe isn't what it used to be. Circa 2013 with publication of the IPCC AR5, RCP8.5 was regarded as the business-as-usual emissions scenario, with expected warming of 4–5°C by 2100. Now there is growing acceptance that RCP8.5 is implausible, and RCP4.5 is arguably the current business-as-usual emissions scenario (Section 7.1). Only a few years ago, an emissions trajectory that followed RCP4.5 with 2–3°C warming was regarded as climate policy success. As limiting warming to 2°C seems to be in reach (now deemed to be the "threshold of catastrophe"),[28] the goal posts were moved in 2018 to reduce the warming target to 1.5°C.[29]

What is accomplished by moving the goal posts and the continued amplification of the catastrophe rhetoric surrounding climate change? Catastrophizing is motivated by a desire to amp up the urgency for action in eliminating fossil fuels. Continued catastrophizing has produced a political battle between two extremes: those who insist on urgent elimination of fossil fuels; and a range of others that are castigated as deniers of climate science because they do not support the rapid elimination of fossil fuels until reliable replacement fuel sources are in place.

While the IPCC Assessment Reports do not support the catastrophe narrative (Section 1.3), there is a catastrophizing lock-in for many organizations that would lose their *raison d'etre* if the climate change threat was diminished. Letting go of apocalyptic rhetoric is difficult for those who have built careers based on climate catastrophism. Climate catastrophe rhetoric now seems linked to extreme weather events, most of which are difficult to identify any role for human-caused climate change in increasing either their intensity or frequency (Section 7.4.1).

With the growing realization that warming in the twenty-first century will not be nearly as bad as previously thought, the narrative is shifting to "but the impacts are worse than we thought." This shift in perception of twenty-first century climate change and its impacts argues for prioritizing adaptation (addressing emergency risks) over emissions reductions (the slower, incremental risk).

15.2.1 Apocalyptic Climate Politics

"Politically correct climate change orthodoxy has completely destroyed our ability to think rationally about the environment." (Economist Richard Tol)[30]

Motivated by international treaties and the UNFCCC Paris Agreement, countries and municipalities are declaring a "climate emergency" that requires urgent and strong climate policies to avoid both local and global catastrophe.[31]

Physicist Robert H. Socolow argues that the 2°C target was not so much a policy goal but rather a political motivation, reflecting "a mindset that is common to the entire exercise: to create maximum pressure for action."[32] Business-as-usual climate policy is based on what has been referred to as the politics of "climate scarcity" whereby there is an upper limit to the level of warming (and thereby CO_2 emissions) that must not be exceeded to avoid dangerous climate change.[33] The politics of climate scarcity is associated with the politics of energy and material scarcity, blaming climate change on extravagant lifestyles and requiring a long period of belt-tightening if we are to survive the crisis.

The failure of the world's governments to make much headway in reducing emissions is blamed on several factors. The main impediment to progress is blamed on fossil fuel companies, who wield power through political influence via financial contributions and propaganda.[34] Capitalism is being blamed because manufacturers, farmers, and others use fossil fuels to produce food and equipment needed by the economy and general population. Democracy is being blamed, since democratic decision-making is too slow and sometimes people do not make the "right" decision.[35] Arguments are being made for degrowth, which is the idea that economic growth is environmentally unsustainable and should be halted at least in wealthy countries.[36] Increasingly, apocalyptic climate politics is becoming a useful cudgel to approach other social issues.

With the failure of most countries to significantly reduce CO_2 emissions, activists and governments are using "random gambits to kneecap fossil fuel production."[37] Attempts to restrict the supply of fossil fuels include restricting permits for fossil fuel production, canceling oil and gas pipelines, getting organizations to divest their funds from fossil fuel companies, and restricting access to loans and other financial resources for fossil fuel companies.

Attempts to limit CO_2 emissions from the demand side by imposing carbon taxes have been politically very unpopular.[38] The politics of scarcity is not an easy sell, particularly when framed in terms of anti-democracy, anti-capitalism, and degrowth. Making energy less abundant and/or increasing its price is politically toxic unless there is an urgent, short-term need for austerity.

While the vast majority of people believe that climate change is a real problem,[39] many fear a future without cheap, abundant fuel and continued economic expansion far more than they fear climate change. Climate politics business-as-usual (the apocalyptic version) expects people in developed countries to exercise energy and material restraint for the altruistic motives of "saving the climate," while at the same time slowing down development in Africa by not supporting access to their own energy resources.

The politics of alarm and fear and scarcity aren't working as well as its advocates have hoped. Further, the lock-in of apocalyptic climate politics is becoming increasingly disconnected from physical and political realities.

15.2.2 Framework for a Post-Apocalyptic Politics

"We address ourselves not to their humanity but to their self-love, and never talk to them of our own necessities but of their advantages." (Adam Smith, *An Inquiry into the Nature and Causes of the Wealth of Nations* 1776)[40]

There is growing support for a climate politics that harnesses enlightened self-interest, rather than focusing on austerity.[41] This plays to the values of the central objectives of human flourishing and thriving.

There are three major policy issues that fall under the climate umbrella:

1. The desire for clean, abundant, and cheap energy.
2. Concerns about vulnerability to extreme weather and climate events.
3. Concerns about rising atmospheric concentrations on CO_2 and its impact on the climate.

Issues #1 and #2 are primarily dealt with by national and subnational entities, and can be expected to receive widespread political and economic support since they support local self-interests. Issue #3 is politically controversial since international policies have attempted a top-down approach that impacts #1 and #2 by the emphasis on rapid reduction of fossil fuel emissions, with the specter of energy scarcity and a redirection of funds away from development and adaptation.

Focusing on issues #1 and #2 is a quieter kind of climate politics, which does not require the apocalyptic rhetoric and plays well into local strategies that people are enthusiastic about. Willing actions are more effective politically and have higher moral legitimacy because of the absence of coercion. A focus on issues #1 and #2, while concurrently supporting flourishing and thriving of the global population, can act to de-escalate the political controversies associated with the climate change issue.

Another strategy for post-apocalyptic politics is to eliminate the mismatch in timing of the costs versus benefits of energy and climate policies.[42] Any near-term emissions-reducing policies should also provide short-term benefits, with the long-term goals of reducing CO_2 emissions taking on secondary importance. Clean energy that increases abundance and reduces costs is an example. The reduction of methane emissions from fossil fuel production is another example that has short-term benefits (Section 14.2). Adaptation strategies also provide near-term benefits (Chapter Thirteen).

Once we acknowledge that we do not currently know how to stabilize atmospheric concentrations of CO_2 on the timescale of decades, we can search for new and more effective approaches for issues #1 and #2 and other ancillary issues under the climate umbrella, all the while focusing on supporting human flourishing and thriving in the twenty-first century.

And finally, we need to recognize that what has been cast as a global "crisis" is for the most part thousands of local vulnerability emergencies that are revealed by extreme weather events.

15.2.3 Politics of Climate Uncertainty

"We must accept finite disappointment, but we must never lose infinite hope." (US civil rights leader Martin Luther King, Jr.)[43]

Futures are unknown—uncertainties are ubiquitous, diverging interests and perspectives introduce ambiguities, and we should expect to be surprised. Policy scientist Daniel Sarewitz has characterized climate change as "inherently open, indeterminate, and contested."[44]

Uncertainties can create fear and anxiety, resulting in attempts at premature closure of the knowledge base for policy making. Exaggerated certainty about the knowledge base is a defensive mechanism that displaces anxieties around uncertain futures.[45]

As a consequence of the exaggerated sense of knowledge and control surrounding climate policy, some highly uncertain issues that should remain open for political debate are ignored in the policies of targets and deadlines. Premature foreclosure of uncertainties in mainstream climate change politics is suppressing the interests and needs of marginalized communities, cultures, and environments. Failure to embrace uncertainty and ambiguities surrounding climate change has been characterized as an invisible form of oppression that is foreclosing possible futures.[46]

In formulating a politics of climate uncertainty, the challenge lies in rejecting fear in favor of reframing uncertainty as "an invitation to plural hopes, and respectful recognition of difference, rather than singular fears."[47]

The politics of uncertainty is based upon an appreciation of uncertainty and its diverse framings. As such, the politics of uncertainty can provide the seeds for imagining diverse, alternative futures. By opening up space to re-imagine futures, uncertainties can be confronted in positive ways: not as threats or sources of fear, but as sources of possibility and opportunity. The spaciousness of uncertainty provides room for creativity, curiosity, entrepreneurship, discovery, and innovation.[48]

The politics of climate uncertainty abandons the ideal of negotiated destinations such as the 2°C target. The politics of uncertainty leads to framing climate change as an ongoing predicament.[49] Accepting climate change as an ongoing predicament avoids relying on human prescience. Even if human-caused climate change is somehow eliminated, natural climate variability and inevitable surprises will provide ongoing challenges that require continuing adaptation by communities.

As a framework for which a better future may be realized, the politics of uncertainty is about humility (not hubris) regarding what is known, hope (not fear) regarding what is possible, and allowing for greater diversity of values.[50]

15.3 Climate Pragmatism

"We can only see a short distance ahead, but we can see plenty there that needs to be done." (Alan Turing, English mathematician and author of *Computing Machinery and Intelligence*)[51]

After reaching this point in the book, readers may or may not agree as to whether warming is dangerous or whether urgent action to reduce CO_2 emissions is needed. However, I hope that the reader is better informed as to the uncertainties and the various values in play surrounding these judgments. It is often far easier to agree on specific solutions/actions/policies than it is to agree on the causes of problems (Section 12.2). Most people can agree on the overall strategies of innovation, building resilience, and no regrets policies.

The failure of the UNFCCC Kyoto Protocol to limit CO_2 emissions spawned several policy analyses that pointed towards a more pragmatic approach to the climate change issue. In 2007, UK social scientists Gwythian Prins and Steve Rayner published an article entitled "Wrong Trousers: Radically Rethinking Climate Policy."[52] The article responded to the failure of the Kyoto Protocol by providing some principles for a viable climate policy. The general framework is to harness enlightened self-interest to drive a process designed to generate a range of possible solutions, which can be evaluated, changed, and refined as the goal of climate security is pursued. Using a wicked problem framework, Prins and Rayner argue that better climate policies may emerge from a more oblique approach which recognizes that successful climate policy does not necessarily focus instrumentally on the climate. Prins and Rayner coined the term "silver buckshot," in contrast to a "silver bullet" solution. They emphasize social learning, which includes constant course corrections and improvements.

They characterize climate change as a multi-level governance problem and recommend that problems be dealt with at the lowest possible levels of decision-making. Recommendations are made for investing in technology research and development and adaptation.[53]

In 2010, the so-called Hartwell paper was published,[54] authored by 14 natural and social scientists from Asia, Europe, and North America. Also invoking the wickedness characterization, the paper posits that climate change has been misrepresented as a conventional environmental problem that is capable of being solved. The Hartwell paper argued that decarbonization will only be achieved successfully as a benefit contingent upon other goals which are politically attractive and relentlessly pragmatic. The paper recommends expanding access to energy for the poor, quickly reducing non-CO_2 climate forcings, and adapting to the changing climate.[55]

The term "climate pragmatism" was coined in a follow-on paper in 2011 that provided an American context to the Hartwell paper.[56] Pragmatism has a deep intellectual heritage in the United States, originating in the late nineteenth century from philosophers Charles Sanders Pierce, William James, and John Dewey.[57] Pragmatism values pluralism over universalism, flexibility over rigidity, and practical results over utopian ideals.

Energy innovation, resilience to extreme weather, and no regrets pollution reduction—each of these goals has diverse justifications that are independent of climate change policies. Climate pragmatism prioritizes no-regrets policies by avoiding options with controversial, uncertain, or immeasurable benefits. Climate pragmatism emphasizes the exercise of strategic judgment, placing a premium upon understanding what we do not know. Climate pragmatism reduces the role for public policy in context of more oblique approaches.[58]

The pragmatic approach does not depend on reaching global agreement on an alleged perfect solution that requires moral and political coercion. Climate pragmatism recommends moving away from centralized top-down approaches, such as international treaties and accords, in favor of breaking the problem into smaller, human-relevant problems and solutions and prioritizing them by the importance of the problem and feasibility and costs of the solution.[59]

The bottom-up approach to decision-making addresses smaller problems with incremental policies at the lowest possible level of decision-making.[60] More flexible, multi-level approaches include policy actors in local governments, firms, trade associations, nongovernmental organizations, and scientific and technical organizations. Policy actors at lower levels have much more flexibility in responding to market signals, abandoning courses of action that aren't working, and learning by doing.

15.4 Wicked Science for Wicked Problems

"There are really two kinds of optimism. There's the complacent, Pollyanna optimism that says, 'Don't worry—everything will be just fine,' and that allows one to just lay back and do nothing about the problems around you. Then there's what we call dynamic optimism. That's an optimism based on action." (Ramez Naam, technologist, and science fiction writer)[61]

At this point, the momentum for a transition away from fossil fuels is being driven more by geopolitics, macroeconomics, and technological innovation than it is by the rhetoric of catastrophic climate change. It is difficult to foresee any climate research over the next decade that would change this trajectory.

If climate science is to be relevant in decision-making beyond international agreements aimed at reducing emissions, different frameworks are needed. Three different solution-oriented directions that involve climate science are evolving: (1) big data approaches, (2) climate tech; and (3) wicked science.

Big data approaches are an outgrowth of the information technologies that have driven Big Tech. We now have the capability of collecting and analyzing massive amounts of data. Climate-relevant data are being collected from diverse sources that include smart sensor networks, cell phones, and constellations of small satellites, with an emphasis on geospatial and hyperlocal data. Big data approaches, when interpreted by machine learning and artificial intelligence approaches, have great potential for providing information and knowledge for managing the risks of climate change. Big data analytics are most useful when they are interpreted by human intelligence and combined with theory-driven and/or small data approaches.[62]

Another trend that comes from the culture of Big Tech is "climate tech." Climate tech is defined as technologies that focus on reducing/sequestering greenhouse gas emissions or addressing the impacts of global warming. Climate tech entrepreneurs are developing diverse products ranging from carbon accounting software to agricultural technology to electric vehicle batteries to nuclear fusion projects. There is a substantial migration of engineers underway, away from Big Tech and towards the new opportunities in climate tech. Climate tech is currently regarded by many to be a generational investment opportunity and the most exciting space in tech.[63]

And finally, we have the evolution of "wicked science."[64] Wicked science is a process that is tailored to the dual scientific and political natures of wicked societal problems (Section 3.4). As such, wicked science is massively transdisciplinary, including natural sciences and engineering along with social sciences and humanities. Wicked science uses approaches from complexity science and systems thinking in a context that engages with the political roles and perspectives of decision makers, planners, and other stakeholders. Wicked problems and the strategies devised to address them cannot be defined by scientific experts alone, but include the experiential and operational knowledge of a range of stakeholders.

Two recent papers by atmospheric/climate scientists have articulated something similar to wicked science for the climate sciences. A paper entitled "Usable climate science is adaptation science" emphasizes that the localness of adaptation implies much greater uncertainty in the relevant climate science.[65] Climate science for adaptation is more about characterizing uncertainty for robust decision-making (Chapter Twelve). Usable climate science requires that scientists engage in co-production of usable science with stakeholders, with a willingness to learn to understand how the human factors are manifest in a particular setting. A paper entitled "Small is beautiful: climate-change science as if people mattered" addresses strategies for grappling with the complexity of local situations.[66] The strategies include expressing climate knowledge in conditional form in terms of scenarios developed via the storyline approach (Chapters Eight and Nine), and working with local communities to make sense of their own situations.

Combining and integrating knowledge from diverse disciplines and other sources to provide insights, explanations, and solutions to wicked problems is a substantial challenge. For the solution orientation of wicked science to be meaningful, we need an overarching philosophy for navigating wicked problems.[67] We need to acknowledge that control is limited, the future is unknown, and it is difficult to determine whether the impact you make will be positive. We need to accept that climate change will continue to disrupt natural systems and human wellbeing; this acknowledgment helps avoid the urgency trap. By acknowledging that there is no road back, we can focus on the road ahead.[68]

The road ahead can be facilitated by broader, transdisciplinary thinking about the climate change problem and its solutions. This requires moving away from the consensus-enforcing and cancel culture approach of attempting to restrict the dialogue surrounding climate change and the policy options. We need to open up space for dissent, disagreement, and discussion about scientific uncertainty and policy options, so that multiple perspectives can be considered and broader support can be built for a range of policy options.

Wicked scientists are willing to become embroiled in political debates and thorny social problems. To be effective, we need to break the hegemony of disciplinary researchers, particularly those who are strident political activists, as being regarded as experts for solutions to the wicked problem of climate change. While the IPCC has operated via a loose cooperation between multiple disciplines,[69] genuine transdisciplinary understanding and collaborations, across disciplines and with a broad range of stakeholders, is needed for meaningful contributions to wicked problems.

This book encompasses my own philosophy for navigating the wicked problem of climate change. As such, this book provides a single slice through the wicked terrain. By acknowledging uncertainties in the context of better risk management and decision-making frameworks, in combination with techno-optimism, there is a broad path forward for humanity to thrive in the twenty-first century and beyond.

Notes

1 Alan Kay, "Learning vs. Teaching with Educational Technologies.," EDUCOM, 1984, https://eric.ed.gov/?id=EJ293879.
2 Philip Oltermann, "German Cities Impose Cold Showers and Turn off Lights amid Russian Gas Crisis," The Guardian (Guardian News and Media, July 28, 2022), https://www.theguardian.com/world/2022/jul/28/german-cities-impose-cold-showers-and-turn-off-fountains-in-face-of-russian-gas-crisis.
3 AP, "Hungary Declares 'State of Emergency' over Threat of Energy Shortages," euronews, July 13, 2022, https://www.euronews.com/2022/07/13/hungary-declares-state-of-emergency-over-threat-of-energy-shortages.
4 "Almost Half of UK Adults Fear Falling into Fuel Poverty before the Year Ends," Scottish Housing News, August 3, 2022, https://www.scottishhousingnews.com/articles/almost-half-of-uk-adults-fear-falling-into-fuel-poverty-before-the-year-ends.
5 Michael Shellenberger, "The West's Green Delusions Empowered Putin," Common Sense (Bari Weiss, March 1, 2022), https://www.commonsense.news/p/the-wests-green-delusions-empowered.
6 Ted Nordhaus, "Russia's War Is the End of Climate Policy as We Know It," Foreign Policy, June 5, 2022, https://foreignpolicy.com/2022/06/05/climate-policy-ukraine-russia-energy-security-emissions-cold-war-fossil-fuels/.
7 TCS Wire, "Trudeau Moves Forward with Fertilizer Reduction Climate Policy," The Counter Signal, July 23, 2022, https://thecountersignal.com/trudeau-moves-forward-with-fertilizer-reduction-climate-policy/.
8 Anna Holligan, "Why Dutch Farmers Are Protesting over Emissions Cuts," BBC News (BBC, July 29, 2022), https://www.bbc.com/news/world-europe-62335287.
9 Keean Bexte, "Ireland Debates 30% Emissions Cap on Farmers," The Counter Signal, July 26, 2022, https://thecountersignal.com/ireland-debates-30-emissions-cap-on-farmers/.
10 Michael Shellenberger, "Green Dogma behind Fall of Sri Lanka," Michael Shellenberger, July 9, 2022, https://michaelshellenberger.substack.com/p/green-dogma-behind-fall-of-sri-lanka.
11 Vijaya Ramachandran, "Rich Countries' Climate Policies Are Colonialism in Green," Foreign Policy, November 3, 2021, https://foreignpolicy.com/2021/11/03/cop26-climate-colonialism-africa-norway-world-bank-oil-gas/.

12 Abrahm Lustgarten, "Barbados Resists Climate Colonialism in an Effort to Survive the Costs of Global Warming," ProPublica, July 27, 2022, https://www.propublica.org/article/mia-mottley-barbados-imf-climate-change.

13 Fiona Harvey, "African Nations Expected to Make Case for Big Rise in Fossil Fuel Output," The Guardian (Guardian News and Media, August 1, 2022), https://www.theguardian.com/world/2022/aug/01/african-nations-set-to-make-the-case-for-big-rise-in-fossil-fuel-output.

14 Fiona Harvey, "UN Climate Talks End in Stalemate and 'Hypocrisy' Allegation," The Guardian (Guardian News and Media, June 17, 2022), https://www.theguardian.com/environment/2022/jun/17/un-climate-talks-stalemate-hypocrisy-allegation-european.

15 Tess McClure, "Climate Crisis 'Insufficient' to Halt Oil and Gas Exploration, Says New Zealand Government," The Guardian (Guardian News and Media, July 27, 2022), https://www.theguardian.com/world/2022/jul/28/climate-crisis-insufficient-to-halt-oil-and-gas-exploration-says-new-zealand-government.

16 "Sustainable Development Goals," UNDP (UN), accessed August 17, 2022, https://www.undp.org/sustainable-development-goals.

17 Ibid.

18 Roger Pielke Jr., "Opening up the Climate Policy Envelope," Issues in Science and Technology, 2018, https://issues.org/opening-up-the-climate-policy-envelope/.

19 Elisabeth Sifton, The Serenity Prayer: Faith and Politics in Times of Peace and War (New York, NY: W.W. Norton, 2004), 7.

20 John Asafu-Adjaye et al., "An Ecomodernist Manifesto," An Ecomodernist Manifesto, April 2015, ecomodernism.org.

21 Wally Nowinski, "America's Top Environmental Groups Have Lost the Plot on Climate Change," Noahpinion, January 14, 2022, https://noahpinion.substack.com/p/americas-top-environmental-groups.

22 Rafaela Hillerbrand and Michael Ghil, "Anthropogenic Climate Change: Scientific Uncertainties and Moral Dilemmas," Physica D: Nonlinear Phenomena 237, no. 14–17 (August 2008): pp. 2132–2138, https://doi.org/10.1016/j.physd.2008.02.015.

23 Bjorn Lomborg, "Welfare in the 21st Century: Increasing Development, Reducing Inequality, the Impact of Climate Change, and the Cost of Climate Policies," Technological Forecasting and Social Change 156 (July 2020): p. 119981, https://doi.org/10.1016/j.techfore.2020.119981.

24 Mike Hulme, "One Earth, Many Futures, No Destination," One Earth 2, no. 4 (April 17, 2020): 309–311, https://doi.org/10.1016/j.oneear.2020.03.005.

25 Shinichiro Asayama, "Threshold, Budget and Deadline: Beyond the Discourse of Climate Scarcity and Control," Climatic Change 167, no. 3–4 (August 10, 2021), https://doi.org/10.1007/s10584-021-03185-y.

26 Mike Hulme, "Climate Change Forever: The Future of an Idea," Scottish Geographical Journal 136, no. 1–4 (February 14, 2021): 118–122, https://doi.org/10.1080/14702541.2020.1853872.

27 Peter W. Wood, "The Climate Change Conformists," The Spectator World, December 15, 2021, https://spectatorworld.com/topic/the-climate-change-conformists/.

28 António Guterres, "Secretary-General's Remarks to Economist Sustainability Summit Secretary-General," United Nations Secretary General (United Nations, March 21, 2022), https://www.un.org/sg/en/content/sg/statement/2022-03-21/secretary-generals-remarks-economist-sustainability-summit.
29 V Masson-Delmotte, ed., "Global Warming of 1.5°C. An IPCC Special Report on the Impacts of Global Warming of 1.5°C above Pre-Industrial Levels and Related Global Greenhouse Gas Emission Pathways, in the Context of Strengthening the Global Response to the Threat of Climate Change, Sustainable Development, and Efforts to Eradicate Poverty," 2018, pp. 1–616, https://doi.org/10.1017/9781009157940.
30 Richard Tol, "Hot Stuff, Cold Logic," The American Interest, December 10, 2014, https://www.the-american-interest.com/2014/12/10/hot-stuff-cold-logic/.
31 "Summary Report 31 October – 12 November 2021," Glasgow Climate Change Conference (IISD Earth Negotiations Bulletin), accessed August 18, 2022, https://enb.iisd.org/glasgow-climate-change-conference-cop26/summary-report.
32 Robert H Socolow, "Truths We Must Tell Ourselves to Manage Climate Change," Vanderbilt Law Review (Vanderbilt University Law School, November 2012), https://scholarship.law.vanderbilt.edu/vlr/vol65/iss6/2.
33 Shinichiro Asayama, "Threshold, Budget and Deadline."
34 Michael E. Mann, The New Climate War: The Fight to Take Back the Planet (New York, NY: PublicAffairs, 2021).
35 Shinichiro Asayama et al., "Why Setting a Climate Deadline Is Dangerous," Nature Climate Change 9, no. 8 (July 22, 2019): 570–572, https://doi.org/10.1038/s41558-019-0543-4.
36 Matthias Schmelzer et al. The Future Is Degrowth: A Guide to a World beyond Capitalism (London, UK: Verso, 2022).
37 Matthew Yglesias, "Climate Politics for the Real World," Slow Boring, March 16, 2022, https://www.slowboring.com/p/sunrise-movement.
38 Jake Cigainero, "Who Are France's Yellow Vest Protesters, and What Do They Want?," NPR, December 3, 2018, https://www.npr.org/2018/12/03/672862353/who-are-frances-yellow-vest-protesters-and-what-do-they-want.
39 "The Majority of People around the World Are Concerned about Climate Change," The Lloyd's Register Foundation World Risk Poll, May 12, 2022, https://wrp.lrfoundation.org.uk/2019-world-risk-poll/the-majority-of-people-around-the-world-are-concerned-about-climate-change/.
40 Adam Smith, Inquiry into the Nature and Causes of the Wealth of Nations (London, UK: W. Strahan and T. Cadell, 1776).
41 Gwyn Prins and Steve Rayner, "The Wrong Trousers: Radically Rethinking Climate Policy," LSE Research Online (James Martin Institute for Science and Civilization, University of Oxford and the MacKinder Centre for the Study of Long-Wave Events, London School of Economics and Political Science, July 27, 2009), http://eprints.lse.ac.uk/24569/.
42 "The Case for a Sustainable Climate Policy: Why Costs and Benefits Must Be Temporally Balanced," University of Pennsylvania Law Review 155, no. 6 (June 2007): 1843–1857.
43 Martin Luther King and Coretta Scott King, The Words of Martin Luther King, Jr. (New York, NY: Newmarket Press, 2001).

44 Daniel Sarewitz, "Stop Treating Science Denial like a Disease," The Guardian (Guardian News and Media, August 21, 2017), https://www.theguardian.com/science/political-science/2017/aug/21/stop-treating-science-denial-like-a-disease.

45 Mark Fenton-O'Creevy, "Solidarity, Insurance, Emotions and Uncertainty," Pathways to Sustainability (STEPS Centre, September 19, 2019), https://steps-centre.org/blog/solidarity-insurance-emotions-and-uncertainty/.

46 Ian Scoones and Andy Stirling, *The Politics of Uncertainty: Challenges of Transformation* (New York, NY: Routledge, 2020).

47 Ibid.

48 Ibid.

49 Mike Hulme, "Climate Change Forever."

50 Ian Scoones and Andy Stirling, *The Politics of Uncertainty.*

51 Alan M. Turing, "Computing Machinery and Intelligence," *Mind* 59, no. 236 (October 1950): pp. 433–460, https://doi.org/10.1093/mind/lix.236.433.

52 Gwyn Prins and Steve Rayner, "The Wrong Trousers."

53 Ibid.

54 Gwyn Prins et al., "The Hartwell Paper: A New Direction for Climate Policy after the Crash of 2009," LSE Research Online (Institute for Science, Innovation & Society, University of Oxford; LSE Mackinder Programme, London School of Economics and Political Science, May 2010), https://eprints.lse.ac.uk/27939/.

55 Ibid.

56 Daniel Sarewitz et al., "Climate Pragmatism: Innovation, Resilience and No Regrets," The Breakthrough Institute, July 25, 2011, https://thebreakthrough.org/articles/climate-pragmatism-innovation.

57 Catherine Legg and Christopher Hookway, "Pragmatism," Stanford Encyclopedia of Philosophy (Stanford University, April 6, 2021), https://plato.stanford.edu/entries/pragmatism/.

58 Gwyn Prins and Steve Rayner, "The Wrong Trousers."

59 Ibid.

60 Ibid.

61 Mark Tercek, "Q&A With Ramez Naam: Dialogues on the Environment," Big Think, July 1, 2013, https://bigthink.com/articles/qa-with-ramez-naam-dialogues-on-the-environment/.

62 James D. Ford et al., "Big Data Has Big Potential for Applications to Climate Change Adaptation," *Proceedings of the National Academy of Sciences* 113, no. 39 (September 27, 2016): 10729–10732, https://doi.org/10.1073/pnas.1614023113.

63 Michelle Ma, "Move over, Silicon Valley. Engineers Are Quitting for Climate Tech.," Protocol, August 10, 2022, https://www.protocol.com/climate/tech-workers-quitting-climate-jobs.

64 Mark Moritz and Nicholas C. Kawa, "The World Needs Wicked Scientists," *American Scientist* 110, no. 4 (2022), https://doi.org/10.1511/2022.110.4.212.

65 Adam H. Sobel, "Usable Climate Science Is Adaptation Science," *Climatic Change* 166, no. 1–2 (May 6, 2021), https://doi.org/10.1007/s10584-021-03108-x.

66 Regina R. Rodrigues and Theodore G. Shepherd, "Small Is Beautiful: Climate-Change Science as If People Mattered," *PNAS Nexus* 1, no. 1 (March 2022), https://doi.org/10.1093/pnasnexus/pgac009.

67 Coyan Tromp, *Wicked Philosophy: Philosophy of Science and Vision Development for Complex Problems* (Amsterdam, NL: Amsterdam University Press, 2018).

68 Melissa Kreye, "A Philosophy for Working on Wicked Problems," Connect (Extension Foundation, April 1, 2022), https://connect.extension.org/blog/a-philosophy-for-working-on-wicked-problems.

69 Andreas Bjurström and Merritt Polk, "Climate Change and Interdisciplinarity: A Co-Citation Analysis of IPCC Third Assessment Report," *Scientometrics* 87, no. 3 (March 11, 2011): 525–550, https://doi.org/10.1007/s11192-011-0356-3.

INDEX

Printed in the USA
CPSIA information can be obtained
at www.ICGtesting.com
LVHW052305051223
765809LV00033B/189

9 781839 989254